COAL BURNING ISSUES

To Tim

With the hope that this is helpful to the solution of the important Florida problem within your purview.

Warmest regards

Alex

COAL BURNING ISSUES

*A monograph reporting the results of the scoping phase
of an interdisciplinary assessment of the impact of
the increased use of coal*

A. E. S. Green, *Editor*

A University of Florida Book

UNIVERSITY PRESSES OF FLORIDA
FAMU / FAU / FIU / FSU / UCF / UF / UNF / USF / UWF
Gainesville

Library of Congress Cataloging in Publication Data

Main entry under title:

Coal burning issues.

Bibliography: p.
Includes indexes .
1. Coal. I. Green, Alex Edward Samuel, 1919-
II. Alexander, John Franklin, 1943-
TP324.C51485   333.8′22   79-25376
ISBN 0-8130-0667-8

Contents

Contributors

John F. Alexander, Jr., Associate Professor, Urban and Regional Planning, College of Architecture; Acting Director, Center for Wetlands

William E. Bolch, Jr., Professor, Environmental Engineering Sciences, College of Engineering

Eugene F. Brigham, Graduate Research Professor, College of Business; Director, Public Utility Research Center

Barney L. Capehart, Professor, Industrial and Systems Engineering, College of Engineering

Lynne C. Capehart, Legal Research Associate, ICAAS

William L. Chameides, Assistant Professor, Physics, College of Liberal Arts and Sciences

William M. Denty, Jr., Student Assistant, Industrial and Systems Engineering, College of Engineering

Raymond W. Fahien, Professor, Chemical Engineering, College of Engineering

Faramarz Fardshisheh, Adjunct Postdoctoral Research Associate, ICAAS

Louis C. Gapenski, Research Associate, Public Utility Research Center, College of Business

Alex E. S. Green, Graduate Research Professor, Physics and Nuclear Engineering Sciences, Colleges of Liberal Arts and Sciences and Engineering; Director, ICAAS

Harold P. Hanson, Professor, Physics, College of Liberal Arts and Sciences

Robert K. Hutchinson, Graduate Research Assistant, Urban and Regional Planning, College of Architecture

Marc J. Jaeger, Professor, Physiology, College of Medicine

Michael A. Kenney, Graduate Assistant, Environmental Engineering Sciences, College of Engineering

Hong H. Lee, Assistant Professor, Chemical Engineering, College of Engineering

Joseph W. Little, Professor, College of Law

Joel G. Melville, Assistant Professor, Civil Engineering, College of Engineering

Dennis J. Miller, Graduate Assistant, Chemical Engineering, College of Engineering

Joseph D. O'Connor, Graduate Assistant, Nuclear Engineering Sciences, College of Engineering

M. J. Ohanian, Professor, Nuclear Engineering Sciences; Associate Dean for Research, College of Engineering

Daniel E. Rio, Graduate Assistant, Nuclear Engineering Sciences, College of Engineering

Ralph J. Rognstad, Jr., Graduate Research Assistant, Urban and Regional Planning, College of Architecture

Walter A. Rosenbaum, Professor, Political Science, College of Liberal Arts and Sciences

Michael J. Rowe, Staff Engineer, Nuclear Engineering Sciences, College of Engineering

Evelyn H. Schlenker, Postdoctoral Research Fellow, Physiology, College of Medicine

Richard T. Schneider, Professor, Nuclear Engineering Sciences, College of Engineering

Jerome M. Schwartz, Associate in Atmospheric Sciences, ICAAS

Jimmy J. Street, Assistant Professor, Soil Sciences, Institute of Food and Agricultural Sciences

Dennis P. Swaney, Adjunct Assistant in Wetlands Ecological Research, Center for Wetlands.

Karl E. Taylor, Assistant Professor, Physics, College of Liberal Arts and Sciences

Paul Urone, Professor, Environmental Engineering Sciences, College of Engineering

Shreve S. Woltz, Professor, Agricultural Research and Education Center, Institute of Food and Agricultural Sciences

Preface

THIS BOOK is a product of the collective work of a group of faculty members of the University of Florida. Although each writer is a full-time teaching or researching member of a disciplinary department of the university, all participate regularly in interdisciplinary research conducted under the auspices of the Interdisciplinary Center for Aeronomy and (other) Atmospheric Sciences, known as ICAAS. Since 1970 ICAAS has conducted research projects pertaining to atmospheric pollution and related phenomena. These studies are exemplified by a multidisciplinary assessment, now underway with Board of Regents support, of an anticipated large increase in the utilization of coal in Florida.

Although many ICAAS members are physical scientists, the institute's research has not been limited to hard science issues. Rather, ICAAS employs a broad gauge public policy approach, bringing together the hard sciences, life and agricultural sciences, social sciences, economics, medicine, law, and other disciplines with the goal of seeking technical solutions to complex problems that are also socially and politically acceptable.

ICAAS members foresee a period of severe social stress ahead as this nation is forced to switch to coal from oil and natural gas as its primary energy sources. The forces are multiple, posing specific technical, environmental, political, legal, and other kinds of issues that are individually intractable and that in sum pose an extremely formidable barrier to the nation's well-being. ICAAS members believe that a concentrated multidisciplinary effort by academic scholars and researchers can help the nation get over that barrier.

Such efforts have many precedents. After the outbreak of World War II the American academic community rallied quickly to the nation's defense. In addition to specialized training programs, important contributions were made to technological developments, such as radar, missiles, rocket-assisted take-off aircraft, the proximity fuse, weather forecasting techniques, operations analysis methods, and many others which helped win World War II. Beyond doubt, the atomic bomb is the best known World War II product of academic scientists. Although almost all of Japan's war potential was destroyed before it was used, the bomb still is generally credited with avoiding the invasion of Japan, which would have taken many lives.

Today, academicians, like all citizens, are grateful that the nation is not engaged in a shooting war. Nevertheless, the independence of the nation is threatened so severely by the impending energy shortfall that President Carter has called it the moral equivalent of war. The relatively slow development of this "moral war," which truly is an economic struggle for survival, affords the academic community with yet another opportunity to serve the nation. The nature of this war and the general direction that the United States must now take have been outlined in Project Independence (1974), The National Energy Plan (1977), and the Camp David plans (1979). Although wide areas of disagreement about what should be done still remain, both Republican and Democratic administrations have agreed that the essential aims set forth in these plans must be achieved: (1) import less foreign oil; (2) conserve energy; (3) to live within the nation's resources, make coal the major alternative to foreign oil.

How can the academic community help achieve these goals? In general, it is not as well-equipped as large industrial and governmental laboratories to handle major engineering developments, but it can originate new ideas, evaluate and extend innovative technology, and carry out early phases of development work. Perhaps most important, a university faculty, being a respository of experts and scholars of many disciplines, is unusually well-equipped to perform multidisciplinary assessments.

This book singles out one topic, namely *coal burning issues*, as the focus of its attention. The writers have examined the technical, medical, environmental, legal, economic, and public policy issues that must be addressed as the nation increases its dependence on coal. The book does not purport to supply answers; too much work remains to be done such as a more in-depth examination of the apparently most feasible solutions. Instead, it is written with the hope of accelerating examination of a series of critical, long-term strategic and short-term tactical options. This must be done without undue delay because passing time irrevocably closes options.

This book employs no single style. In part it employs quantitative analyses and in part it examines social, political, and legal choices influencing the nation's energy future. No attempt has been made as yet to eliminate all disagreements. For example, the reader may note divergencies among some of the chapters in estimates of coal and other resources. These estimates have traditionally varied by large factors depending upon the perspectives of the estimators and divergencies are characteristic of the technical publications examined in the course of this study. Accordingly, they do not detract from but perhaps enhance this work, which aims to highlight both settled and disputed facts and viewpoints that will influence imminent public policy choices.

In the past, America's academic community has played a vital role in protecting the nation in shooting wars. It seems certain that no less a role is to be played in winning the economic war of survival that now entangles the nation's future. The writers of this book hope that the presentation made here points the way to solutions to some of the coal burning issues. The next goal of the group is to help supply some of the detailed solutions to the energy, environment, and economy problems related to the increased utilization of coal.

Acknowledgments

In such a broad interdisciplinary effort many persons besides those listed as authors have made important direct or indirect contributions. First, it is a pleasure to thank some of the key University of Florida administrators who ten years ago encouraged the establishment of ICAAS to undertake broad comprehensive analyses of public policy issues. These include H. P. Hanson, S. C. O'Connell, R. B. Mautz, E. T. York, G. K. Davis, H. E. Spivey, R. E. Uhrig, and Pat Rambo. We also want to thank R. Q. Marston, J. A. Nattress, R. A. Bryan, and F. M. Wahl, who recently encouraged us to continue our quest and made available funds which made this project possible.

Next we wish to thank the contributors to this report, the faculty, research staff and graduate students whose names appear on page *vi*, almost all of whom not only met very demanding deadlines, but also worked far beyond the call of duty. Particularly our "tiger team," Dennis J. Miller, Michael J. Rowe, Michael A. Kenney, and Daniel E. Rio, made extra efforts needed to cover important issues overlooked in our first draft. In addition, Mr. J. R. Jones, Jr., of our University of Florida Libraries, deserves special thanks for his invaluable assistance in literature searches. Miss Lisa K. Gregory was of assistance in our research on trace elements.

The staff of University Presses of Florida gave valuable guidance in the formulation of the book from a publishing standpoint. In this same vein, we thank our Scientific Advisory Board members: Dr. Woodrow W. McPherson, Graduate Research Professor in IFAS, and Dr. Marvin E. Shaw, Professor of Psychology, who were especially helpful in providing advice and assistance.

The copy editing of this monograph was done by Rita H. Barlow, who came to us in our hour of need and made it possible to meet our publication schedule. The typing of this manuscript was largely accomplished under considerable pressure by Roxie Mays and Olivia S. Berger. Additionally, the efforts of university department secretaries in Physiology, Environmental Engineering Sciences, Nuclear Engineering, and the College of Law should be acknowledged for providing final copy. The drafting services by Wesley E. Bolch and Woodrow W. Richardson and the photographic services by Hans W. Schrader also merit our thanks. We also acknowledge the fiscal services of Grethel Greene and the secretarial services of D. D. Ogle.

The focus of this coal study was indirectly suggested by a request for a proposal formulated by the State Energy Office and issued under the Florida Board of Regents Star Grant Program. We would like to thank the state officials who established this program and Dr. J. S. Dailey and others involved in its administration for their valuable contributions. In addition to the support from the Star Grant Program this work was supported by the Gatorade trust fund, the Francis B. Parker Foundation, the U. S. Department of Energy, and the Colleges and Departments of the listed authors.

CHAPTER 1

INTRODUCTION AND SUMMARY

Interdisciplinary Center for Aeronomy and (other) Atmospheric Sciences

I. INTRODUCTION

The exhaustion of the oil and natural gas of the world, and particularly the United States, has been predicted for some time. While the
day when the wells run dry may move into the future as new discoveries are
made and old wells are rejuvenated with new recovery methods, no knowledgeable person can deny that ultimate exhaustion of stored fossil fuels
is inevitable if consumption continues at present rates. The urgency of
the problem was underscored in 1979 by three events that have had a
major impact in bringing the energy supply situation to the crisis
stage. First, the revolution in Iran supplanted a government friendly
to the United States and that has recently made extensive purchases in
return for oil with one that is antagonistic to this nation. Second,
the Three-Mile Island nuclear reactor accident has had a damping effect
on the growth in number of nuclear reactor electric power facilities,
thus undercutting a main alternative that the nation had been counting on
to the use of oil and gas. Public concerns over storage of radioactive wastes and the proliferation of nuclear technology have now been
augmented by concern as to the safety of current reactors and the
adequacy of training provided reactor operators. Third, OPEC oil price
increases have forced the United States into virtual economic warfare
with many of the oil producing countries; a war that the nation
appears to be losing, if the rising price of gold, the rising interest

rate, and other signs of inflation may be taken as indicators.
The urgency of the need to reduce purchases of foreign oil and,
thereby, help stem the net outflow of dollars to pay for imported
energy is now so acute that almost everybody acknowledges it.

What can the citizens of the United States do? The most obvious
thing, albeit a little old-fashioned, is to take stock of our natural
resources, particularly our renewable resources, and use American common sense to find a way to live within the nation's means and re-
establish national self-respect. Highest priority must be given to
conservation and the use of renewable resources since these actions do
not involve any drain upon "energy savings." Nevertheless, phasing in
a renewable resource mode of life will take time and could be rather
painful. Fortunately, the United States has a goodly supply of coal
that is estimated to amount to 30% of the world supply, enough to sustain the nation for a hundred years or so before a new energy resource
is discovered. Accordingly, coal is a resource whose use must now be
promoted. On the other hand, coal poses dangers of polluted air, dirty
water, mangled earth, crushed bodies and blackened lungs. Images of
these hazards were created a generation or two ago before oil and gas
became so abundant and cheap that they ousted coal as the fuel of
choice in America. The nation is now compelled to return to coal and
find out how to utilize it without resurrecting those bad images.
Modern technology is the key.

This book examines the potential role of coal in the United States
in the coming decades. It looks at coal utilization broadly to include
burning and other direct uses, liquefaction and gasification. To assure
comprehension, the perspectives of scientists, engineers, systems
analysts, medical scientists, lawyers, economists and other experts
have been employed. Particular attention is given the interim role
that coal must assume to meet America's energy needs in the next
generation or two. This will give a breathing space for science to
ascertain whether or not unlimited energy resources such as fusion
machines or breeder fission reactors are feasible, and for society to
decide whether or not they are acceptable. If affirmative answers to

both questions are not forthcoming after a few decades, society must
then use its remaining coal and other fossil fuels to learn to live
entirely on solar energy in direct or indirect forms.

This book has been written with the assumption that society does
not know what the future holds with respect to these long-range alternatives. This will be discomforting to some readers who have firmer
visions about the future than do the scientists and other experts who
wrote this book. This underlines the importance of the collective
point of view about the use of coal expressed by these writers--coal
buys time for this country and the world until the shape of the long-
range future becomes clearer. Thus, this book concerns itself with
the best way to utilize coal from the viewpoint of our energy needs,
our environmental safety and the health of our economy.

II. QUANTITATIVE FACTS ABOUT ENERGY SUPPLIES AND CONSUMPTION RATES

The old saw about history's continually repeating itself has
obvious validity so far as the use of coal is concerned. The curves
of Figure 1 depict how during this past century or so the U.S. has
shifted from a renewable resource, that is, fuel wood, to coal, and then
to petroleum and natural gas. Each shift has been spurred by factors
such as availability, convenience, and economic advantage. Since coal
is still abundantly available it obviously must have some relative
disadvantages that must be coped with. Almost everyone has heard about
Pittsburgh's reputation as "the smoky city," a nickname that no longer
applies, and the descriptions about the coal smoke pollution of London
are even more fearsome. Perhaps fear, both legitimate and apocryphal,
about the bad side effects of burning coal even delayed the onset of
the industrial revolution in England. History recounts that in the
year 1306 King Edward I of England feared it so much that he decreed
punishment of death to people who burned coal. Ultimately, however,
the power of coal was not to be denied even in England, because history
has recorded that about four centuries later that nation used coal to
fuel the industrial revolution that made it a great world power.
Similarly, coal has been a major source of energy used in the United

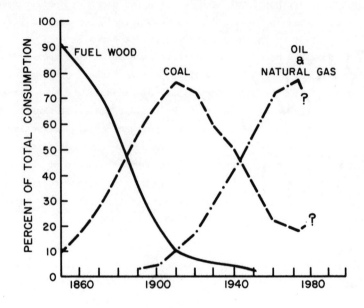

Fig. 1. Percent of Total Energy Consumption

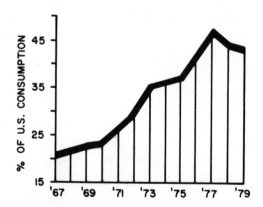

Fig. 2. Imported Oil Use

States during the period illustrated in Figure 1 to build this nation
into an industrial power.   These and other historical coal landmarks
are charted in Table 1.

The world is now on the threshold of shifting back to coal as its
major source of energy.   Convenience and environmental advantage no
longer are the driving forces; this time it is one of necessity.   As is
well known to everyone, petroleum and natural gas are running out and
will by all accounts be depleted as a major source of energy sometime
in the first half of the 21st century.   Also well known is the fact that
escalating American dependence on foreign oil sources has drastically
driven up the price of energy in the United States and played havoc
with not only the nation's but also. the world's economy.   The rapidity
and the extent of the growth of the nation's oil dependency are depicted
in Figure 2.   However, the pattern emerging in Figure 1 from 1970 on-
ward suggests that a new age of coal use is at hand during which our
dependency on petroleum, prrticularly foreign petroleum, could be
driven downward.

Although this book is about coal use, the writers do not want to
leave the erroneous impression that energy conservation is unimportant.
It takes no more than a slight digression to demonstrate that conservation can offer substantial relief that, when coupled with increased
coal use, can help return the United States to a status of energy
independence and sustain it there.   Per capita energy consumption in
the United States is so high by any standard that the very fact suggests
that conservation may be of major quantitative importance.   For example,
as shown in Figure 3 (McPherson, 1965), Americans far exceeded any other
of the world's people in both per capita energy consumption and income
in 1965.   Comparative data for 1974, pictured in Figure 4, show that
although the per capita consumption of energy in the United States again
exceeded that of any other nation, the equivalent per capita income in
many industrialized nations had overtaken or even exceeded that enjoyed here.   Taken together, these factors suggest that conservation
in America might produce substantial energy savings without curtailing
the standard of living.

Table 1: *Some Historical Coal Burning Landmarks*

Time	Country	Person/People	Use or Context
1500 BC	Wales		coal used in funeral pyres
1100 BC	China		heat for homes
950 BC	Israel	King Solomon	mentioned in Bible
300 BC	Greece	Aristotle	noted disagreeable smell
121 AD	England	Hadrian	used in dwellings near wall
852 AD	England	Anglo-Saxon	coal used in rental payment
1306	England	King Edward	decrees use of coal punishable by death
1200	North America	Pueblo	coal used in pottery production
1679	North America	Fra Hennepin	observed black mineral along Illinois river
1694	England	Clayton	first production of coal gas
1750	England		coal fuels industrial revolution and rise of England
1770	United Colonies	Washington	noted coal mine on Ohio river
1771	Pennsylvania		anthracite discovered in eastern Pennsylvania
1840	U.S.A.		coal industry grows 1,000,000 tons mined
1850	U.S.A.	Gesner	coal liquefaction
1890	U.S.A.		electric steam generators expand use of coal
1940	Germany		large-scale liquefaction of coal by Bergins process
1950	U.S.A.		Middle East, Venezuela oil flood world markets
1973	U.S.A.		oil crisis begins
1974	U.S.A.	Nixon*	Project Independence
1977	U.S.A.	Carter*	National Energy Plan
1979	U.S.A.	Carter*	oil crises worsen - Camp David proposals

*proposes major switch to coal utilization

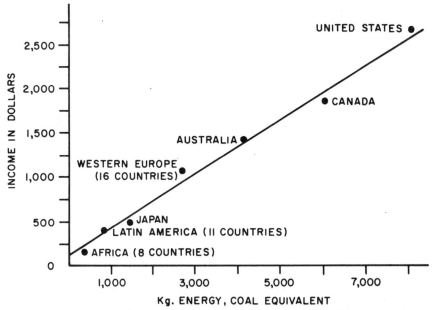

Fig. 3.   *Consumption of Energy Per Capita, Inanimate Sources, and*
          Gross Domestic Product Per Capita, Selected Areas, 1962
                        from McPherson (1965)

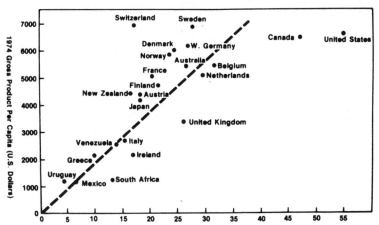

Fig. 4.   *1974 Energy Consumption in Barrels of Crude Oil*
                        Equivalents Per Capita
          Source of Data: U.N. Statistical Yearbook, 1975

This may be illustrated by comparing the fuel efficiency of
American cars to that of imported cars. In 1975 the average domestic
car used about five gallons of gas to travel the same distance as the
average imported car would go on three gallons. Hence, much petroleum
could be saved by merely converting to fuel-saving cars. This will
occur if present federal programs successfully bring the average rate of
fuel use of the domestic fleet down to that of imported cars. The waste
in American transportation apparently has its industrial counterpart if
comparisons between American and West German industrial fuel consumption
data are a reliable guide. For example, the West Germans use about
20% less energy per unit of primary metals produced than does American
industry (SRI, 1975). Other major energy intensive industries of West
Germany have even greater relative efficiencies. Clearly, a role for
better industrial technology in energy conservation is evident. What
happens in industry and transportation seems to represent quite faith-
fully the relative energy voracity of the American way of life. The
huge demand for energy imposed by the average American compared to that
for people from other countries has been noted by Smil (1979) who compares the per capita amounts of energy consumed measured in kilocalories
per day (kcal per day) of the developing world with that of the USA.
This work shows that American per capita energy consumption is exorbi-
tantly large, about 215,000 kcal per day whereas many of the world's
developing countries get along on about 10,000 kcal per day per capita
with little fossil fuel augmentation to solar energy input. In China,
for example, the energy use per capita is approximately 13,500 kcal
per day. Thus, Americans could help alleviate the energy crisis by
moving the per capita energy use downward, possibly to 100,000 kcal per
day, the approximate energy use of other western industrialized nations.
If, for example, we could just lower the rate of growth of energy use
from 3.5% to 2.3% per annum we would save 20 millions barrels of oil
per day by the year 2000.

The objective of this book, however, is not to dwell upon the
energy conservation measures that the writers unanimously subscribe to,
but rather is to examine phenomena associated with a transition to coal

use, particularly in the United States. The use of coal in this country may be what will make it possible for Americans to maintain a high standard of living, leaving the foreign oil to alleviate the plight of poorer people in the world.

Fortunately, the world and especially the United States have large supplies of coal to consume. In gross dimensions, the world's coal reserves are illustrated by the data in Table 2, extracted from Peters and Schilling's (1978) appraisal of world coal resources. The data are presented in metric tons with the energy equivalence of hard coal (HC) rather than in gross weight. Hard coal here encompasses anthracite and bituminous and brown coal (BC) includes sub-bituminous and lignite. Gross supply data also have been adjusted to reflect only supplies that are economically and technically recoverable. Hence, other estimates of coal resources could vary substantially from those shown. Be that as it may, this table indicates that the United States has more recoverable coal resources than any other nation. The last two rows give the U.S. production and export rates in 1975 in million tons.

Table 2. *Economically and Technically Recoverable*
Coal Resources of Leading Coal Countries
in Gigatons (10^9 metric tons)

		HC	BC	Total	Prod	Exp
	World	492	144	636	2593	199
1	USA	113	64	177	581	60
2	USSR	83	27	110	614	26
3	China	99	--	--	349	3
4	Great Britain	45	--	45	129	2
5	West Germany	24	10	34	126	23
6	India	33	--	33	73	--
7	Australia	18	9	27	69	29
8	South Africa	27	0	27	69	3
9	Poland	20	1	21	181	39
10	Canada	9	1	10	23	12

Peters and Schilling also have predicted how rapidly coal will be put into production in the coming years. As shown in Figure 5, the United States is expected to lead the world in this process. Whether or not it does will depend upon whether or not and how the issues examined in this book are resolved. In any event, it seems certain that a substantial shift-over to coal is coming. Already this is manifested by the fact that virtually all new electric power plants purchased in the U.S. are being designed for coal rather than oil or gas.

The fact that shifting to coal opens up the world's last known major resource of non-renewable energy to consumption with an appetite that will inexorably lead to ultimate exhaustion poses a moral and resource issue for humankind that has yet to be squarely faced. What will be the plight of those unborn generations that come along when the resource is depleted? What should we do about the underdeveloped nations who are demanding the right to develop? This necessarily implies moving away from their current equilibrium with renewable energy (Smil, 1979)

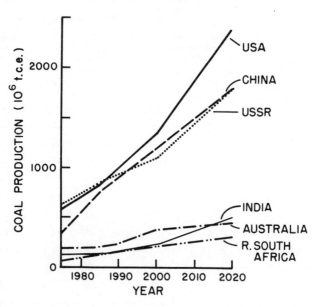

Fig. 5. Survey of the Future Trend in Production

toward a dependence upon stored energy. Certainly if China's billion
people were to move up to America's per capita energy consumption rate
the world's stored energy resources would be wiped out in short order.
This world-wide social dilemma has yet to surface in full force. Unless
a new, inexhaustible energy resource is discovered, consideration of
equity suggests that some sort of balance must be struck in which those
with more, use less and those with less, use more. However, the allo-
cations to each nation of energy resources, property, population limits
and emission limits to control adverse climate changes raises complex
international issues considerably beyond the scope of this work.

Another issue whose complexity makes it beyond the scope of the
present study is the question of monopolistic practices in the energy
field. Such practices by the OPEC oil cartel obviously have contributed
to our energy crisis. Voices are now being raised that major American
companies have established an energy monopoly paralleling the OPEC cartel. However, it is an open question as to whether this cry identifies
a real problem or simply diverts the nation's attention from fundamental
problems: the inevitable exhaustion of the U.S. oil and gas reserves
and the need to use coal to save our economy.

The existing coal supplies of the USA are large enough, even if
used up at a voracious rate, to provide time for the development of even
more plenteous energy resources that may exist, such as nuclear fusion
or breeder fission, or to learn how to live on renewable energy resources better than the developing countries. Referring again to
Figure 5, the reader can see that the actual 1979 coal production is 700
million tons per year in the United States. At that rate, the estimated
American reserve of 177 Gigatons of economically and technically recoverable coal would last 250 years, and even if the consumption rate were to
triple, the supply would last nearly 100 years. This underscores again
the central purpose of this book; namely, to examine the issues and
problems that must be resolved to make coal a safe and environmentally
acceptable energy workhorse for the USA during the coming 50 year or so.

III. ENERGY UNITS AND THE DIMENSIONS OF THE PROBLEM

The units used to describe energy quantities vary greatly, even among professional people. Human energy commonly is measured in consumption of kilocalories (kcal) per capita per day. For example, under moderate exertion the average American male uses 3,220 kcal per day and the average American female 2,320 kcal per day. By contrast, heat engineers describe energy in terms of British Thermal Units (BTU), physicists use the kilojoules (kilowatt-sec), electrical engineers use kilowatt-hours, petroleum engineers use barrels of oil, gas engineers use millions of cubic feet, and mining engineers use tons of coal, either in short-tons (2,000 pounds) or metric tons (1,000 kilograms). All describe the same basic quantity, but in terms that have no readily apparent conversion equivalences, such as the trivial conversion of, say, feet to inches. This communication hitch has been made even worse by the growth in use of a number of new words to describe energy resources in global terms. Thus, the quad = 10^{15} BTU, the Gigaton = 10^{9} tons of coal equivalent (GTC), have come into use and now compete with older terms such as millions of barrels of oil (MBO) and trillions of cubic feet of natural gas (TCF) to describe energy resources.

To supply a common base for communicating about energy quantities, the information in Tables 3 through 5 is provided. Table 3 merely defines the units; Table 4 displays some of them in a fashion that makes conversion among micro-units relatively easy; and Table 5 displays others in a fashion that does the same for macro-units. The BTU is used to tie Tables 4 and 5 together.

As a quantitative overview of the dimensions of the United States problem in terms of these units, Table 6 gives a detailed picture of the U.S. energy supply and demand from 1960 to 1978. Readers may find it helpful to mark these tables for future reference. Right now this nation's coal production capability far exceeds the current demand and production could go to as high as 2 Gigatons by the year 2000. Thus, coal could replace oil and natural gas as our main energy source. Whether or not it should or would is the main concern of this book.

Table 3: *Definition of Energy Units*

Common Abbreviation	Meaning
kcal	Kilocalorie (also denoted Cal)
kwatt-hr	Kilowatt-hour
kjoule	Kilojoule (a kilowatt-sec)
BTU	British Thermal Unit
BOE	Barrel of Oil Equivalent
CFG	Cubic Feet of Natural Gas equivalent
TCE	Metric Ton of Coal equivalent (TCE = 1.032 short ton equivalent)
GTC	Gigaton of Coal equivalent (metric)
quad	10^{15} BTU C = 10^{16} BTU Q = 10^{18} BTU
MBO	million barrels of oil
TCF	trillion cubic feet gas

Table 4: *Energy Unit Conversion Factors (Micro Units)*

One of these energy units	Equivalent Number of these units			
	Kcal	Kwatt-hr	K-joule	BTU
Kcal =	1	11.627×10^{-4}	4.185	3.968
Kwatt-hr	860	1	3600	3413
K-joule	0.238	27.76×10^{-5}	1	0.948
BTU	0.252	29.29×10^{-5}	1.055	1

Table 5: *Energy Unit Conversion Factors (Macro Units and BTU's)*

Energy Units:	Equivalent number of these units:				
	BTU	BOE	CFG	TCE	quad
BTU =	1	1.72×10^{-7}	9.79×10^{-4}	3.59×10^{-8}	10^{-15}
BOE =	5.8×10^6	1	5.68×10^3	.208	5.8×10^{-9}
CFG =	1.02×10^3	1.76×10^{-4}	1	3.67×10^{-5}	1.02×10^{-12}
TCE =	27.8×10^6	4.79	27.2×10^3	1	27.8×10^{-9}
quad =	10^{15}	1.72×10^8	9.79×10^{11}	3.59×10^7	1

1 quad = 35.9 MTC = 172 MBO = 0.970 TCF = 41.5 M Short Tons Coal(MST)

Table 6. *U.S. Energy Supply and Demand**

	1960	1970	1972	1974	1976	1978
Energy Supply (quads) [a]	46.01	70.91	74.28	75.64	76.93	79.89
Oil [b]	20.39	30.38	32.94	34.17	35.26	38.04
Natural gas	12.82	22.52	23.26	22.20	20.47	20.24
Coal	11.12	15.05	14.49	14.47	15.85	15.11
Nuclear	0.01	0.24	0.58	1.27	2.11	2.98
Hydro, other	1.67	2.72	3.01	3.53	3.24	3.47
Energy consumption (quads) [c]	44.08	66.82	71.63	72.35	74.16	78.01
Transportation	11.23	16.76	18.34	18.40	19.39	20.59
Residential, commercial	14.37	23.77	25.84	25.57	27.18	29.30
Industrial, other	18.48	26.29	27.45	28.39	27.59	28.13
Oil consumption (MBD)	9.80	14.70	16.37	16.65	17.46	18.73
Oil imports	1.62	3.16	4.52	5.89	7.09	7.86
U.S. reserves [d]	31.6	39.0	36.3	34.2	30.9	27.8
Gasoline consumption (MBD)	4.13	5.78	6.38	6.54	6.98	7.41
Natural gas consumption (TCF/yr)	11.97	21.14	22.10	21.22	19.95	19.41
Residential	3.10	4.84	5.13	4.79	5.05	4.97
Industrial	5.77	9.25	9.63	9.77	8.60	8.14
Electric Utilities	1.73	3.93	3.98	3.44	3.08	3.22
U.S. reserves [e]	262.3	290.7	266.1	237.1	216.0	200.3
Coal production (MST)	415.5	602.9	595.4	603.4	678.7	653.8
Underground mines	284.9	338.8	304.1	277.3	294.9	243.5
Surface mines	130.6	264.1	291.3	326.1	383.8	410.3
Coal Use Electric Utilities	173.8	318.3	350.2	390.3	447.0	480.1
Coke	81.0	96.0	87.3	89.7	84.3	71.1
Exports	36.5	70.9	56.0	59.9	55.4	39.8
Electricity production (quads)	8.40	16.51	18.74	19.94	21.51	23.76
From coal (per cent)	53.5	47.3	44.1	44.5	46.4	44.3
From oil	6.1	12.2	15.6	16.0	15.7	16.5
From natural gas	21.0	25.0	21.5	17.1	14.5	13.8
From hydro	19.4	16.7	15.6	16.1	13.9	12.7
From nuclear	0.0	1.5	3.1	6.1	9.4	12.5
Population (millions)	180.0	203.8	208.2	211.4	214.7	218.1
Automobiles (millions)	61.7	89.2	97.1	104.9	110.4	117.1
Gross national product (billions of 72 dollars)	736.8	1,075.3	1,171.1	1,217.8	1,271.0	1,385.7
Energy consumption/GNP [f]	59.8	62.1	61.2	59.4	58.3	56.3

[a] includes U.S. imports, U.S. exports and changes in inventories

[b] includes natural gas liquids

[c] includes distribution loss from electrical generattion

[d] proved reserves in billion barrels of oil

[e] proved reserves in trillion cubic feet

[f] in 1000 of BTU/per 72 dollars

* Adapted from National Journal, July 21, 1979, page 1200.

The following section of this chapter consists of summaries of Chapters 2 to 18 of this book. Chapters 2 to 8 largely consist of descriptions of our resources, present and future coal technologies and current coal issues. Chapters 9 to 14 describe known or possible environmental problems. Finally, Chapters 15 to 18 discuss and describe possible assessment techniques which might be used in public policy decisions related to coal burning issues.

In addition to the references cited in this chapter we list below a brief bibliography representing very recent works on energy in general and on coal utilization in particular.

References

Executive Office of the President, The National Energy Plan, Energy Policy and Planning, U.S. Government Printing Office, Washington, D.C., 1977.

McPherson, W. W., Economic Development of Agriculture, "Input Markets and Economic Development," Chapter 6, pp. 99-117; Iowa State University Press, Ames, Iowa, 1965.

Peters, W., and Schilling, H. D., An appraisal of world coal resources and their future availability, World Energy Resources, pp. 1985-2020.

Smil, Vaclav, Energy flows in the developing world, American Scientist, 67, p. 522, 1979.

Recent Bibliography

Berkowitz, N., An Introduction to Coal Technology, Academic Press, New York, 1979.

Crane, A. T. et al., The direct use of coal, prospects and problems of production and combustion, Congress of the United States, Office of Technology Assessment, Washington, D.C., 1979.

Davis, R. M. et al., National coal utilization assessment, a preliminary assessment of coal utilization in the South, Oak Ridge National Laboratories Report No. ORNL 1 TM-6122, 1979.

Department of Energy, U. S. Fuel use act, final environment impact statement. Washington, D.C.,

Lansberg, H. H. et al., Energy: The next twenty years, A report sponsored by The Ford Foundation, Ballinger Publishing Company, Cambridge, Mass., pp. 273-408, 1979.

Leonard, J. W., Coal, The World Book Year Book, World Book-Childcraft International, Inc., Chicago, Ill., pp. 566-582, 1979.

Loftness, R. I., Energy Handbook, Van Nostrand Reinhold G., New York, 1978.

McNeal, W. H., and Nielsen, G. F. et al., Keystone Coal Industry Manual, McGraw-Hill Mining Publications, New York, 1977.

Stobaugh, R. and Yergin, et al., Energy Future, Report of the Energy Project at the Harvard Business School, Random House Publishing, New York, 1979.

Surles, T., Gasper, J., and Hoover, J. J. et al., An assessment of national consequences of increased coal utilization, U. S. Department of Energy, Washington, D.C., 1979.

Chapter 2, Coal Availability and Coal Mining, by Ohanian and
Fardshisheh, summarizes the U.S. coal resource base by geographic location, access (i.e., deep or shallow mines), and classification. A
general assessment of demand projections is based on the assumption
that electrical power generation be the primary user of coal over the
next two to three decades.

Coal is the most abundant non-renewable energy resource in the
United States which has approximately 28% of the world's total coal
supply. U.S. coal, which represents 70% of the non-renewable energy resource in the country is located in three distinct regions: Eastern,
Central and Western. Bituminous coal comprises 43% of the resource
base, sub-bituminous 27%, lignite 28% and anthracite 1%. The sulfur
content of Western coal is lowest (generally 1% or less) with Central
coal highest at 2-4% and Eastern coal at most 2% or less. However, the
heat value of coal per unit weight is, in general, lower for the
Western coal reserves.

A summary discussion of modern mining techniques and the factors
that give rise to the use of a particular mining methodology are also
presented. The role played by technological development through the
evolution of machinery and equipment used in the mining system is discussed. The environmental issues arising from coal mining are examined.
Since coal crushing, washing, drying, etc. is often required, it is
considered an integral part of the mining, and coal preparation is also
covered briefly. Since safety plays a significant role, the problems
related to safety are discussed. A related topic which is of paramount
importance, namely the occupational hazards associated with coal mining,
is reviewed. It is seen that since the enactment of the 1969 Federal
Coal Mine Health and Safety Act, mine fatalities have been reduced
appreciably. However, no greatly measurable reduction in the rate of
disabling injuries has occurred. Figures are tabulated which show
miners are likely to suffer from occupational health and safety hazards
at a rate higher than other occupations.

Chapter 3, <u>An Energetics Analysis of Coal Quality</u>, by Alexander, Swaney, Rognstad and Hutchinson uses energy circuit language, a method of systems analysis, to determine the quality of energy embodied in coal. Three distinctly different approaches are employed to calculate coal's energy quality.

The first method employed to determine the energy embodied in coal uses a theoretical model of the coal formation process. As a starting point a detailed description of the formation of the various ranks of coal is given. From photosynthesis, atmospheric CO_2 is transformed to plant biomass and then to peat via microbial decomposition in water. Then through catagenesis (autoclaving geological processes), peat is transformed to lignites to bituminous and anthracite forms. The geological and ecological energies involved in these tranformations are evaluated.

The second method used to determine coal quality compares the energetics of coal burned as an electrical generating fuel to the use of wood and oil in electrical power production. These two fuels were chosen as a basis for comparison because the energy value of wood is easily traced to the energy flows of natural systems and the energy in oil is the largest component of the contemporary fossil fuel energy base.

The third approach to estimating the energy quality of coal is based on the energetics of converting coal to higher quality synthetic gas. It is postulated that this will give an upper range to the energy quality of coal, while conversion to electricity will approximate the lower limit.

It appears that 1.36 - 2.2 thermal calories of coal are required to do the same amount of work that one calorie of oil or gas can accomplish. In effect, this means that the magnitude of the world's coal reserves when expressed in thermal units, should be reduced by 26 - 55%.

Chapter 4, <u>Coal Transportation</u>, by B. Capehart, O'Connor and Denty discusses the important links in the overall processes of supplying coal to ultimate users. Coal is shipped by a variety of modes and may even require multiple modes, were 65% by rail, 11% by water, 12% by truck and 1% by other modes (slurry pipelines, conveyors, etc.) Eleven percent was used at mine mouth plants. Future expansion of the coal transportation network is clouded by present federal policy which appears to emphasize short-term expansion of natural gas use, rather than coal. If the coal transportation system is to be expanded, the rail system which can most easily be expanded must carry the major burden of growth since it is the most easily expanded transport system. Expansion of the coal transportation network will result in major environmental and social problems attendant on both construction and operation. Air, water and noise pollution accompany the different modes in varying degree. Social impacts such as community disruption and safety also occur.

Operation and future construction of transportation modes are greatly affected by federal regulations and regulatory boards. The extent of regulations concerning the different modes varies from almost no control to, in some cases, a complete roadblock to expansion. Expansion of coal transportation will require the examination of major public policy decisions. Environmental, social and economic problems are all interrelated, and must be considered in a systems approach. Complex cost/benefit studies will be necessary to help determine these public policies. Finally, economic comparisons of the various modes of transportation are needed in order to make optimum economic decisions and expansion plans. This is also a complex area since economic decisions are not independent of environmental, social and regulatory requirements.

The future of the coal transportation system is still very uncertain. Until questions of future demand are answered more completely, few major investments in expansion are likely to be made.

Coal-oil mixtures, considered for pipeline transportation, also provide a rapid means of reducing oil and gas consumption with only minor retrofitting of oil and gas burning facilities.

Chapter 5, Coal Burning Technology, by Schneider and Rowe is devoted to combustion aspects of the direct use of coal. The discussion
centers primarily on concentrated use such as by electric utilities, or
large industries and dilute use, as in residential coal burning which
probably will be an environmental problem in populated regions. Concentrated techniques provide for economy of scale for the removal of harmful
pollutants. The important parameters in coal combustion are the heat of
combustion, the burning temperature, the oxidant control, burning times
and the coal quality. These parameters control the energy output as well
as waste removal state and gaseous by-product production.

Historically, the first type of coal burner was the "fixed bed"
burner where fixed bed refers to the location of the coal burning zone.
Here heat rises to an overhead boiler to produce steam. In these installations the stoker for feeding and coal and ash removal involves the
majority technology.

Suspension burners in which coal dust is injected into a fire box
surrounded by boiler tubes represented the next improvement in coal
burning techniques. Improvements of injection technology led to the
present day "cyclone" furnace. Suspension burners have a very high temperature which offers good heat transfer and high efficiency. Unfortunately, they produce large amounts of gaseous NO_x.

"Fluidized bed" burners derive their name from the liquid-like properties of the combusting material. With improved heat transfer techniques,
fluidized bed burners can be operated at lower temperatures thus producing much less NO_x, yet producing the same amount of steam as a high
temperature suspension burner. Furthermore, chemical techniques in the
fluidized bed can be used to remove SO_x. Other techniques for the direct use of coal which are still in the development stage include magnetohydrodynamic (MHD) reactors and gas turbine-steam turbine hybrid systems.
In the latter the gas turbine could be supplied by an MHD nozzle or water-
gas combustor. The gas turbine exhaust would then feed a conventional
boiler to produce steam. The established techniques and experimental
techniques are all described and assessed in the chapter on coal burning technology.

Chapter 6, <u>Synthetic Fuels from Coal,</u> by Miller and Lee, is devoted
to coal conversion, the processes of making clean burning liquid and
gaseous fuels from coal. The basic concept of coal conversion chemistry
is to enrich raw coal with hydrogen in some form to produce hydrocarbon
products with relatively large hydrogen content. This can be accomplished
by the reaction of coal with hydrogen or steam, or by removing excess
carbon from coal to give a hydrogen-rich product and a carbonaceous by-
product. The liquid fuels produced in the liquefaction process range
from light gasoline to heavy tars. Liquefaction and gasification reactors involve many overlapping processes and many reactors produce both
gases and liquid products.

Several commercially operating gasification processes are now in
use and many more are in the pilot plant stage. These gasification processes can be categorized in terms of reactor type and products formed.
The raw synthesis gas produced in most gasifiers can be used either as a
fuel or as a feed stock to make synthetic natural gas, methanol, hydrogen, ammonia or liquid hydrocarbon fuels, via the Fischer Tropsch synthesis.

Coal conversion to synthetic fuel has the great potential to accommodate cleaning up processes and could provide an important part of the
clean gaseous and liquid energies of fuels needed for the next generation
or two. The technology is available for producing both gaseous and liquid fuels from coal and only economic considerations, the capital intensive nature of coal conversion plants and the uncertainty as to the
OPEC cartel controlled price of foreign oil has inhibited our large-scale
development of coal conversion in America.

The possibilities of converting oil shales, tar sands and heavy petroleum residuals into liquid or gaseous synthetic fuels also constitutes
an important issue. The relative economics of these synfuels vs. coal-
derived synfuels will determine the synthetic fuels which have the greatest short-range and long-range promise in the United States. The environmental impact of the expanded coal mining and disposal of large amounts
of sulfur and ash produced in coal conversion is another important issue
related to coal conversion.

Chapter 7, <u>Technological Innovations</u>, by Hanson <u>et al</u>. summarizes
some technological advances which might ameliorate, postpone, or even
eliminate the energy shortage.

<u>Integrated Utility Systems</u> (IUS) as a concept originally involved
the tactical use, as dictated by economic circumstance, of gas, oil,
coal and solid waste. However, because of the cost and unavailability
of petro-fuels, the IUS concept has evolved into one in which coal is
augmented by processed solid waste. Nevertheless, such units have certain advantages, including financial, which can make them the unit of
choice.

<u>Electrically Propelled Automobiles</u>, an old technology, permits the
energy burden of travel to be shifted from gasoline to coal. Batteries,
which have been greatly improved, are charged by coal-produced electricity thus freeing the car owner from the dependence on petro-chemicals.
Further, by charging at night, the load on the power plant becomes
better distributed.

<u>Air Pollution Control Technology</u> could make extensive burning of
coal acceptable to society by proper control of effluents and residues.
There is continuing research in all phases of the technology of pollution control. Particulates and SO_2 have been given greatest attention,
but NO_x control is also under development.

<u>Coal Cleaning</u> makes it possible to eliminate noxious or undesirable
by-products <u>before</u> the coal is actually burned. Techniques have been
developed for removing incombustible ash, pyritic sulfur, and water from
unprocessed coal.

<u>Off-Shore Power Plants</u> originally proposed for nuclear power plants
might be applied to coal-fired units to isolate and buffer the power
facility. Since almost half of today's power demand exists within a
200 mile strip along our various coasts, the concept might find extensive applications.

<u>Coal Plant Siting Techniques</u> in general provide the opportunities
of coal plants to maximize operational efficiency, minimize electric
transmission losses and environmental impacts.

Chapter 8, _Water Resources,_ by Melville and Bolch, suggests that two of the primary constraints on the utilization of coal energy will be water availability and the potential for pollution of water that would otherwise be available for other uses. An analysis of the available freshwater reveals a breakdown of 95% groundwater, 3.5% lakes, swamps, reservoirs, and river channels, and 1.5% solid moisture. The groundwater stresses are emphasized and are shown to be the most threatening.

Surface water pollution is easily detectable. The most menacing characteristic of groundwater pollution is that damage is probably irreversible or at least corrective time scales could be measured in terms of years or even generations.

Suggested actions for reduction of water resource constraints are listed.

One of the key issues in the increased use of coal will be the allocation of water resources to competitive users. The competing water demands are enumerated and discussed. The differences between point-source and non-point source pollution are discussed as they apply to coal burning facilities.

The possibility of utilizing to good advantage the water content of lignite, which is usually a problem, to facilitate coal conversion is briefly discussed.

Chapter 9, <u>Atmospheric Pollution,</u> by Urone and Kenney, considers the increased emissions and possible degradation of ambient air quality arising from the increased use of coal. The combustion of coal leads to major emissions of sulfur and nitrogen oxides, particulate matter, and volatile trace hazardous and toxic substances. Ambient air quality standards prescribe health and welfare endangering levels when surpassed. The amount of emissions with the attendant allowable calculated environmental impact for expanded coal use will depend upon the base line quality or expected quality of the air basin involved.

To estimate the air quality impact of increased coal use is not a simple matter of taking ratios of smoke stack emissions or air measurements to the amount of increased use. Among the factors that must be considered are the type and properties of the coal to be used. Heating value, sulfur, nitrogen, ash, and trace element content must be considered. The local micrometeorology and plume dispersion patterns are important in determining how much of the pollutants will be found at ground level under normal as well as worst-case conditions. Finally, the location and the degree of emissions control required will be determined by both the quality of the area's ambient air and by the potential impact on any environmentally protected areas in the vicinity. Proper proportioning of all the above factors requires an indepth knowledge of each of the mechanisms involved as well as detailed knowledge of the substances emitted when coal is burned. In addition, the enriching emission process for trace toxic substances forms a subtle, long-term and long-range threat to man and the environment.

Many key issues must be addressed including (1) quantification of increases in emissions, (2) possible insult to air quality standards, (3) long-term impacts of trace toxic substances, (4) reactions of sulfur dioxide, (5) synergistic effects to acid rain, visibility degradation, climate change and other environmental questions.

Chapter 10, <u>Air Pollutant Dispersion Modeling</u>, by Fahien, summarizes the status of three major approaches which may be used for the
quantitative prediction of the impact of increased coal burning on air
quality. Such calculations require not only a knowledge of the emissions
to be expected from coals of a certain chemical composition but also a
quantitative method to relate emission rates to air quality. To do the
latter requires the use of a dispersion model, an important link between
emission rates and dose-response or other health effect studies.

Dispersion models are classified as follows: (a) Gaussian models,
which assume that the concentration of a pollutant from a point source
follows the normal error curve; (b) transport models, which are based on
the law of conservation of mass; and (c) stochastic models, which are
based on the laws of probability.

The Gaussian model has very limited validity and its use requires
knowledge of parameters ("dispersion coefficients") which cannot be predicted accurately. As a result, errors of several hundred percent are
not uncommon. Nevertheless, it is widely used and forms the basis for
the EPA-recommended "off-the-shelf" models.

The transport models are more rigorous--especially when chemical
reactions are involved--but usually require knowledge of eddy diffusivities or "K" values which also cannot be accurately predicted. Since
they are less limited than the Gaussian models, they are mathematically
more complex and usually require more computer time. Stochastic models
are the most rigorous and most adaptable, but these are in a develop-
mental stage and require meteorological data that are not always available. The actual selection of a model for use in a "preconstruction
review" is for practical purposes limited by law to EPA-recommended
models. Examples of previous modeling studies are discussed.

Most compelling is the great need to develop reliable yet practical
methods for the quantitative prediction of the dispersion of the air
pollutants emitted in coal burning.

Chapter 11, <u>Atmospheric Modifications</u>, by Taylor, Chameides and Green, is concerned with the atmospheric impact of the release of pollutants from combustion of fossil fuels, particularly coal. Such releases can cause global perturbations by changing the average composition of the earth's lower atmosphere. Whether such perturbations are significant depends upon the magnitudes of the combustion sources as compared to other natural sources and the rapidity with which these pollutants are scavenged before they are dispersed throughout the atmosphere.

In this chapter we focus on the potential impacts of the continued or accelerated release of CO_2, NO_x, SO_2 and aerosols upon global climate and other important environmental parameters. Our present understanding of the atmospheric budgets of these pollutants indicates that anthropogenic emissions of CO_2 have already led to an increase in global CO_2 levels, while NO_x and SO_2 levels may be affected in the coming decades. While the climatic perturbation implied by a global CO_2 increase appears to be the most significant global pollution problem we presently face, the consequences of increases in NO_x and SO_2 upon the environment (i.e., acid rain) are also of concern. In the case of atmospheric particulates, the oxidation of S compounds to produce $SO_4^=$ aerosols in both the troposphere and stratosphere may ultimately lead to a significant degradation in visibility. Some simple techniques for monitoring visibility are described. One important method of monitoring aerosols and the deterioration of visibility on a global scale is the application of remote sensing with space technology. This technology is also applicable to the global monitoring of CO_2, SO_2, NO_x and the earth's radiation budget. The concluding section of the chapter illustrates this rapidly advancing technology.

Chapter 12, Solid Waste and Trace Element Impacts, by Bolch, considers the potential environmental impacts of the solid wastes and trace
element releases due to coal utilization for electric power. Coal may
contain a wide spectrum of trace elements including As, Cd, Ce, Cr, Cu,
Hg, Mn, Pb, Se, Sr, V, Zn, and naturally occurring radioactivity, especially the uranium and thorium series. A recent monograph by Torrey reviews the potential impacts from trace elements, and a companion monograph focuses on the recovery of these waste products as beneficial resources. Both monographs were published before the impact of the Resource Conservation and Recovery Act was reflected in the proposed hazardous waste regulations. Some of the topics not emphasized by Torrey
are more adequately covered in the recent impact statement on the Fuel
Use Act (DOE, 79).

During the development of the environmental regulations of the last
half of this decade, there has been an increasing emphasis on the less
obvious pollutants from fossil fuel power plants. The return to coal, a
defined national policy, results in the consideration of the ultimate
fate of trace elements in coal as a significant coal burning issue.

This chapter summarizes the trace elements in coal and fly ash, including the radioactive components of uranium and thorium. The fractionation of coal ash within a typical power plant is presented and the association of element with various fractions is discussed. A brief review
of potential health effects is presented. It is suggested that environmental transport and dose-to-risk models be developed and presented in
order to place coal burning on an environmental cost scale with other
energy sources. A brief discussion of the impact of mining, coal cleaning, storage and coal conversion are discussed. Lastly, the importance
of new laws, namely, The Toxic Substance Control Act (TOSCA) and its
companion, The Resource Recovery and Conservation Act, is discussed.

Chapter 13, Agriculture, by Woltz and Street, describes the environmental impact of increased coal usage on agriculture and natural ecosystems. The impact of coal residues on agricultural and forestry environments depends on the partitioning of these materials between bulk solid waste (90%) and released emission products (10%). Main environmental concerns are sulfur oxides released as atmospheric SO_2 and acid rain from sulfuric acid aerosols. These subjects have been well documented qualitatively but not so well quantitatively. The magnitude of the effects is assayed indirectly through appraisal of visible acute damage to plants. Settlement of specific damage claims for SO_2 effects requires a consideration of the pertinent features of etiology, environment and degree of susceptibility of the plant populations at risk.

Ecological effects of SO_2 are recognized in terms of genetic adaptation which takes place in individual plant species in response to airborne SO_2. Also, the makeup of populations by species in exposed areas represents a biological adaptation that occurs. Changes in genetic and species makeup of plant populations may or may not be desirable. Acid rain and acid mist from airborne sulfates and nitrates have been shown to have adverse effects on native and cultivated vegetation, soils, and aquatic ecosystems. These effects have increased in recent years. Trace elements from coal burning may impact adversely on agriculture and ecosystems but the effects will most probably be of a much lower order of magnitude than those of sulfur oxides.

There is a potential benefit from disposing of coal solid waste residue on agricultural and forestry soils insofar as the chemical and physical alterations of the soil are not detrimental to the production of quality food and fiber. Due to the high variability of the chemical and physical nature of coal ash and soil a compatibility must be ascertained before indiscriminatory use of coal ash on soils can be permitted. Preliminary studies indicate that certain plant essential elements contained in coal ash are readily available to plants grown on ash-amended soils. However, there are possible unfavorable changes in the soil upon addition of coal ash which must be investiaged.

Chapter 14, Air Pollution Health Effects, by Schlenker and Jaeger, deals with the possible health effects of air pollution resulting from large-scale coal utilization. The pollutants (primary ones such as SO_2, particulates, NO_2, CO, trace elements, and secondary ones such as O_3 and aerosols) are evaluated according to their physical, chemical and biological properties. To understand the health impacts of these pollutants pertinent epidemiological studies and controlled laboratory studies are presented in which animals and humans were exposed to various pollutants. Each technique has inherent advantages and disadvantages. Animal studies show primarily species-specific responses to exposures, but such studies are invaluable mirrors of biomedical and morphological changes which may occur as a result of exposures. For ethical and legal reasons, studies using human subjects are restricted in length of exposure and concentration of pollutant. However, the most important information obtained from such studies is the acute response by normal and sensitive (such as asthmatic) human beings to well-defined levels of single pollutants or combinations of pollutants. Epidemiological studies allow one to evaluate the effects of air pollution on large numbers of people over a lifetime. Such studies, unfortunately, are the most difficult to conduct.

Epidemiological studies can be subdivided according to the health effect considered. These subdivisions are mortality rates, morbidity rates, and cancer rates. Confounding factors such as location and number of monitoring sites, smoking habits of the population (another form of self-induced pollution, elaborated upon in this chapter), geographical seasonal variations, socio-economic factors, occupational exposures, and migration of individuals make interpretation of epidemiological studies difficult. Taking all these factors into consideration, in many cases the measurable relationship of air pollution to health is decreased. There is evidence, however, that sulfur-dioxide, sulfates and particulates, major products of coal combustion, have some detrimental effects on the health of children and adults; particularly sensitive subjects react to a larger extent.

Chapter 15, Quantitative Public Policy Assessments, by Green and Rio, reports the overall approach and principal results of a study (ICAAS, 1978) which attempted to carry out an integrated interdisciplinary assessment of air pollution abatement alternatives in the Tampa area of Florida, a region where coal is a major source of electric power. ICAAS's initial concept of public policy decision methodologies (PPDM) for regulating air pollution began in connection with a proposed Air Quality Index (AQI) Project (ICAAS, 1970). This AQI project was designed to be a broad socio-technical research program leading to the establishment of a quantitative scale for air quality. From this AQI program plan, we developed our first practical cost/benefit analysis approach (ICAAS, 1971). These two overall systems approaches were implemented in the ICAAS-FSOS study (ICAAS-FSOS, 1978) which involved a chain of component studies within the framework of two types of public policy decision methodologies on sulfur oxide pollution. One methodology, the Disaggregated Benefit/Cost Analysis (DB/CA), was essentially an advanced form of economic analysis in which the distributional aspects of B/C are considered (i.e. the question of who gets the benefits and who pays the costs). The second methodology, the Quantitative Assessment of the Level of Risk (QALR), is a non-economic analysis that by-passes many of the difficult problems of translating all important decision factors into monetary terms.

This chapter concentrates on a few specific but vital facets of these system approaches. In particular we discuss quantitative characterizations of ambient air quality from air pollution health effects viewpoint. We also describe a "factor of safety" approach to quantitative dose response relations based on the use of an air quality index (AQI). These are followed by the application of the QALR-PPDM used in the ICAAS-FSOS study which also uses an AQI. We also describe the essence of the DB/CA-PPDM used in the ICAAS-FSOS study. Finally, we describe some recent dose response results for plants and materials which can be used in PPDM's.

Chapter 16, <u>Financing Capacity Growth and Coal Conversions in the</u>
<u>Electric Utility Industry</u>, by Brigham and Gapenski, considers first the
demand for electric power which is expected to increase greatly from
1980 to 1995. To meet this demand, and also to finance the conversion
of existing oil and gas fired plants to coal, the utility industry must
raise and invest unprecedented large sums of money.

At the start of the 1960s, the electric power industry was the
epitome of financial strength, with a virtually unlimited ability to
raise funds. Today, however, the average company is so weak financially that it simply cannot meet its capital requirements. This deterioration was caused by a combination of economic and political
factors--inflation, both general and in fuel prices, is the root cause
of the industry's problem. Cost increases have outstripped productivity
gains, which has squeezed profits, necessitating rate increases. However, since utility prices are set by regulatory commissions, time lags
are inherent in obtaining rate relief. If costs rise but prices can be
increased only after a regulatory delay, then obviously profit margins
are squeezed, rates of return on invested capital decline, and the company's financial position suffers.

The situation is masked by the fact that the companies now have
excess capacity that arose from the sudden, sharp reduction in growth
after 1973. Since this excess capacity has permitted the companies to
survive and meet current power demands, the public has not yet suffered
to any significant extent. However, excess reserves will soon be used
up, so that if construction programs have not been started up well in
advance, power shortages will follow, accompanied by severe economic
problems. It is possible, however, to avoid capacity shortfalls. What
is needed is for utility commissions across the country to realistically analyze the situation and then to allow the utility companies to
charge prices that cover the cost of providing service, including the
cost of the capital invested in the plant that provides the service.

Chapter 17, <u>Coal and the States: A Public Choice Perspective</u>,
by Rosenbaum, discusses the Coal Burning Issues problem from the perspective of a political scientist. The development of a national coal
policy will depend heavily upon the states for implementation. The
generous discretionary authority exercised by the states in implementing
federal coal policy, together with traditional state powers affecting
coal use, means the states will be major actors in any national coal-
management program. In general, state activities will affect coal utilization through the siting of mines, the siting and design of power
generating plants, the location of coal logistical facilities, the
enforcement of air and water pollution standards, and much else.

Formulating state coal policy confronts the state governments with
competing, and sometimes conflicting, policy objectives and policy
priorities. Among the major policy choices that must be resolved by
the states are: (a) the relative importance of environmental protection
among other policy objectives in coal use; (b) the priority to be
accorded economic growth in coal development; (c) the distribution of
social costs and benefits from coal development--the "distributive
equities"; and, (d) the relative priority to be given state or regional
interests compared to national ones in choosing coal policy goals.

While the federal government cannot, and should not, attempt to
resolve all these matters at the state level, federal coal policy can
constructively assist the states in resolving these issues. In particular, the federal government should restrain massive new coal utilization to a few decades at most, should create a target growth rate figure
for the U.S., should emphasize conservation of energy in the coal sector
by dampening demand for new electric generating facilities and should
encourage great public involvement in coal policy formulation among the
Western states where the environmental risks of coal development are
especially acute.

Chapter 18, <u>Federal Regulatory and Legal Aspects</u>, by Little and L.
Capehart, examines federal laws which simultaneously promote and constrain the increase in coal use as an alternative fuel to natural gas
and petroleum. Congressional desire to encourage or require coal use is
evident as early as 1974 with passage of the Energy Supply and Environmental Coordination Act which set mandatory coal conversion requirements
for power plants with conversion capability. More recent laws have expanded federal authority to require the use of coal by major fuel
burning installations.

Congress has recognized that an increase in coal use will be possible only if federal assistance is available to weak links in the coal
production-transportation-combustion chain. To date, some limited
financial help has been authorized for developing new underground mines,
for rehabilitating the rail system, and for buying air pollution control
equipment. At the same time, Congress has been standing firm on environmental laws, trying to protect the health of the public by preserving
air and water quality standards. Clean air legislation has come in for
the biggest attack because the control of emissions from coal combustion
places large costs on any coal burning facility. The waste disposal
laws may also create additional financial burdens if coal wastes are
termed hazardous.

Reform of other federal legislation may be appropriate if the goal
of increased coal use is to be met. Regulation of the transportation
industry should be examined to determine inequities in the laws which
favor one mode over competing modes. Examples are unequal federal
subsidies to various modes, and varied formulae for determining rate
structures.

Changes in mining laws have been advocated. The mineral leasing
program of the federal government has been criticized by almost everyone.
The mine safety legislation has been accused of decreasing productivity
without a proportionate increase in worker safety. Changes in federal
law may be necessary to achieve the goal of increased coal use, but such
changes should be carefully examined to determine the side effects as
well as the expected result. Furthermore, proposed changes should be
considered in light of interrelating laws and problems.

CHAPTER 2

COAL AVAILABILITY AND COAL MINING

M. J. Ohanian and Faramarz Fardshisheh

I. THE CHARACTERISTICS OF COAL

Coal deposits have quite variable characteristics, depending upon the original topography and water movement during its formation. The various types of coal are traditionally classified by a number of para- meters, including heating value, ash content, moisture content, sulfur content, and the division of the organic portion of the coal into fixed carbon and volatile matter. The four main ranks of coal are anthracite, bituminous, sub-bituminous, and lignite. Table 1 lists some character- istics of coal in the United States.

The American Society for Testing and Materials (ASTM) rank is based on the degree of lithification and metamorphism of plant material (DOE 1979). According to this system, rank is determined primarily by the percentage of fixed carbon and the heat value (BTU content) of the coal, calculated on a mineral-matter-free basis. A more detailed discussion of the energy quality and other properties of coal based upon the evolu- tionary history of coal formation as interpreted from an energetic analysis perspective is given in Chapter 3.

As described in Chapter 1, coal is the most abundant non-renewable energy resource in the world, and it comprises the major fraction of the estimated energy reserves of coal, petroleum and natural gas. The United States' share of these recoverable coal reserves is estimated to be about 28% or 4,900 quads (Peters and Schilling, 1978). (The present rate of U.S. energy consumption is ∿78 quads, of which 18% is supplied by

33

Table 1: *Characteristics of Coal (USEPA, 1977, Torrey, 1978)*

Type of Coal	BTU/lb.	Fixed C	Moisture Content	Ash	Sulfur
1) Anthracite 96% found in Pennsylvania	13,000–15,000	85–90%	2–5%	Varies from 5–20% by weight for all types	Varies from 0.2–7% by weight for all types
2) Bituminous found in East & Central U.S.	12,000–15,000	45–78%	5–15%		
3) Sub-bituminous found in Western U.S.	8,000–11,000	37–45%	18–35%	---	---
4) Lignites North Dakota and Montana	6,000–7,500	25–30%	40%	4–8%	0.2–1.4%

coal.) Table 2 summarizes, by general geographic location, the world's estimated recoverable reserves of coal. The recoverable reserves are that fraction of the total estimated geological resources (the resource base) which is considered to be technically and economically recoverable under present conditions. Typically, the presently recoverable portion of the resource represents 1/10 to 1/15 of the total resource base. In time, as energy costs increase and new extraction technology may be developed, increasing fractions of the total resource base may become available.

II.   GEOGRAPHIC DISTRIBUTION OF COAL IN THE UNITED STATES

The specific geographic locations of the continental U.S. coal resources are shown in Figure 1. Generally these fall into three regions: Eastern, Central and Western. In Table 3 the estimated identified resources by state to a depth of 3,000 ft. are given for each of the three regions. These resources total 1,580 billion tons of which 685 billion tons are bituminous coal, 424 billion tons sub-bituminous coal, 450 billion tons lignite and 21 billion tons anthracite. In addition to

Table 2. *Estimated Recoverable Reserves of World Coal*
(Peters and Schilling, 1978; Loftness, 1978)

	quads	%
Western Hemisphere		
United States	4,933	28.0
Canada	260	1.5
South America	304	1.7
	5,497	
Eastern Hemisphere		
Africa	948	5.4
Europe	3,555	20.0
U.S.S.R.	3,053	17.3
China (PR)	2,746	15.5
Other Asiatic	1,114	6.3
Oceania	764	4.3
	12,180	
TOTAL:	17,677	100.0

Note: Total world geological resources of coal are
estimated to be 282,000 quads of which 71,000
quads (25%) are in the U.S.

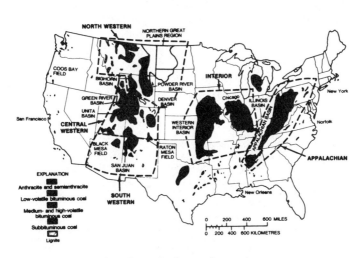

Fig. 1. *U.S. Coal Supply Regions*

Table 3: *Estimated Identified Coal Resources
in the United States (Jan., 1972)
(Million Short Tons)*

	B[(a)]	SB[(a)]	L[(a)]	A[(a)]	Total
Eastern					
Alabama	13,342		2,000		15,342
Kentucky-East	32,421				32,421
North Carolina	110				110
Maryland	1,158				1,158
Ohio	41,358				41,358
Pennsylvania	56,759			20,510	77,269
Tennessee	2,572				2,572
Virginia	9,352			335	9,687
West Virginia	100,628				100,628
Total:	257,700		2,000	20,845	280,545
Central					
Arkansas	1,638		350	430	2,418
Illinois	139,124				139,124
Indiana	34,573				34,573
Iowa	6,509				6,509
Kansas	18,674				18,674
Kentucky-West	32,421				32,421
Michigan	205				205
Missouri	31,014				31,014
Oklahoma	3,281				3,281
Total:	267,439		350	430	268,219
Western					
Alaska	19,413	110,668			130,081
Arizona	21,246				21,246
Colorado	62,339	18,242		78	80,659
Montana	2,299	131,855	87,521		221,675
New Mexico	10,752	50,671		4	61,427
North Dakota			350,630		350,630
Oregon	50	284			334
South Dakota			2,031		2,031
Texas	6,048		6,824		12,872
Utah	23,541	180			23,721
Washington	1,867	4,190	117	5	6,179
Wyoming	12,705	107,951			120,656
Total:	160,260	424,041	447,123	87	1,031,511
Grand Total:	685,399	424,041	449,473	21,362	1,580,275[(b)]

(a) B = Bituminous   (12,000 - 15,000 Btu/lb)
     SB = Sub-bituminous (8,000 - 11,000 Btu/lb)
     L = Lignite   (6,000 - 7,500 Btu/lb)
     A = Anthracite   (13,000 - 15,000 Btu/lb)

(b) There is an additional 688 million short tons distributed
     among California, Idaho, Nebraska, Nevada, Louisiana and
     Mississippi

(c) Identified resources: identified deposits that may or may
     not be evaluated as to the magnitude and quality and which
     may or may not be economically and technically recoverable

(d) The above are for 0 - 3,000 ft. overburden

(e) Source: U.S. Geological Survey

these known (but not necessarily all recoverable) resources, there are
1,643 billion tons of hypothetical resources (i.e., geologically predictable but undiscovered), for a total U.S. coal resource base of 3,223
billion tons, which compares to the resource base of 71,000 quads in
Table 2, using an average heat content of 11,000 Btu per pound of coal.
As noted in Table 2, of the total resource base, 4933 quads or 225 billion tons of coal is presently considered recoverable. Of this amount,
about 30% can be mined at the surface, the balance is underground.

In Table 4a the percentage distribution of coal by type in the three
major geographic regions is shown. Bituminous coal predominates in
the Eastern and Central regions and sub-bituminous coal and lignite in
the West. While about 65% of all the coal in the United States is
found in the West, at present almost 71% of the coal production is in
the East.

It is important to note that, while Western coal has lower sulfur
content (see Table 4b), it also generally has a significantly lower heating value. Therefore, as noted by Schmidt and Hill (1976), the low-sulfur
advantage is mostly compensated by the requirement of burning more of
it to make up for the lower Btu content. In effect, then, the net quantity of low sulfur coal in the United States is reduced.

Table 4a: *Percentage Distribution of Coal by Type*

	$B^{(a)}$	$SB^{(a)}$	$L^{(a)}$	$A^{(a)}$	Total
Eastern	16.3	–	0.13	1.3	17.73%
Central	17.0	–	0.02	0.03	17.05%
Western	10.0	27.0	28.3	0.006	65.3%
Total:	43.3	27.0	28.45	1.34	

(a) B   = Bituminous
   SB = Sub-bituminous
   L   = Lignite
   A   = Anthracite and semi-anthracite

Table 4b: *Percentage Distribution of Coal by Sulfur Content*

	0 - 1.0%	1.0 - 2.0%	2.0 - 3.0%	3.0 - 4.0%	< 4.0%
Eastern	5.0	4.5	5.3	1.5	–
Central	–	1.5	3.0	12.4	7.0
Western	57.0	2.8	–	–	–
Total:	62.0	8.8	8.3	13.9	7.0

Note: The regional totals in Tables 3a and 3b do not
agree exactly because the data sources are not
identical

Source: Adapted from (Loftness, 1978)

III. DOMESTIC DEMAND

The domestic demand for coal, by consuming sector, is summarized
in Table 5 for the period 1976 through 1979. After a 3.5% growth in
demand during 1976-77, demand leveled off for two years and now is on
the upswing again with a projected growth of 6-13% in 1979. (In the
first half of 1979 coal consumption was about 13% higher than during
the same period in 1978.)

Table 5: *Domestic Demand for Coal (Monthly Energy
Review, Sept., 1979) (million short tons)*

	1976	1977	1978	1979[a]	1983[b]
Electric Utilities	448.4	477.1	481.3	515–550	615–685
Coke Plants	84.7	77.4	71.4	77–86	70–78
Other Industry	61.8	61.6	60.5	58–60	85–125
Residential/Commercial	8.9	9.2	10.4	9–10	
Total:	603.8	625.3	623.6	659–706	770–888
Production	684.9	697.2	660.2	750–800	812–940

[a] Estimated from data for first half of 1979

[b] The Energy Daily, October 3, 1979; Source: National Coal Assoc.

The major user of coal is the electric utility sector which has increased its share from 74% of coal demand in 1976 to 78% today. With the
planned further increases in coal utilization by the utility sector, this
trend will continue and is projected to reach 80% by 1983. The demand
for coal by the other consuming sectors has generally stayed level and
no major shift in these trends is expected. The electric utility sector
will continue to dominate coal consumption for the next few decades.

As is shown in Table 5, domestic coal production has more than kept
up with the demand, with the excess capacity (mostly bituminous and lignite) being exported.

IV. MINING

Mining methods for the extraction of coal from coal bed (seams) are,
in general, categorized into two groups: surface (strip) mining and underground (deep) mining. The two methods are applicable under quite different geological settings, and there are several mining techniques associated with each method. A techno-economic study taking into account all
of the factors including safety considerations ultimately helps to determine the method and the technique to be used for a given mine.

In this section a summary discussion of modern mining techniques and
the factors that give rise to the use of a particular methodology are presented. The information which is implicit in the description should enable one to envisage the qualitative implications of a relatively high
rate of growth in the extraction of coal per annum (note that this rate
for 1979 could be as high as 13%). Since safety plays an important role,
the problems related to safety are also pointed out.

A. Surface Mining

Nearly all surface mining consists of stripping away the overlying
soil and rocks (overburden) and extracting coal from the coal seam thus
exposed. This procedure is referred to as strip mining. In hilly terrain where the slope of the overburden is too steep, auger mining is utilized. In this case a huge drill with cutting heads up to 7 feet tall
and capable of driving into the coal seam up to 200 feet is employed.

Surface mining is usually limited to coal seams within 200 ft. of the earth's surface. Until the last decade surface mining was considered sound for coal seams where the overburden-to-seam thickness ratio was not higher than 10:1. Nowadays, coal within 150 feet of the surface may be economically recoverable even when the overburden-to-seam thickness ratio is as high as 30:1 (The Direct Use of Coal, 1979). At present 60 percent of the coal produced in the United States comes from surface mines.

Most strip mining follows the same basic steps, but the equipment used may vary to produce a safe and optimum performance. The steps and equipment used are:

1. Topsoil removal and leveling using bulldozers, loaders, and trucks.

2. Overburden removal, using blasting (if necessary) and scooping out with bucket wheel excavators, stripping shovels, bulldozers, or draglines. Walking draglines can be as tall as a 20-story building and have a bucket up to 200 cubic yards and boom lengths of up to 375 feet. They can remove up to 3500 metric tons of overburden per hour. Sometimes up to 40 tons of overburden may be removed for every ton of coal mined.

3. If the coal cannot be scooped up directly, it is first blasted by explosives. A coal-digging machine or a shovel is used to load the coal into giant trucks. The trucks usually take the coal to a nearby processing facility or a conveyor belt leading to one. After crushing and cleaning, the coal is stored at the tipple for transport to the market.

There are three techniques used in strip mining: (1) Area mining, (2) Open pit mining, (3) Contour mining.

Area mining (Fig. 2), the major strip mining technique used on Western and some Midwestern coal lands, is practiced where the terrain is relatively

level and the coal seam is parallel to the surface. It involves the development of open pits in a series of long narrow strips (usually about
100 feet wide by a mile or more long) (The Direct Use of Coal, 1979).
The mining begins with the cutting of a trench across the end of the
strip using an excavator or a dragline. The top soil is stored for reclamation and the remaining overburden is piled alongside the cut on the
side that is away from the mining area to form a ridge called a spoil
bank. A loading shovel then loads the coal (blasted out, if needed) to
short haulage trucks. The dragline or excavator is moved to the next
strip to dig a new trench parallel to the first while transferring the
overburden to the mined-out cut. This cycle is then repeated to the
limit of the mine boundaries.

Open-pit mining is similar to area mining except that it involves the
preparation of a large rectangular area, say 1000 by 2000 feet. The
overburden is moved around in the pit to uncover the coal seams for
scooping. This technique is primarily used for the very thick seams
which occur frequently in the Western mines.

The contour strip mining technique (Fig. 2) is used where a seam "out-
crops" (is exposed) all around a hill. The technique consists of removing the overburden above the bed by starting at the outcrop and proceeding along the contour of the bed in the hill side. After the seam is exposed and removed by the first cut, additional cuts are made into the
hill until the ratio of overburden to seam thickness brings this technique to its limit. Here, auger mining may be employed to extract the
coal which remains within the depths along the high wall of the hill.
If a seam lies near the top of the hill, earthmoving equipment may simply
remove the hill top and thus expose the coal.

    Restoration of the Land: The primary environmental problems caused
by strip mining result from fertile topsoil being buried under piles of
rock, which tend to give off acids when exposed to moisture. Thus rainwater running down the bare slopes carrying acids and mud with it can
cause irreparable damage to the neighboring areas and rivers. In 1977
the U.S. Congress passed a law requiring all the land being used for
strip mining after 1978 to be reclaimed as nearly as possible to its

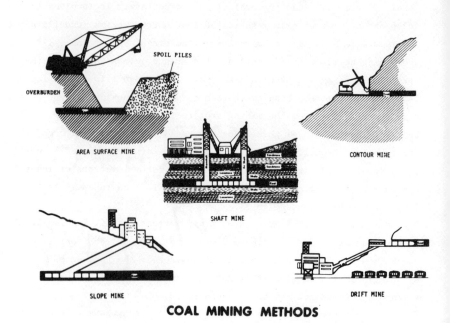

COAL MINING METHODS

Fig. 2.   Surface and Underground Mining Methods (Coal Data, DOE, 1978)

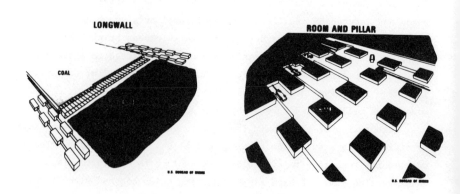

Underground Mining Systems

Fig. 3.   The Two Underground Mining Systems (Coal Data, DOE, 1978).

original condition. The steps involved in reclamation are:

1. Backfilling the overburden, generally simultaneously with removing it from a fresh area, and grading and compacting it, usually
with bulldozers.

2. Topsoil is returned and spread.

3. The area is revegetated.

B. Underground Mining

Underground mining is considerably more hazardous and complex than
surface mining. Underground miners must work with overburden above them
and tackle problems involving safety such as roof support, lighting,
ventilation, methane gases, drainage, equipment access, and coal transport. Underground mining requires more human labor than surface mining
although all operations are mechanized.

In most cases, miners start by digging two access passages from the
surface to the coal seam, one to be used by the miners and the other
for hauling out the coal, and both serving for circulating the air.

Underground mines consist of three types depending on the angle at
which the passages are dug: (1) Shaft mine (passages straight down),
(2) Slope mine (passages at a slant), (3) Drift mine (passages along the
seam). The selection of location and type of passages depends on a particular site. From these passages, parallel entries are driven into the
coal to provide corridors for haulage, power, ventilation, etc. Cross-
corridors then reach to the sides of the mine.

The two main systems of underground mining used are: (1) Room and
pillar system, and (2) the longwall system.

The Room and Pillar System involves leaving pillars of coal standing in
a mine to support the overburden (Fig. 3). The deeper the mine the
bigger the pillars must be relative to the mined-out areas. In some
cases less than 50% of the coal can be removed. In such cases it may be
prudent to mine the pillars as the equipment retreats towards the main
corridors and the roof allowed to collapse. There are two methods used
for the Room and Pillar system: conventional mechanized mining and continuous mining. In the last quarter century *the continuous method* has

increasingly replaced *the conventional mechanized one. The conventional mechanized* method involves five main steps: (1) A huge chain saw protruding from the bottom of a self-propelled vehicle cuts a deep slit at the base of the coal face. The undercut slit may be 6 inches high, 10 feet deep, and perhaps 20 feet across the face. (2) A self-propelled vehicle with a long auger attached to a movable boom drills holes in the face of the coal. (3) Each hole is loaded with explosives or compressed gas and blasted. (4) A machine loads the coal onto a conveyor or a shuttle car. (5) A bolting machine drills holes in the roof and inserts anchor bolts which firmly attach the roof to stronger overlying layers of rock. Bolts are generally required on a 4 x 4 foot array。 This is one of the most dangerous activities in the underground mine. Exposed areas are rock-dusted to prevent coal dust explosions. In all seams frequent methane testing is required at the face. Some machines must be hooked to a water supply for a spray to control dust.

The continuous method was developed to combine the first four operations of conventional mechanized mining and thus further increase productivity. The continuous miner gouges the coal from the face and automatically loads it onto a shuttle car or a conveyor belt. Haulage of coal away from the face is a major constraint; this together with withdrawal of the machinery to allow for bolting of the roof and extending of the ventilating duct hampers operation and limits the continuous miner's activity to 20-30% of the available time.

The Longwall System: This system is different from the room and pillar system and has long been used in Europe and is now gaining increasing popularity in the U.S. Corridors 300 to 700 feet apart are driven into the coal and interconnected. The longwall of the interconnection is mined in slices. The miners move a cutting machine back and forth across the face plowing or shearing off the coal which falls onto a conveyor belt and is transferred to the main corridor. The roof is held up by steel jacks while the cutter makes a pass across the face. To make a new pass, the roof jacks are advanced with the shearer. The roof is then allowed to collapse in the mined out area. Practically all the coal can be extracted by this process。

Ventilation is a major problem in modern mines because the
highly productive equipment produces high levels of dust and methane.
Dust is reduced by water sprays at the working face. Methane must be
diluted with air to less than flamable levels to avoid the possibility
of underground explosions. In a few mines, methane is removed from
the coal seam before the mining operation begins.

C. Coal Preparation

Most mining companies have a preparation plant as an integral part
of the mine-to meet the customer's needs in terms of size, moisture,
mineral concentration, heat content, etc. In some cases the coal is
only crushed. About half of the coal produced in the U.S. is cleaned
to reduce ash, sulfur and such mineral impurities. Above and beyond
the environmental concerns, the reduction of the inert materials improves
the heat content of coal and reduces deposits in furnaces and also re-
lieves the load on filtering and particulate removing equipment. Coal
washing also reduces the sulfur and the trace elements.

Crushing is performed on nearly all bituminous coal to provide a
product of uniform size and also to increase the exposure of impurities
for removal. In 1974, of the 374 million tons of coal which was cleaned,
a net production of 267 tons of processed coal and 107 million tons of
refuse resulted.

The steps involved in coal preparation are as follows:
1. Crushing to reduce impurities and provide a product of desirable
   size.
2. Sorting to ensure uniformity of size using screens or fluid classification.
3. Washing to remove impurities often using the jigging technique.
   The devices separate the impurities by means of specific gravity.
   A bed of raw coal is stratified in water by pulsations that move
   lighter coal particles to the top and heavier refuse to the bottom.
   This drops to a refuse bin. Another process is the dense-medium
   technique where coal is immersed in a heavy fluid in which the coal
   floats and the refuse sinks. The dense-medium cyclone is a variant
   of this technique that uses centrifugal force to assist in the sep-

aration. Froth flotation and pneumatic techniques are also used
for cleaning purposes.

4. Drying and dewatering. Washing leaves the coal wet and, in order
to reduce the weight for shipping and alleviate problems of handling and combustion, dewatering or drying is utilized. Dewatering
is effected with screens, filters, and centrifuges of various
types. Drying on the other hand uses hot air to remove the water;
and, in addition to being relatively expensive, it can cause air
pollution problems.

5. Water clarification is the process of removing enough suspended
particles of coal from the water to permit it to be recycled.
For this purpose settling tanks are generally used.

D. Mining Safety and Health

Since 1900, more than 100,000 fatalities have occurred in mining
accidents in the U.S. The number of disabilities and injuries are much
more (J. Leonard, 1979). The improvements in mine safety have greatly
reduced the death rates from mine accidents since the enactment of the
1969 Federal Coal Mine Health and Safety Act. In the early 1900's,
about 3-5 miners per 1000 were killed in mine accidents annually; this
rate has dropped to about 0.5 per 1000 today. The largest improvement
has been in the more hazardous underground mines. The underground mines'
fatality rate now is 0.35 deaths per million employee-hours, which is
about half of 0.84, the figure for 1969 (Lansberg et al., 1979). However, no appreciable reduction in the rate of disabling injuries has
occurred and the total number of injuries has been increasing. The
Federal Act particularly addressed some work-related health and safety
hazards arising from explosions and dust control.

Miners are likely to suffer from occupational disease, injury, and
death at a rate higher than other occupations. Table 6 delineates this
fact for the energy sector.

The threat to personal safety is greatest in the underground mines
and caused by roof, rib, and face falls. Underground coal mining is
more dangerous than surface mining with an injury frequency rate of 4:1.
The data in Table 7 elucidate on this statement. This data also allows

a glance comparison with an overall industry average figure.

Table 6: *Occupational Hazards by Energy Sector*
for 1975 (Surles et al., 1979)

Sector	Deaths	Injuries
Mining	164	10,237
Processing	6	340
Transportation	214	3,330
Electrical Generation	15	509

Table 7: *Coal Mining Accident Rates (EPA, 1977)*

Type of Mining	Disabling Injuries per Million-Employee-Hours
Underground	35.0
Surface	10.0

Overall Industry Average	9.8

(All Member Companies of National Safety Council)

The percentage and types of coal mining accidents are presented in
Table 8.

Table 8: *Types of Coal Mining Accidents (EPA, 1977)*

Type of Accident	Percentage
Roof, rib, and face falls	50
Fires and Explosions	10 - 12
Coal Haulage	10 - 15
Underground Total ∿	80
Fall of high wall, equipment Misoperation, electrical system malfunction	20
Surface Total	20

Technological development has been able to partially remedy some safety
aspects, regretably not without some ensuing problems. For example,
the continous mining method raised the productivity greatly and lowered
fatality (mainly due to reduction of work force by 70%) but has caused
a higher incidence of respiratory diseases collectively called black
lung diseases. Also long wall mining appears to be more productive and
safer in terms of fatalities, but not in injuries. Therefore, labor-
saving technology may or may not improve safety. Training of the mining
employees for safe procedures including the safe operation of the equipment with effective control and inspection can greatly reduce hazards.

Occupational health is a major problem in the coal mining sector.
Black lung disease, a non-clinical name for a variety of respiratory
illnesses of which coal workers' pneumoconiosis (CWP) is the most promi-
nent, is the primary problem. Between 1970 and 1977, 420,000 Federal
compensation awards were made costing the government 5.5 billion dollars. Due to the 1977 black lung legislation (amendments to the 1969
Act) mining industry will pay more of the compensation in the future.
At present about 10% of the mining work force show X-ray evidence of
CWP and perhaps twice that number show other black lung diseases.
Exposure to nonrespirable dust and other toxic substances is neither
measured nor regulated. Hearing loss is another hazard.

References

Congress of the United States, Office of Technology Assessment, The direct use of coal,
   prospects and problems of production and combustion, Washington, D.C., April 1979.
Department of Energy, U.S., Energy Information Administration, Monthly Energy Review, DOE/EIA-
   0035/9/79, Washington, D.C., September 1979.
Department of Energy, U.S., Fuel use act, final environmental impact statement, Washington,
   D.C., April 1979.
Environmental Protection Agency, U.S., Office of Research and Development, A summary of
   accidents related to non-nuclear energy, EPA-600/9/77-012, Washington, D.C., May 1977.
Environmental Protection Agency, U.S., Office of Research and Development, Potential radio-
   active pollutants resulting from expanded energy program, EPA-600-7-77-082, Washington,
   D.C., August 1977.
Lansberg, H. H., et al., Energy: The next twenty years, A report sponsored by the Ford
   Foundation, Ballinger Publishing Company, Cambridge, Mass 273-408, 1979.
Leonard, J. W., Coal, The World Book Year Book, World Book-Childcraft International, Inc.,
   Chicago, Ill., 566-582, 1979.
Loftness, R. I., Energy Handbook, Van Nostrand Reinhold G., New York, 1978.
Peters, W. and H. D. Schilling, An appraisal of world coal resources and their future avail-
   ability, World Energy Resources 1985-2020, World Energy Conference, 1978.
Schmidt, R. A. and G. R. Hill, Coal: Energy keystone, Annual Review of Energy, Volume 1,
   Annual Inc., Palo Alto, California, 1976.
Surles, T., J. Gaspar, and J. J. Hoover, An assessment of national consequences of increased
   coal utilization, U.S. Department of Energy, Washington, D.C., 1979.
Torrey, S., Trace contaminants from coal, Pollution Technology Review No. 50, Noyes Data
   Corp., N. J., 1978.

CHAPTER 3

AN ENERGETICS ANALYSIS OF COAL QUALITY

John F. Alexander, Jr., Dennis P. Swaney, Ralph J. Rognstad, Jr., and Robert K. Hutchinson

I. INTRODUCTION

Coal and lignite make up 89% of the world's recoverable fossil reserves when compared in thermal energy units (Hubbert 1971). Currently many nations are seriously considering the transition from an oil-to coal-based economy. Public policy and private investment decisions are being made by directly comparing coal reserves to oil reserves in heat equivalent energy content. In this chapter the question of the energy quality of coal and the possible net energy yields from coal-use scenarios are examined and evaluated. Energy circuit models and concepts developed by H. T. Odum (1971) are presented in a holistic assessment methodology to guide decisions regarding utilization. The energy quality and net energy of coal are addressed from three distinctly different points of view: (1) a theoretical model is presented which shows the energy quality of coal as a function of the degree of ecological and geological energies concentrated in the formation process; (2) the energy quality of coal is compared to that of alternative fuels, wood and oil, by evaluating the cost and efficiency of converting each fuel to a common-end product—electricity; and (3) the quality of coal is compared to the quality of synthetic gas by evaluating the coal to gas conversion process.

A. Coal Classification

Scientists and engineers have long recognized significant
differences between coal and other fossil fuels and among various types
of coal. This fact is born out by the predominantly qualitative coal
classification schemes summarized in this section. It should be noted
that qualitative classification systems are necessary in order to
estimate and understand world energy reserve potentials but are not
adequate for comparisons between resources (e.g., between coal and other
fossil fuels).

Coal has been the subject of several conflicting schemes of
classification, including those based on content of energy, volatile
matter, oxygen, ash residue and on physical properties such as swelling,
caking, coking, and grindability. Three useful measures emerge from the
confusion: rank, type, and grade.

The *rank* of a particular coal species is primarily dependent on the
carbon content of its dry, ash-free portion. Rank increases from
lignites through the classes of bituminous coals to anthracite, which
has the highest percentage of carbon. Because carbon content increases
with the exposure of the coal bed to the coal-forming forces, rank is an
excellent measure of degree of coalification of a coal species.

Coal *type* is a classification of the parent plant materials found
in coal based on the relative proportions of organic minerals, or
macerals, which are derived from the specific plant parts from which the
coal was formed. Vitrinite, a maceral formed from the woody parts of
ancient trees, is the predominant maceral. Others include alginite
(from algae), sporinite (from plant spores), resinite (from resins), and
sclerotinite (from fungi). The proportions of these components can be
used to infer the environmental conditions present in the parent
ecosystems of coal.

Finally, coal *grade* is a measure of the relative purity of coal
from contaminants such as ash, sulfur, and other materials. Most ash
residue results from inorganic materials mixed with layers of peat at an
early stage of formation. Sulfur content depends upon the degree of

anaerobic decomposition in the swamp as well as the predominant plant species present during the deposition of the parent material. Thick seams of low-ash, low-sulfur coal are high grade and represent either long periods of uninterrupted peat deposition or shorter but more highly productive periods. Poorer grade coal may indicate interference with the ecological process of the parent swamp, due to local meterological or geological catastrophes. Riverine washouts, mud inundations, and siltation processes are represented as disruptions and intrusions in the coal seam. Coal grade, then, is a measure of the economic value of coal and is dependent upon both the rank and type of coal. The economic worth of coal, can thus be seen as a function of the ecological conditions prevalent over the period of deposition of coal parent materials, and the subsequent geological conditions of its transformation (Neavel 1977).

II. METHODS

A. Description of Modeling Language and Symbols

The symbols used in the systems diagrams were established by H. T. Odum (1971) and are part of an energy circuit language. The language combines several techniques which simultaneously illustrate the energetics of a system and provide insight into its mathematical description. Energy circuit language contains a hierarchy of symbols, which allow the diagraming of several levels of complexity in one model (Fig. 1).

Energy circuit models are constructed by arranging production function and storage symbols within group symbols to show systems and to facilitate comprehension of the concepts embodied in the model. Standard convention also calls for arranging the model so that subsystems that concentrate dilute energy are at the left, while energy flows increase in quality toward the right of the diagram. Heat sinks are used to show the loss of degraded energy from the system. The creative use of these symbols can provide several levels of hierarchy. The highest level is formed by arranging the diagram by energy concentration or quality, the intermediate level by the use of group

ENERGY SOURCE

The circle indicates a source of energy from outside the system under consideration.

HEAT SINK

The arrow pointing downward, seemingly into the ground, symbolizes the loss of degraded energy (that is, energy that cannot do any more) from the system.

ENERGY STORAGE

The tank indicates a storage of some kind of energy within the system. The symbol could indicate energy stored in an elevated water tank, in the manufactured structures of a building, or in any way that makes it (energy) more valuable.

ENERGY FLOW

Lines indicate a flow of energy or energy containing materials.

ENERGY INTERACTION OR TRANSFORMATION

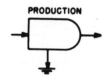

The pointed block is used to show the interaction of two or more types of energy required for a process. For example, on a farm sunlight would interact with water, soil, machinery, and stored structure to produce food.

PRODUCTION

The bullet-shaped symbol indicates the processes, interaction, storages, etc. involved in producing high-quality energy from a dilute source like sunlight. It is used in large-scale diagrams to denote entire ecological systems such as forests, swamps, estuaries, and farms.

SELF-MAINTAINING ACTIVITY

The hexagons indicate subsystems that have self-contained internal storages, energy flows, and interactions. Such subsystems can also be considered consumers because they require high-quality energy for their operation and maintenance but do not produce it directly themselves. Examples include organisms, industries, and towns.

Fig. 1 Energy circuit language symbols

symbols, while the level of dynamic and kinetic detail is described with
mathematical symbols.

B. Energy Quality and Embodied Energy

Energy quality theory is built on the laws of thermodynamics. The
law of energy conservation provides the basic underpinnings for all
energy circuit diagrams. In a first-law energy diagram, the average
heat equivalent stored or flowing per unit of time is used to evaluate
the model. When evaluated in heat equivalents, all systems of man and
nature conserve energy, that is, the sum of all the flows into a system
minus the energy flowing out must equal the net changes in the energy
storages. The system models in this chapter meet this first law
requirement. Unfortunately, heat equivalent measures provide little
information regarding the potential value of an energy flow or storage
for performing work or controlling a system. This is especially true
when comparing resource potential, such as among various ranks of coal
or between coal and oil.

To assign a value to a particular energy resource such as coal the
law of energy degradation must be used. This law states that in all
useful processes some energy must be degraded and thus lose its ability
to do work. The concept is incorporated into energy circuit language by
requiring heat sinks or energy degradation pathways on all energy flows,
processes, and storages. The quantitiy of energy dissipated by each
energy concentration process is also evaluated in heat equivalents in
energy circuit diagrams, even though these energy pathways can do no
useful work.

The concept of energy quality places an embodied energy cost on the
concentrated flows and storages that result from the energy convergence
property of systems. Energy quality factors for systems are defined to
be the ratio of the heat equivalent energy produced or upgraded by a
system to the quantity of energy ultimately required to power the
system.

The maximum power principle makes it possible to relate the value
calculated on the basis of energy consumed to the value based on the
effect of energy flow. The maximum power principle proposes that

systems that maximize their flows of energy will survive in competition
(Lotka 1922). The theoretical relationship between the energy embodied
in a resource and the energetic effect of properly using the resource is
grounded in the idea that the energy-consuming system must use its
energy flows and storages effectively to survive in competition. Hence,
in the long run, it is thought that energy cost is related energy
effect.

Solar energy is used as the basis for energy quality calculations
in this chapter. By constructing an energy model, of coal formation,
evaluated in heat equivalents, the energy quality of the various ranks
of coal were calculated based on the quantity of solar energy required
to power the system as suggested by Odum (1978).

C. Calculation of Energy Ratios

For purposes of comparison, several useful ratios are calculated
for the energy conversion processes in this chapter. These are the
investment ratio, the energy yield ratio, an efficiency ratio, and a
calculation of energy quality, all of which are based on the work of H.
T. Odum and are summarized in Fig. 1.

The net energy of a resource such as coal is the energy yielded
from the source beyond the energy required in extracting, processing,
and/or converting the source to some usable form of energy. Odum and
Odum (1976) define the net energy of a process to be the difference
between the output of the process and the feedback (F) from the main
economy to the process. The energy yield ratio is calculated by
dividing the feedback into the output. This ratio is often referred to
by other authors as the net energy ratio, in that a process with a yield
ratio of x produces x times the energy required in running the process.
The efficiency ratio is similar to standard engineering efficiency
calculations and is obtained by dividing the output by the input. The
investment ratio is related to the potential profitability of a process
and is calculated by dividing the input into the feedback. It should be
noted that all flows are first converted to a unit of common energy
quality, usually coal equivalents.

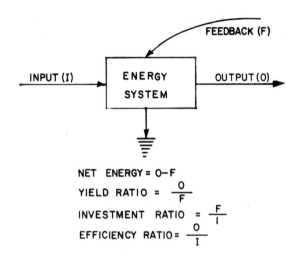

Fig. 2 Block diagram used in defining energy ratios
and net energy calculations.

III. RESULTS

A. The Energy Quality of Coal Based on Formation

Figure 3 is an energy circuit language representation of the
evolutionary history of coal formation. Both biological and physical
processes are involved in the transformation of simple, low-energy
molecules into the finished product of coal. As seen in the figure, the
radiant energy of sunlight provides the initial energy needed to start
the process. Photosynthesis has been the most significant biochemical

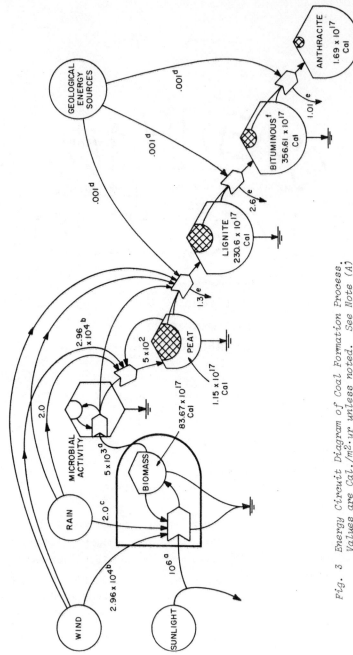

Fig. 3 Energy Circuit Diagram of Coal Formation Process.
Values are Cal./m².yr unless noted. See Note (A)

process on earth (with the possible exception of the replication of
nucleic acids), being responsible for the transformation of atmospheric
CO_2 into high-energy, carbon-rich sugars, which fuel other synthetic
processes within plant tissues. In addition, the process has radically
altered the composition of the atmosphere, resulting in a change from
the reducing conditions of the primeval earth, to the oxygen-rich
environment extant today.

After eons of life in the seas, where plants provided a base of
food for higher life forms and the detritus (which would later result in
petroleum deposits), vegetation made its way onto land, where, over
geologic time, forest ecosystems evolved in warmer, rainier, and more
humid conditions than are present today. During the Carboniferous era,
approximately 250-300 million years ago, vast forests covered the land,
populated by Sigillaria, Lepidodendrons, and Sphenopterids, genera
related to modern ferns and mosses (van Krevelen 1961). Primitive
insects populated the forests, remains of which are occasionally found
in a coal seam. The debris of plant life that fell on dry land or was
subjected to grazing rapidly decomposed and was recirculated as nutrient
material to other plants. Lowland trees, such as the Calamites, stood
in humid swamp forests, accumulating organic deposits around them as,
generation upon generation, they added to the debris at the bottom of
the swamp. Modern forests existing in these conditions are among the
most productive in the world. Cypress and hardwood hammock ecosystems
in south Florida exhibit annual gross productivities greater than 52,000
$Cal/m^2 \cdot yr$ (Lavine et al. 1979).   Subject only to the depredation of
microbes and safe from the biochemical decomposition occurring in the
open air, layers of organic material were transformed into peat deposits
(see Fig. 3).

Peat is a complex organic stew, 70-90% water, which results from
microbial decomposition of vegetation in water. Both aerobic and
anaerobic activity play a part in the production of peat, aerobic
bacteria breaking down some of the vegetable material shortly after
deposition and anaerobic activity continuing long afterward as other
layers accumulate above. The geological and biological processes
associated with the early stages of peat transformation are called

diagenetic processes. In northern latitude bogs of the Soviet Union and
northern Europe, where cold conditions limit initial aerobic
decomposition, large deposits of peat are found. Marshes of Florida,
with high rates of productivity and consequently high rates of organic
matter deposition, also possess large amounts of peat. Current global
stocks of peat are estimated at 23-120 x 10^{19} metric tons (Table 1),
some of which is burned as fuel, but the relatively high water and
oxygen content of peat make it a poor source of energy.

As peat beds are buried by further accumulation of organic material
and by sedimentary deposition processes driven by wind and rain,
anaerobic activity diminishes due to accumulations of toxins and to
gradually increasing temperatures. These deeper peat layers are
compressed to about 10% of their surface thickness. Sealed off from the
surface, peat begins to undergo a new phase of transformation.

Catagenesis, which refers to the geological processes occurring in
the transformation of peat to lignites and then to bituminous and anthracite is essentially an autoclaving process (Tissot and Welte 1978).
The organic material of the coal bed, compressed by the weight of
sediment above and previously altered by microbial activity, is heated
and pressurized over geologic periods of time by the geologic forces
which act over those intervals.

Large-scale convective processes originating deep beneath the
earth's surface result in local subsidence, or sinking, of the surface.
Deep burial by hundreds of meters of rock or seawater provided the
pressure and temperature changes necessary to compress the organic
matter in peat and to squeeze out quantities of water and volatile gases
(methane, CO_2, and other materials). The complex macromolecules
present in the material are fused together into networks of hexagonal
carbon chains (aromatic hydrocarbons characteristic of coal), that are
free of oxygen, hydrogen, and trace elements in proportion to the degree
of catagenesis. The succeedingly small proportions of these elements up
the evolutionary scale of coal are listed in Table 1 and are represented
by the shaded storage tanks in Fig. 3.

Other impurities, including ash and some sulfur compounds produced
anaerobically in the presence of clays or ferric compounds may remain.

Table 1. Characteristics of Organic Materials and Coal Ranks

Fuel	Estimated Global Abundance	H_2O Content (% wt)[1]	Ash Content (% wt)[1]	Carbon Content (% DAF[2] wt)	Oxygen Content (% DAF wt)	Hydrogen Content (% DAF wt)	Nitrogen Content (% DAF wt)	Sulfur Content (% DAF wt)	Calorific Content (Cal/kg-DAF)
Terrestrial Plant Material	1823[m](dry) [5469-fresh][3]	66[l]	4[e]-25	54[i]	37[i]	6.0[i]	2.8[i]	0.3[i]-0.6[p]	4600 (dry)[L] [1530 fresh][3]
Wood	1381[m,n] (dry) [2302-3069 fresh][3]	40-55[b]	1-2.9[b]	50[i]-52[b]	40.5[b]-43[i]	6.0[i]	0.1[b]-1.0[i]	0.60[p]	4630-5278 (dry)[b] [2987[3]-3345[k] fresh]
Peat	23.4c-120d	70-95[f]	3-25[f]	59[g]-60[a]	33.9[d]	4.9[d]-6[g]	0.9[d]-2.5[p]	.05[d]-0.6[p]	1955.6[k] [fresh]
Lignite	5013.3[c]	40[e]-60[f]	4.2[a]	21-69[i]	16.2-27.8[d]	4.6-5.3[d]	0.9-1.6[d]	0.2-1.1[d]	3800[i]-4600[k]
Sub-bituminous Coal	1555.7[c]	14-31[a]	3.7-7.0[a]	69[i]-80[h]	13.0-20.5[h]	5.1-6.2[h]	---	.59-.88[h]	5600[j]
Bituminous Coal	3877.6[c]	1-12.2[a]	3.3-11.7[a]	82[i]	13[i]	5[i]	0.8[i]	.51-6.84[h]	6950[j]
Anthracite Coal	24.3[c]	0-2.5[a]	---	92-98[a]	2.5[i]	2.5[i]	---	.55-1.0[h]	6950[j]

Note: Values should be taken as Representative of a range for each parameter.

(1) As coal rank is based on the carbon or calorific content of the dry, ash-free portion, these figures are not necessarily correlated with rank, but are included for purposes of comparison.
(2) Dry, ash-free portion
(3) Fresh weight value estimated by including water content (Col. 2) in the calculation.

[a]Van Nostrand Scientific Encyclopedia, pp. 570-597, 1976
[b]Tillman, D. A., pp. 68-71, 1978.
[c]Proceedings of the World Energy Conference, pp. 294-299, 1976 (Courtesy of Leonard Westenstrom, DOE).
[d]van Krevelin, p. 172, 1961.
[e]Darrah, W. C., p. 151, 1960.
[f]Schopf, J. M., p. 588, 1976.
[g]Stutzer, O., p. 179, 1940.
[h]Flueckinger, L., R. Dutcher and A. Cameron, 1972

[i]Rosler, H. H., and H. Lange, pp. 335-336, 1972.
[j]Westerstrom, L. (DOE), personal communication
[k]Singer, S. F., App. III, 1979
[l]Odum, E. P., pp. 39-56, 1971
[m]Whittaker, R. H. and G. E. Likens, p. 306, 1975 (agricultural biomass not included.
[n]Whittaker, R. H. and P. L. Marks, pp. 80-81, 1975.
[o]Chalmers, B., p. 157, 1963.
[p]Brady, N. C., p. 365, 1974.

These impurities must be washed away from the fuel after mining, at some cost to the economic system. Other organically bound forms, such as ester sulfates, are the residues of the original plant structure, and when burned with the coal can result in significant environmental degradation.

Lignitic, or brown coal, although the most primitive rank of coal, represents a significant departure from peat. It is substantially richer in carbon and consequently in fuel energy, than peat (Table 1). Lower concentrations of other elements make it a cleaner fuel than peat, but not too much cleaner. Burning low-rank coal may require cleaning precautions before and after the actual combustion process that are not needed with coals of higher rank, which is why coals of higher rank are favored.

Higher coal rank is produced by longer or more intense exposure to the diagenetic conditions of coalification. Older coal beds (seams) often produce higher quality coal, all things being equal, than seams of recent formation. As the material of the bed ages, more water is driven out. Volatile material continues to escape. The coal is compressed, becomes harder, shinier, and more dense, resulting in sub-bituminous and bituminous coal (see Fig. 3). This coal, driven further from oxidation, will produce more energy when burned than would any of its parent materials. Luckily, world reserves of this fuel are more plentiful than are those of its antecedent forms (see Table 1).

Anthracite coal is still further evolved (Fig. 3). This form is only one step away from the purity of graphite and is nearly pure carbon, roughly 95% by weight (Table 1). The energy content of this fuel, though high, is less than that of its immediate predecessors, due to the loss of the energy content when the last of the volative materials boiled off. The value of anthracite lies in its purity, and hence, its cleanliness. Global stocks of anthracite are rather low (Table 1).

Summarizing the processes of coalification:

(1) Coal is formed as a result of two general processes--a
    biological set and a geological set.

(2) Photosynthesis is the basis for coal formation, concentrating the dilute energy of the sun into the concentrated form of carbon bonds.

(3) Long periods of uninterrupted production, deposition, and accumulation in wet environments subject to microbial activity and burial (i.e., diagenesis) set the stage for the long-term process of coalification.

(4) Catagenetic processes of geology, including sedimentation and geological subsidence, result in the altered conditions of temperature and pressure which are necessary to transform organic deposits into coal structures.

(5) The grade of coal (i.e., purity and energy content) generally increases with both rank (degree of coalification) and type (condition of parent material) and thus with the energy embodied in coal production. This is due to the inherent lower energy quality of wood in comparison to coal.

B. The Energy Quality of Coal Based on Use

The second method used to determine the energy quality of coal is based on the comparison of several resources of varying quality used in similar processes. In the coal formation process diagramed in Fig. 3 plant biomass is concentrated from its relatively dilute energy content to higher energy fossil fuels such as bituminous and anthracite coals. The purpose of the previous section was to show that the process of biologically and geologically transforming biomass to anthracite was in effect increasing the energy quality of the material in each step by embodying large quantities of natural energy. Therefore, it should be possible to get more work done per thermal Calorie of anthracite than of wood. This section illustrates the effect of higher levels of embodied energy (increased energy quality) by comparing the use of wood and coal in the production of electric power. It will be shown that the cost of converting wood to electric power is higher than the cost of using coal to generate electricity, and it is suggested that this is due to the inherent lower energy quality of wood in comparison to coal.

Figure 4 is the energy circuit model of wood to electric power
conversion system used for this analysis. Wood was selected as a
comparative point because of the availability of relatively good
estimates for the amount of solar energy embodied in wood and because
wood represents the first step in the coal formation process.

Dilute solar-based energy flows such as sun, wind, and rain are
converged by the natural production system of the forest to produce
wood. As a rule of thumb, it takes approximately 1000 Calories of
sunlight to produce one Calorie of plant sugar. In other words, roughly
1000 Calories of solar energy are embodied in one Calorie of wood (Odum
and Odum 1976).

The conversion of wood into electricity requires fossil fuel based
energies in addition to the natural energies that are embodied in the
wood. The fossil energies required include both the direct use of fuels
and the indirect use of fossil based energy flows (e.g., capital, labor,
and goods and services).

Figure 5 diagrams the conversion of wood, coal, and oil to electric
power in 1000 megawatt generation facilities. The evaluation of the
wood plant is based on an operating ten megawatt plant in Burlington,
Vermont (Bryan 1979). The coal and oil evaluations were based on data
from Westinghouse Electric Corporation for 1000 megawatt power plants.
The coal plant uses a conventional steam boiler and a stack gas scrubber
while the oil-fired plant is of the combined cycle type. Table 2 summarizes the energy yield ratio for the three power plants shown in Fig.
5. The energy quality for wood and oil with respect to coal (based on
each fuel's ability to generate electricity) is also listed in Table 2.
The yield ratio was calculated by dividing a power plant's output by the
feedbacks required to construct and operate it. The energy quality
factor was calculated based on an equivalent output in all three
electrical generation processes. That is, electricity, which has a
constant heat content, served as a standard for determining the energy
quality of these three fuels.

First, the energy value of an amount of electricity, expressed in
coal equivalents (CE) was determined. This was done by adding the
annual fuel inputs (14.6 trillion Cal/yr of coal) to the required

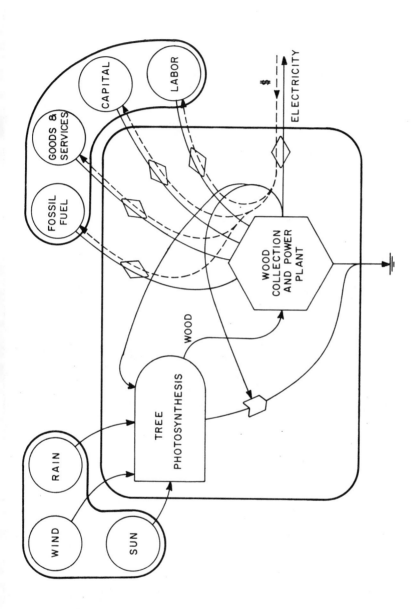

Fig. 4 An energy circuit diagram of the interaction of solar based natural sources and fossil fuel based energy sources to produce electric power from wood.

(a)

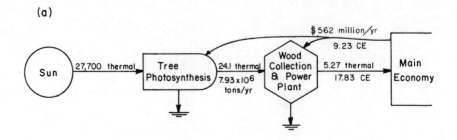

(b)

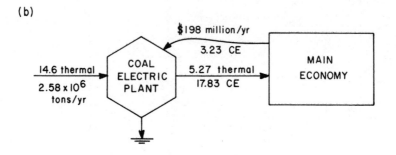

(c)

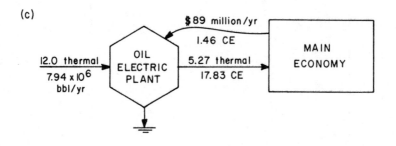

*Fig. 5   Energetics of Electricity Generation Using Various Fuels.
Values are 10¹² Cal/yr unless noted.   See Note (B)*

Table 2.   Yield Ratios and Energy Quality Factors for Various Electrical Generation Plant Fuels.

Fuel	Yield Ratio	Energy Quality Factor
Wood	2.2	0.36
Coal	5.5	1.00
Oil	12.2	1.36

feedback energies (3.23 trillion Cal/yr CE) giving a total energy requirement of 17.83 trillion Cal/yr CE. The electric output of this power plant is 5.27 trillion Cal/yr thermal. Dividing the total energy requirement (in coal equivalents) by the plant's electric output (in thermal units) gives a quality factor of 3.39 for electric power. In other words, 3.39 Calories of coal are equivalent to one Calorie of electric energy.

Now that the energy quality factor of electricity (expressed in coal equivalents) is known, it is a relatively simple matter to calculate the energy qualities of wood and oil, since both are also used to generate electricity. These calculations were accomplished using the formula:

$$Q = (O-F)/I$$

where - Q is the energy quality ratio of the fuel input, expressed in coal equivalent units per unit of thermal output
- O is the amount of electricity generated, expressed in coal equivalents
- F is the energy value of all feedbacks, expressed in coal equivalents
- I is the input energy, expressed in thermal units.

The results of these calculations are that wood has an energy quality of 0.36 and oil has an energy quality of 1.36 when compared to coal. That is, 0.36 Calories of coal can do the work of one Calorie of wood, and one Calorie of oil is equivalent to 1.36 Calories of coal.

C. The Energy Quality of Coal Based on Transformation

A third method for comparing coal to other fuel reserves is through conversion. Calculation of the amount of energy (in coal equivalents) required to produce a Calorie of gas or oil would pin down a maximum difference in the quality of the two resources.

To facilitate comparison of coal to gas a 250 million cubic feet per day coal-gas plant was analyzed using data from the American Gas Association. The results of the analysis are present in Fig. 6. A feedstock and fuel input of 27.84 x 10^{12} Cal/yr of coal plus 3.50 x 10^{12} Cal/yr CE of feedbacks produced 14.28 Cal/yr (thermal) of synthetic gas. Using the methodology outlined in the previous section and detailed in the notes, an energy quality for synthetic gas of 2.2 was calculated. That is, 2.2 Calories of coal will deliver one Calorie of gas and pay all costs.

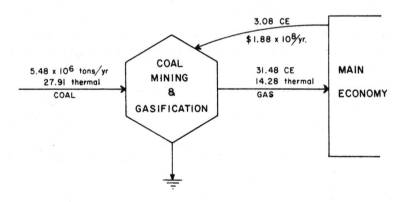

Fig. 6   A 250 MMcfd coal gas plant.   Values are
          10^{12} Cal/yr. unless noted.   See Note (C)

IV. Discussion

A. Some Preliminary Limits on Energy Quality of Coal

In our comparison of coal to oil by converting them both to a common-end product, electricity, a quality factor of 1.36 for oil was found. This means that 1.36 Calories of coal were required to do the work of one Calorie of oil. This represents a low estimate of the quality of coal with respect to oil. An upper limit was estimated by analyzing the direct conversion of coal to synthetic gas. This upper limit was found to be 2.2; that is, 2.2 Calories of coal can be converted to one Calorie of gas. While gas is probably higher quality than oil (because it is easier to transport and use, and burns cleaner) for placing a preliminary range on the quality of coal with respect to petroleum fuels, it appears that from 1.36 to 2.2 thermal Calories of coal will be required to do the work of one thermal Calorie of oil or gas.

B. Thoughts on the Energy Value of Global Coal Reserves

Estimates of the world's recoverable fossil fuel resources are 4.8×10^{19} Calories of coal and lignites; 2.80×10^{18} Calories of petroleum liquids; 2.53×10^{18} Calories of natural gas; 4.3×10^{17} Calories of tar sand oil; and 2.7×10^{12} Calories of shale oil (Hubbert 1971). In these estimates, coal makes up 89% of the total global energy reserves when compared in terms of the ability to produce heat. When compared on a basis of common energy quality, however, the magnitude of the coal reserve is effectively reduced, because most coal is of a lower energy quality than oil or gas. The energy quality factors for petroleum products developed in this paper suggest the amount of useful work which can be derived from the world's coal reserve is actually somewhere between 45% and 74% of current estimates.

The significance of this conclusion is that coal cannot be substituted for oil or gas on a Calorie for Calorie basis. This fact could be an important determinant when calculating the impact that the world's projected rate of energy consumption will have on coal reserves if major shifts to a coal based economy are made.

Notes to Figures

(A) Notes to Figure 3.

[a] Average annual insolation = 10^6 Cal/m^2·yr (E. P. Odum 1971).

[b] Average annual production of kinetic energy of wind, assumed to be equal to dissipation equals 2.96×10^4 Cal/m^2·yr (Monin 1972 and Swaney 1978).

[c] Chemical energy value of rain Cal/m^2·yr (Costanza 1976 and Swaney 1978).

[d] Estimate of heating contribution of geological subsidence to a 30 cm coal seam with specific heat equals 0.3145 Cal/kg-°C.

[e] Low estimate of energy losses associated with the degassing of volatiles assuming a 30 cm coal seam, and the coalification model of Juntgen and Kein (Tissot and Welte 1979, p. 219). Free energy values of volatile production over typical generation times are shown.

[f] Includes sub-bituminous.

(B) Notes to Figure 5a.

1. Efficiency of net production of wood estimated as 0.1% of solar insolation (Odum et al. 1976).

2. Output of one thousand megawatt power plant with a load factor of 70% for one year is:
   1000 MW x 0.70 x 860,000 Cal/MW x 8760 hr/yr = 5.27×10^{12} Cal/yr (thermal)

3. Green wood collected from forest and delivered estimated at $13.50 per ton (Bryan 1979) and 3.04×10^6 Calories (thermal) per ton (Singer 1979).

4. Calculation of Calorie to dollar ratio. (Sources: U.S. Dept. Commerce 1979, EIA Review 1978, National Energy Plan II 1979.)
   - U.S. fossil energy consumption (1978) = 19.66×10^{15} Cal
   - U.S. natural energy input (CE)   = 15.19×10^{15} Cal
   - Total U.S. energy use   = 34.85×10^{15} Cal
   - U.S. Gross National Product (1978)   = 2.13×10^{12}
   - 1978 Cal/$GNP = 16,400 Cal/$

5. The cost of operating a one thousand megawatt power plant in 1976 (Odum et al. 1976), increased by the change in the consumer price index from 1976 to 1978, and converted to cost equivalents:
   (4.91×10^8) x (1.146) x (16,400 Cal/$) = 9.23×10^{12} Cal/yr (CE)

6. The energy quality of wood is:
$$Q = \frac{(17.83 - 9.23) \times 10^{12} \text{ Cal/yr}}{2.4 \times 10^{12} \text{ Cal/yr}} = 0.36$$

Notes to Figure 5b.

7. Input for a year to a one thousand megawatt electric plant with a load factor of 70% and a thermal efficiency of 36%:
$$\frac{1000 \text{ MW x 0.70 x 860,000 Cal/MW x 8760 hrs/yr}}{0.36} = 14.6 \times 10^{12} \text{ Cal/yr (thermal)}$$

8. Electric load from the plant for one year:
   14.6×10^{12} Cal/yr (thermal) x 0.36 = 5.27

9. Total cost per year for coal plant (Westinghouse 1979):
   700 MW x $51/MWH x 8760 hrs/yr = $313 million/yr

10. Fuel cost per year for coal plant (Westinghouse 1979):
    14.6×10^{12} Cal/yr (thermal) x $7.94/$10^6$ = $116 million/yr

11. Feedback cost equals total cost (9) minus fuel cost (10):
    $313 - $116 = $197 million/yr

12. Feedback energy equals feedback cost (11) converted to Calories of coal equivalents (see note 4 above): $197,000,000/yr x 16,400 Cal/$ = 3.23 x 10^{12} Cal/yr (CE)

13. Output in coal equivalents equals input plus feedback energies:
(14.6 + 3.23) x 10^{12} = 17.83 x 10^{12} Cal/yr (CE)

14. Energy quality factor of coal compared to electricity is output (in coal equivalents) divided by output (in thermal units):
$$17.83/5.27 = 3.38$$

Notes to Figure 5c.

15. Input for a year to a one thousand megawatt oil fired combined cycle power plant with a load factor of 70% and a thermal efficiency of 44% is:
$$\frac{1000 \text{ MW} \times 0.70 \times 860,000 \text{ Cal/MW} \times 8760 \text{ hrs/yr}}{0.44} = 12.0 \times 10^{12} \text{ Cal/yr (thermal)}$$

16. Total cost per year for oil plant (Westinghouse 1979): 700 MW x $51/MWH x 8760 hr/yr = $313 million/yr

17. Fuel cost per year for oil plant (Westinghouse 1979):
12.0 x 10^{12} Cal/yr (thermal) x $18.65/$10^{6}$ Cal = $224 million/yr

18. Feedback cost equals total cost (16) minus fuel cost (17):
$313 - $224 = $89 million/yr

19. Feedback energy equals feedback cost (18) converted to Calories of coal equivalents (see note 4 above):
$89,000,000/yr x 16,400 Cal/$ = 1.46 x 10^{12} Cal/yr (CE)

20. Output in coal equivalents equals input plus feedback energies:
(12.0 + 1.46) x 10^{12} = 13.46 x 10^{12} Cal/yr (CE)

21. Energy quality factor of oil compared to electricity, and expressed in coal equivalents per thermal unit:
$$Q = \frac{(17.83 - 1.46) \times 10^{12} \text{Cal (CE)}}{12.0 \times 10^{12} \text{ Cal (thermal)}} = 1.36$$

(C) Notes to Figure 6.

22. A 250 MMcfd coal gas plant delivers 155.2 x 10^{9} Btu per day through conventional residential appliances (American Gas Association 1979, p. 540). Thermal Calories delivered are
$$\frac{155.2 \times 10^{9} \text{ Btu} \times .252 \text{ Cal/Btu}}{250 \text{ MMcfd}} = 156.4 \text{ Cal/cf (thermal)}$$

91.3 billion cf/yr x 156.4 Cal/cf = 14.28 x 10^{12} Cal/yr (thermal)

23. Calculation of feedback from main economy: (American Gas Association 1979, p.540.)

Capital Charge	$9.36/$10^{6}$ Cal output of gas
Fuel Cost	$2.88/$10^{6}$ Cal output of gas
Operating & Maintenance	$3.84/$10^{6}$ Cal output of gas
Credit for By-products	$2.88/$10^{6}$ Cal output of gas

$$\frac{(\$9.36 + \$2.88 + \$3.84 - \$2.88)}{10^{6} \text{ Cal}} \times 14.28 \times 10^{12} \text{ Cal/yr} = \$1.88 \times 10^{8}\text{yr}$$

$1.88 x 10^{8}/yr x 16,400 Cal/$ (CE) = 3.08 x 10^{12} Cal/yr (CE)

24. Based on $7.50/ton sub-bituminous western coal at 5.09 x 10^{6} Cal/ton and fuel cost of $2.88/$10^{6}$ Cal output of gas, the amount of coal needed for the process is:
$$\frac{\$2.88/10^{6} \text{ Cal} \times 14.28 \times 10^{12} \text{ Cal/yr}}{\$7.50/\text{ton}} = 5.48 \times 10^{6} \text{ tons/yr}$$

5.48 x 10^{6} tons/yr x 5.09 x 10^{6} Cal/ton = 27.91 x 10^{12} Cal/yr (thermal)

25. Gas delivered in terms of coal equivalents is:
27.91 x 10^{12} Cal/yr (CE) + 3.08 x 10^{12} Cal/yr (CE) =
30.99 x 10^{12} Cal/yr (CE)

References

American Gas Association, A comparison of coal use for gasification versus electrification, pp. 531-542, 1980 Department of Energy Authorization, Hearings before the Subcommittee on Energy Development and Applications of the Committee on Science and Technology, U.S. House of Representatives, 96th Congress, U.S. Government Printing Office, Washington 1979.

Brady, N. D., The nature and properties of soils, 8th ed., MacMillan Co., New York, 1974.

Bryan, R. W., Getting serious about wood fuel; public utility begins conversion, For. Ind., May 1979.

Bureau of Economic Analysis, Survey of current business, U.S. Dept. of Commerce, June, 1979.

Casagrande, D., and K. Siefert, Origins of sulfur in coal: importance of the ester sulfate content of peat, Science, 195:675-676, 1977.

Chalmers, B., Energy, Academic Press, New York, 1963.

Considine, D. M., ed., Van Nostrand's scientific encyclopedia, 5th ed., Van Nostrand Reinhold Co., New York, 1976.

Cook, E., Man, energy, society, W. H. Freeman & Co., San Francisco, 1976.

Costanza, R., Embodied energy basis of economic and ecologic systems, Ph.D. diss. Un. Fla., 1979.

Darrah, W. C., Principles of paleobotany, The Roland Press Co., New York, 1960.

Flueckinger, L., R. Dutcher, and A. Cameron, Statistical evaluation of coal compositional data, J. Geol., 80, 237-247, 1972.

Executive Office of the President, National energy plan II, Washington, D.C., 1979.

Hubbert, M. K., The energy resources of the earth, Sci. Am., 242(3):60-7-, 1971.

Krasilov, V. A., Paleoecology of terrestrial plants, John Wiley & Sons, New York, 1975.

Laboratory of Architecture and Planning, EIA Review, MIT, Cambridge, October, 1979.

Lavine, M. J., T. J. Butler, and A. H. Meyburg, Energy analysis manual for environmental benefit/cost analysis of transportation actions, Volumes I and II, Center for Environmental Research Cornell University, Project 20-11B, National Cooperative Highway Research Board, NRC, 1979.

Lotka, A. J., Contribution to the energetics of evolution, Proc. Nat. Acad. Sci., 8:147-150,1922.

Monin, A. S., Weather forecasting as a problem in physics, MIT Press, Cambridge, 1972.

Neavel, R. C., Coal science and classification, pp. 77-79, Scientific problems of coal utilization, NTIS, Springfield, Virginia, 1977.

Odum, E. P., Fundamentals of ecology, W. B. Saunders Company, Philadelphia, 1971.

Odum, H. T., Environment, power, and society, John Wiley & Sons, New York, 1971.

Odum, H. T., et al., Net energy analysis of alternatives for the United States, pp. 253-302E, Middle and long-term energy policies and alternatives, part I, Hearings before the Subcommittee on Energy and Power of the Committee on Interstate and Foreign Commerce, House of Representatives, 94th Congress, U.S. Govt. Printing Office, Washington, 1976.

Odum, H. T. and E. P. Odum, Energy basis for man and nature, McGraw Hill, New York, 1976.

Odum, H. T., Energy analysis, energy quality, and the environment, pp. 53-87, M. W. Gilliland, ed., Energy analysis: a public policy tool, AAAS Selected Symposium No. 9, Westview Press, Inc., Boulder, Colorado, 1978.

Rosler, H. J., and H. Lange, Geochemical tables, Elsevier Publishing Co., Amsterdam, 1972.

Schopf, J. M., Definitions of peat and coal and of graphite that terminates the coal series (graphocite), J. Geol., 74(5):584-592, 1976.

Singer, S. F., Energy, W. H. Freeman & Co., San Francisco, 1970.

Skinner, B. J., Earth resources, Prentice Hall, Englewood Cliffs, New Jersey, 1976.

Stutzer, O., Geology of coal, University of Chicago Press, Chicago, 1940.

Swaney, D. P., Energy analysis of climatic inputs to agriculture, Master's thesis, Un. Fla., 1978.

Tillman, D. A., Wood as an energy resource, Academic Press, New York, 1978.

Tissot, B. P., and D. H. Welte, Petroleum formation and occurence, Springer-Verlag, Berlin, 1978.

Van Krevelen, D. W., Coal, Elsevier Publishing Co., Amsterdam, 1961.

Whittaker, R. H., and P. L. Marks, Methods for assessing terrestrial productivity, pp. 55-118, H. Lieth and R. H. Whittaker, eds., Primary productivity of the biosphere, Springer-Verlag, New York, 1975.

Whittaker, R. H., and G. E. Likens, The biosphere and man, pp. 305-328, H. Lieth and R. H. Whittaker, eds., Primary productivity of the biosphere, Springer-Verlag, Berlin, 1978.

Woodwell, G. M., The carbon dioxide question, Sci. Am., 238(1):34-43, 1978.

Westinghouse Electric Corporation, Coal gasification combined cycle: a commercial assessment, pp. 945-1021, 1980 Department of Energy Authorization, Hearings before the Subcommittee on Energy Development and Applications of the Committee on Science and Technology, U.S. House of Representatives, 96th Congress, U.S. Government Printing Office, Washington, 1979.

CHAPTER 4

COAL TRANSPORTATION

Barney L. Capehart, Joseph D. O'Connor and William M. Denty, Jr.

I. INTRODUCTION

Transportation plays an important role in the utilization of coal as an energy source. Historically, transportation costs have comprised a large percentage of the delivered price of coal. Primary methods of coal transportation are railroads, waterways, highways and slurry pipelines. In addition to these, a significant amount of coal is utilized at the mine mouth for electrical generation and subsequent transmission through high voltage transmission lines. Table 1 shows estimates of current and future coal transportation requirements. As can be seen, a substantial increase is anticipated in transportation requirements for each mode. Although the relative percentage transported by each mode is not expected to change significantly, factors such as extensive slurry pipeline or synthetic fuel development could have a major effect on future coal transportation. Implementation of President Carter's proposed $88 billion synthetic fuel program (July 1979) probably would alter present transportation forecasts.

Coal transportation is greatly affected by the geographical distribution of coal. In the past the majority of coal transportation has originated in the eastern and midwestern coal fields. However, rapid expansion   is anticipated in the mining of low sulphur western coal reserves and this is expected to create a corresponding need for the expansion of western coal transportation. Table 2, taken from a report by Gundwaldsen et al. (1977), shows estimates of the expected utility de-

Table 1: *Projected Coal Transportation Mode (Million Tons)*

MODE	1975 ACTUAL	1985 EEI	1985 BOM	1985 NEP	2000 EEI	2000 BOM
Rail	418	503	637	780	608	1,023
Water	69	83	106	132	101	170
Truck	79	95	120	144	115	193
Mine-mouth Use	74	89	113	132	107	181
Pipeline and other	8	9	12	12	11	19

EEI = Edison Electric Institute
BOM = Bureau of Mines
NEP = National Energy Plan

Source: Congressional Research Service: National Energy Transportation, Volume III - Issues and Problems, March, 1978.

Table 2: *Estimated Utility Demand for Coal by Region (Million Tons)*

REGION	1975	Year 1985	2000(L)	2000(H)
Eastern	332.4	475	555	575
Western	87.8	340	600	580
TOTAL	420.2	820	1155	1155

Source: Gunwaldsen et al., (1977)

L: Low Synfuel Production (4.5 Quads Total)
H: High Synfuel Production (9.0 Quads Total)

mand for eastern and western coal. These estimates require an increase
of over 350% in western coal transportation by 1985 and over 650% by the
year 2000. However, the 1977 Clean Air Act Amendments may substantially
lower these projections since all new power plants are required to use
stack gas scrubbers and this reduces the need for low-sulfur western
coal.

There are many concerns related to the expansion of both eastern
and western coal transportation. In the eastern United States heavy
demands are already placed on rail and waterway transportation. As a
result there is considerable concern over the ability of these modes to
handle increased coal transportation requirements as well as the shipments of other commodities. With respect to western coal transportation, questions have been raised about the ability of the financially-
troubled rail system to finance the necessary expansion to western
mines. In addition there is concern over the effect coal slurry pipelines will have on these railroads and also on scarce western water reserves.

II. RAILROAD TRANSPORTATION

Railroads are the principal transporters of coal in the U.S. In
1977 they moved 437 million tons of coal or approximately 74.2% of the
coal that was transported in this country. In addition, coal accounts
for approximately 30% of the total tonnage shipped by rail and represents over 13% of gross rail revenue (Welty, 1978).

Table 1 shows a number of estimates for future rail transportation
requirements for coal. These estimates vary considerably, depending on
the source. It is generally agreed, however, that transportation requirements for coal will increase significantly. The largest increase
is expected to be in the transportation of western coal to accommodate
the increase expected in western mining.

Transportation of coal by rail may be in the form of single car,
multiple car, or unit train shipments. Single hopper cars with a capacity of approximately 100 tons are used primarily for shipments from
small mines, or to small consumers, and are hauled in combination with

other freight. The main disadvantage of single car movement of coal is
its low efficiency in terms of manpower, equipment utilization, and turn
around time. Energy consumption for rail shipment of coal is estimated
to be 560 BTU per ton-mile (Gunwaldsen et al., 1977).

In the past, multiple car shipments comprised the majority of coal
shipments by rail. These shipments usually originate from mines of
medium to large size. By loading, unloading, and shipping a number of
hopper cars together, the overall efficiency of the shipping process is
increased. Unit trains, which now carry approximately 50% of the coal
moved by rail, are made up solely of coal cars and are an extremely efficient means of transportation (Welty, 1978). They average 100 hopper
cars in length, are pulled by up to six locomotives, and shuttle (at
speeds of 50 to 60 mph) exclusively between coal mine and consumer.

In addition to the economy of direct routing, loading and unloading
practices have a major effect on overall efficiency of unit trains. For
conventional shipments of coal, these practices can account for over 40%
of the total shipping time. Unit trains, however, invariably reduce
processing time requirements. The average unit train can be loaded or
unloaded, while moving, in as little as 3 to 4 hours. As a result,
travel comprises the majority of a unit train's operating time, as compared to only 12% for conventional methods (Campbell and Katell, 1975).
It should be noted, however, that rapid loading and unloading of unit
trains is a necessity for economical operations. Accomplishing this
requires expensive loading and unloading facilities which are practical
only for very large customers (Szabo, 1978).

Current capacities and characteristics of eastern railroads vary
markedly from those of western railroads. Presently, eastern and southern railroads are the major rail transporters of coal, accounting for
over 80% of rail coal revenues in 1974 (Office of Technology Assessment,
1978). Although often in need of upgrading, the track system in the east
is highly developed and already serves most eastern coal mines. In general, eastern railroads are characterized by very poor financial health;
however, the major coal carrying railroads are reported to be in better

financial condition than the rest (Congressional Research Service, 1978).
In the west, the rail systems are much less developed, but are often in
better financial shape. Another difference is the route lengths. The
Office of Technology Assessment (1978), estimated the average western
coal route to be 650 miles, as compared to 235 miles in the east.

Current constraints on expansion of railroads are primarily economic. The poor financial health of many of the railroad companies in this
country has raised numerous questions regarding the ability of the railroads to meet expected demand. Typical concerns are whether the railroads can attract needed capital for expansion, whether they can acquire
necessary rolling stock quickly enough, and whether they can upgrade and
construct the track necessary to meet this demand.

The most serious environmental and social problems associated with
rail transportation are dust emissions, noise, decreased property values
and community disruption. Dust emission during loading, unloading, and
shipping can be quite large, ranging from .05 to 1% of the total tonnages
shipped (Szabo, 1978). Noise from horns, brakes, whistles and
track/wheel interaction not only has a detrimental effect on the health
and well-being of individuals in the community but also has a devastating
effect on property values. Also affecting property values is the decrease in a community's aesthetic appearance resulting from unsightly
rail systems and support facilities.

One of the most serious problems is community disruption, especially
with respect to unit trains. As long unit trains pass through towns, at
relatively slow speeds for reasons of safety, considerable congestion and
traffic delay can occur. Accidents resulting from car-train collisions
and derailments are almost certain to increase. Because of the length of
unit trains, sections of towns may be cut off from emergency services.
This is particularly true in small towns with a railroad track bisecting
them. Stobaugh and Yergin (1979) reported that the Sierra Club was involved in a law suit to force a more thorough environmental impact study
of one proposed coal route allowing as many as 48 unit trains a day to
pass through several small western towns.

Interstate transportation of coal by railroads is regulated by the Interstate Commerce Commission (ICC). The ICC has the power to review and revise freight rates, to authorize abandonment or expansion of rail lines, and to require annual renegotiation of shipping contracts. The methods used by the ICC to make regulatory decisions have been criticized as imposing economic disabilities on the railroad companies. Reforms have been suggested in rate structures incorporating the industry's rate of return on investment and long term shipping contracts permitting safe investment in facilities expansion or improvements. Examination of the adverse impact of regulation on the expansion of the rail transportation system is sorely needed.

Rates charged by the railroads are customer tailored to the individual situation and are dependent upon a number of factors. An important consideration is whether unit train shipments are employed. Because railroad companies regard unit train service as a long term, low risk investment, unit train shipping rates are the lowest per ton-mile of all rates. Other influential factors are the annual tonnages involved, distance of the route, frequency of the shipments, and shipper ownership or lease rights to the cars.

Table 3, taken from a report by the Office of Technology Assessment (1978), lists estimated freight costs and rate charges of four hypothetical rail lines as well as the necessary track upgrading and expansion needed in each case. The reported costs range from .0057 to .012 dollars per ton-mile, which falls into the range reported by Reiber et al. (1976).

Table 4 shows estimates by Desai and Whitten (1978) of capital requirements needed for rail expansion through 1985. The largest capital investments, approximately 70% of the total 6.1 billion dollars, will be needed for western expansion. The poor financial situation of the railroads often raises questions about their ability to attract enough investors. Congressional Research Service (1978), concluded that the increased opportunity for profit should generate enough investment to allow railroads to raise the required capital. A related question is

Table 3: *Estimated Cost and Rates of Four Hypothetical Rail Lines*

	Wyoming to Texas	Montana to Wisconsin	Utah to California	Tennessee to Florida
Average Route Length (miles)	1,584	1,055	684	938
Annual Cost[a] Dollars/ton	8.60	6.00	6.80	9.20
Simulated Rate[b] Dollars/ton	9.10	8.40	7.30	10.50
Dollars/ton-mile	.006	.008	.010	.011

[a] 1977 dollars at 6.5% real cost of money.

[b] 1977 dollars with 12.5% nominal return on investment.

Table 4: *Capital Investment Requirements Through 1985 (1976 dollars)*

Type of Investment	East (Millions)	West (Millions)	South (Millions)
Rolling Stock			
Locomotives	600	992	450
Hopper Cars	266	1,400	230
Track and Signaling	154	1,645	120
Yard and Maintenance Facilities	51	219	18

whether this capital can be generated quickly enough to offset possible
equipment shortages. Banks (1977) concluded that, due to the lead time
in developing new mines, this should not be a serious problem.

III. WATERWAY TRANSPORTATION

Waterway transportation of coal was the primary mode of transportation for over 11% of coal produced in 1976. Although only 70 million
tons moved by all water, approximately 146 million tons were shipped by
a combination of water and other modes. Of this total, 107.5 million
was shipped on the inland waterway system, 33.1 million on the Great
Lakes, and 5.3 million by tidewater (Bureau of Mines Mineral Yearbook,
1976). In addition, coal is a very important commodity to waterway
shippers, being second only to iron ore on the Great Lakes and petroleum
products on the inland rivers (Whitten and Desai, 1978).

Shipment of coal on the Great Lakes is primarily by single vessels
whose capacities range from 20 to 60 thousand tons. On the inland waterways, open hopper barges are used with capacities ranging from 900 tons
(standard size) to 1400 tons (jumbo size). These barges are connected
to form tows which may range from 40 jumbo barges on the open waters of
the lower Mississippi to 4-6 standard barges on smaller rivers. These
barges have average shipping distances of 350 miles and travel approximately 90 to 200 miles a day (Larwood and Benson, 1975). In a survey
by Hood and White (1977), the average energy requirement for a barge
transportation of coal was reported to be in the range of 550 to 650
BTU/ton-mile.

Table 5 shows present and estimated future tonnages of coal
moving on major rivers of the inland waterway system. As shown by
the table, the majority of these shipments are made on the Ohio and
Monongahela rivers. In the past, the inland waterway system has
served primarily eastern and midwestern coal fields. However, as western coal mining expands, combination rail/barge shipments are expected to increase. These shipments will use unit trains to transport the coal from western mines to rivers of the Mississippi River
system, where it will be transferred to barges and shipped to consumers

along the rivers (Stacy, 1978).

Table 5: *Projected 1985 Coal Traffic Origination
          on Selected Waterways*

RIVER	1985 TONS (THOUSANDS)	1975 TONS (THOUSANDS)
Mississippi	25,800	12,900
Ohio	71,900	42,100
Tennessee	9,200	4,600
Kanawha	11,400	5,700
Green	31,600	15,800
TOTAL	149,900	81,100

Source: Whitten and Desai (1977)

As western coal mining is expanded, transportation of coal on the
Great Lakes is also expected to increase. According to the report by
the Congressional Research Service (1978), most of this increase will
originate in the northern Great Plains where it will be shipped by rail
to ports along the Great Lakes such as Duluth. From these ports it will
be shipped via the Great Lakes to cities such as Chicago or Detroit.

A major constraint affecting coal transportation on the inland waterways is the lock and dam system. The capacity of any waterway route
is determined by the smallest lock. Stacy (1978) states that certain
locks in the upper St. Louis region are already operating at capacity.
A second constraint which affects both inland waterway shippers and
Great Lakes shippers is the possibility of a waterway user charge. This
charge, currently being considered by Congress, would impose a fee on
waterway shipments to cover the cost of maintaining and upgrading waterway shipping routes.

Major environmental issues associated with water transportation are
those relating to the construction of locks and dams and the dredging
of shipping lanes and harbors. Construction of locks, dams and harbors
may create significant environmental problems. As a result, environmental impact studies would be necessary which would delay construction for

three or more years. Dredging needed to keep shipping lanes and harbors open can also be environmentally damaging and is subject to stringent regulations. Whitten and Desai (1978) stated that such restrictions are a limiting factor in the amount of harbor and channel maintenance to be undertaken on the Great Lakes.

There is little regulation of waterway shipments of coal. Larwood
and Benson (1975) reported that most water transportation of coal is by
privately owned carriers exempt from ICC regulation. Presently the major regulatory issue is the waterway user charge. Related policy issues
are the effect the charge would have on the competitiveness of water
transportation and whether the charge should be in the form of a waterway toll based on maintenance cost of individual segments of the waterway or in the form of a fuel use tax levied equally on all waterway
users.

The economics favoring water transportation of coal are based primarily on two factors. First, waterborne shipments of coal can take
advantage of the navigable waterway systems and hence do not require
the large capital investment in route construction required by railroad
lines or pipelines. Second, the federal government currently maintains
the locks, dams, channels, and harbors of the waterway system. This
greatly reduces the operational expenses of waterway shipping companies.

Rates charged by waterway shippers are individually negotiated between shippers and barge companies. Whitten and Desai (1978) report
that these rates are dependent upon the season of the year, competition
of other modes, and amount of congestion on the waterway. Rates charged
for waterway shipment of coal have been traditionally lower than those
charged for shipments by rail, ranging from 3 to 8 mills per ton mile
(Desteese et al., 1978).

In a report on the utilization of coal in the northeastern United
States, Lewis (1978) stated that utility companies are planning to utilize coal shipments on the Great Lakes to lower substantially the f.o.b.
price of western coal. He estimated that through combined rail/barge
shipments, utilities in New England save up to 4 dollars per ton, compared to the price of Appalachian coal shipped by rail.

IV. SLURRY PIPELINE

The technology of coal transportation by water-based slurry pipelines can be divided into three stages: coal preparation, slurry transmission, and slurry dewatering at the terminal end. To prepare the
coal for transmission, first it is ground to a specified size. Water
is added during and after the grinding until the slurry is 45% to 55%
coal by weight. The slurry is then stored in mechanically agitated
tanks to keep it in suspension.

The slurry is pumped through a pipeline system by positive displacement reciprocating pumps located 50 to 150 miles apart. Holding ponds
are often constructed along the way into which the slurry may be emptied
if the system is shut down. Once the slurry reaches its destination, it
is dewatered. The dewatering may be accomplished by natural settling,
vacuum filtration, or centrifugation. Once the dewatered coal has un-
dergone additional drying, it may be used there or may undergo further
transportation by another mode. The separated water may be treated to
enable it to be used as cooling water, to be discharged, or possibly
even to be recycled (Office of Technology Assessment, 1978).

The energy requirement of a coal slurry pipeline is dependent on the
definition of the boundaries of the system. The Office of Technology
Assessment (1978) estimated an average energy consumption of 610 BTU per
ton mile. However, this figure does not take into account the added
moisture content of the coal after deslurrification. Szabo (1978) estimated the "coal quality loss" for the Black Mesa pipeline to be as high
as 500,000 BTU per ton.   Banks (1977), in his analysis of pipeline
energy requirements, stated that this loss could be reduced by a factor
of 3 to 4 times by using methanol or oil as the slurry medium instead of
water.

Oil appears to be an attractive alternative to water since the coal-
oil slurry can be used directly in coal burners without separation. The
problem in the use of oil, however, is the higher pumping costs associated with high viscosity of the oil-coal slurry. A compromise can be
reached with the use of a coal-oil-water slurry which can be used directly in existing burners. Such a slurry is being developed by CoaLiquid,

Inc., using 50% coal, up to 40% oil, and 10 to 20% water. The slurry
has about the same heating value as fuel oil, but at a cost about 20%
lower.

A new technology in coal slurrying is developing, using methanol as
the liquid in the slurry. In the "methacoal" slurry system, methanol is
produced from coal at the mine via steam gasification and methanol synthesis. About 85% of the coal heating value is retained in the methanol. The methanol slurry has the advantages of both the water and the
oil slurry systems. Pumping costs are comparable to water, but the high
degree of separation is not required, as methanol can be burned with
coal. In addition, methanol is a high quality fuel which can be separated from coal and shipped elsewhere for use in transportation or
heating.

Finally, methanol produced from coal is a clean-burning sulfur-free
fuel, making minimal its environmental impact. Methacoal slurries may
be an intelligent alternative in coal energy transportation.

Table 6 lists the lengths, location and capacities of existing and
proposed slurry pipelines. To date, only two such pipelines have been
built: the Cadiz, Ohio, pipeline and the Black Mesa pipeline. Presently
only the latter is operational, the Ohio pipeline having been driven out
of business in 1963 by lowered rail rates. Interest in slurry pipelines
is primarily in the west due to the long transportation distances to
eastern markets, the large volumes involved, and expanding western coal
fields. Opposition to western slurry pipelines arises primarily from
residents of semi-arid sections of the west fearful for their water
supply, and the railroads who claim it will hinder their future economic
growth. Recently, interest in coal slurry pipelines for the east has begun to increase.

Currently a major constraint to the construction of coal slurry pipelines is the need for eminent domain legislation. Without the power of
eminent domain, pipeline companies may encounter stringent opposition,
particularly from railroads, or exorbitant prices when negotiating
right-of-way across tracts of land (Congressional Research Service,
1978).

Table 6: *Location, lengths, and capacities of*
Existing and Proposed Slurry Pipelines

Pipeline	Length (miles)	Capacity (tons x 10^6)	Location From/to
Black Mesa	273	4.8	Arizona/Nevada
Alton, Nevada Power Co.	183	11.6	Utah/Nevada
Gulf Interstate N.W. Pipeline	1,100	10.1	Oregon/Wyoming
San Marco	900	15.0	Colorado/Texas
Wytex	1,260	22.0	Montana/Texas
ETSI	1,378	25.0	Wyoming/Louisiana
Cadiz	108	1.3	Ohio/Ohio
Florida Gas	1,500	15-45	Kentucky/Florida

Pipeline construction may have a variety of environmental effects. The most significant environmental effect would be habitat disruption. Extensive grading, ditching, noise, and increased human activity result- ing from pipeline construction could seriously upset biological communi- ties, along with farming and other interests.

Major operational impacts of the coal slurry pipeline mainly con- cern water-related issues. At the pipeline's origin, the major impacts are the effects of the pipeline water requirements on local water sup- plies. At the terminal end of the pipeline, the major issue is how to use or dispose of the water once it is separated from the slurry. An- other concern is the effect on the environment of a pipeline spill. The seriousness of such a break would be dependent upon its location, the volume released, and the liquid used in the slurry (Szabo, 1978).

The reduction of western water supplies by coal slurry pipelines is a politically sensitive issue. Because of the scarcity of water and its importance to agriculture, water rights are jealously guarded. Although the water requirements of the pipeline may not be environmentally dam- aging, the water rights may not be available.

Regulations affecting coal slurry pipelines may be divided into those pertaining to water usage, land use, transportation, and environ- mental protection. State laws, particularly in the west, could handicap

the pipelines' ability to obtain water rights. However, any attempt by
the federal government to interfere in water appropriations is likely to
meet with stringent opposition. Issues pertaining to transportation regulations are whether pipelines should be classified as common carriers,
what restrictions should be placed on pipeline ownership, what type of
rate structure should be implemented, and how contracts should be negotiated. If eminent domain privileges are granted to slurry pipelines,
they may be required to serve as common carriers and provide many of the
same unprofitable services required of railroads (Office of Technology
Assessment, 1978). Environmental regulations would come into effect in
both construction and operation of a slurry pipeline. Particularly powerful is the National Environmental Policy Act (NEPA) which requires an
investigation of the environmental impacts resulting from the pipeline if
federal cooperation is involved in the construction of the pipeline
(Stacy, 1978).

The economic competitiveness of slurry pipelines is due primarily to
large economies of scale. Davis and Associates (1976) reported the capital investment required for a proposed 25 MTPY (million ton per year)
pipeline to be 0.03 dollars per annual ton mile as compared to 0.065 dollars per annual ton mile for a 10 MTPY pipeline. Such economies of scale
are due to the insensitivity of such capital costs as right of way,
trenching, labor, and steel to the diameter of the pipeline. In addition,
Gunwaldsen et al. (1977) stated that operating costs also decrease with
increased capacity.

The range of costs for slurry pipeline transportation of coal is
quite wide and is highly dependent upon individual circumstances. The
Office of Technology Assessment (1978) reported estimates ranging from
.005 to .012 dollars per ton mile. Banks (1977), in a review of slurry
pipeline issues, stated that because of the capital intensiveness of
pipelines their long term costs are less succeptible to the effects of
inflation.

V. TRUCK TRANSPORTATION

Trucking is by far the most flexible mode of transportation. The

public highway system already offers a multitude of shipping routes and
expansion is far easier than for railroads and barges. Also the threat
of congestion is less because coal travels primarily on back roads,
avoiding city traffic (Congressional Research Service, 1978). Capacity
is readily adaptable. Increases may be achieved by buying additional
trucks or by contracting with private trucking firms.

Coal transportation by trucks is primarily intrastate, with the average shipment traveling between 50 to 75 miles (Congressional Research
Service, 1977). Although the ton-mile cost of 5 to 8 cents is considerably higher than other modes of transportation, trucking is more economically feasible for hauls of less than 50 miles (Szabo, 1978). In
addition, where mines yield small annual tonnages, trucks are particularly useful and more economical than railroads. However, trucks are
an inefficient mode of transportation, requiring between 2518 and 2800
BTU's per ton mile according to one study (Congressional Research Service, 1977). In some instances, terrain dictates that trucks be used.

Trucks are utilized primarily as collectors and distributors moving
the coal to railroad tipples and barge-loading facilities or to mine-
mouth generators. Trucks were the sole means of transport for 66.4 million tons in 1974 or about 11% (Congressional Research Service, 1978).
This percentage is expected to level off at 12% by 1985 (U.S. Executive
Office of the President, 1979). Trucks account for some portion of
transportation in about 75% of all coal shipments.

Certain environmental and social problems arise from the transportation of coal by trucks. Truck transport often increases the concentration of particulates in the air, primarily coal dust and dust stirred up
along unpaved country roads. To minimize coal dust emissions, (1) covers could be placed over the truck bed, (2) the coal could be washed at
the mine, or (3) the shipment could be misted to reduce the amount of
coal dust in the air. Watering or the placement of oil along the shipping route would help to reduce the road dust problem.

Truck transportation of coal increases the noise level along the
shipping routes. Stricter standards on noise and emissions are required
to alleviate this problem.

Increased truck usage for transporting coal has resulted in great damage to public road systems, even in instances where legal load limits are obeyed. Maximum gross weight for interstate highways is 40 tons. Since the truck itself is approximately 1/4 of the total weight, the maximum legal load on public roads is 30 tons. These limits are occasionally ignored, resulting in even greater damage. It has been estimated that a 55,000 pound truck equals 2500 automobiles in terms of road damage (Congressional Research Service, 1977). Another study revealed that a 3-axle coal truck running on a sound public highway causes $8,000 worth of damage while paying only $1,350 in licensing taxes.

Some states allow trucks of excess weight to run without penalty. Others are unrestricted in the number of axles. Although state regulations limit tonnage, inspection and enforcement of the laws is inconsistent. The burden of road repair is placed upon the taxpayer (Stacy, 1978).

At the federal level, the Interstate Commerce Commission (ICC) regulates only interstate coal shipments by trucks. However, in the future, state agencies will be required to enforce federal regulations on all federally-financed roads.

VI. TRANSMISSION LINES FROM MINE-MOUTH POWER PLANTS

Some electric utilities locate their power plants directly at the coal source and transmit the electricity over high voltage transmission lines. These mine-mouth generators, which have been supplying approximately 12% of the nation's energy over the last three years, avoid the transportation costs incurred by rail, barges, and trucks. However, many estimates for 1985 expect this percentage to drop slightly to between 11% and 11.5% (Szabo, 1978).

Transmission lines connect the power plants to the load centers where the high voltage electricity is transformed to a lower voltage before distribution. The implementation of a power grid -- several power plants interconnected by transmission lines -- would more efficiently use the output of the stations by shifting electricity during peak de-

mand periods from areas with excess capacity to those which are lacking
(Stacy, 1978).

Many of the environmental and social problems of high voltage
transmission lines center around the transmission towers which are aes-
thetically undesirable and remove areas of land from productive use.
Between 110 and 115 feet are cleared beneath the towers and there are
miles of service roads. Solutions to the problem include the implementation of multiple-use right-of-way corridors convenient for and used
by more than one utility. Also, government regulations may require
transmission lines to be placed underground, as is already done in urban and congested areas.

State regulatory agencies have varying powers over electric companies from state to state. Plant sites and right-of-way locations are
usually overseen by state agencies. Federal control is derived from the
need for the electric utilities to meet federal air and water pollution
laws and laws passed by the Federal Power Commission.

Capital costs for high voltage transmission lines are greater than
those for other modes of coal transport. However, the lesser dependence
of transmission lines upon petroleum fuels and other inflationary pressures may make them a more feasible alternative in the future (Stacy,
1978).

VII. OTHER MODES
In addition to the major modes of coal transportation, certain
other methods are available and have been used on a limited scale.
Among these are the conveyor belt and the pneumatic pipeline. These
types have accounted for 1% of coal transportation over the last few
years and estimates say they will continue at this rate through 1985
(U.S. Executive Office of the President, 1979). They may be used in
cases where other modes are not feasible.

The overland conveyor belt is simply a belt upon which coal is
loaded and then transferred to its destination. Its range is usually
from one to ten miles and it is capable of carrying coal over steep inclines (Desteese et al., 1978). Normally, belts are at least partially

covered to prevent dust emissions and rain saturation of the coal. An
estimate for the energy requirements of the conveyor is .36% of the coal
haul. The permanent belt structure is visible in the environment, particularly at overpasses, fill areas, and trestles. Right-of-way corridors range from 50 to 200 feet in width depending upon the landscape
(Szabo, 1978). Conveyors are found to be more economical than trucks
for all vertical lifts and distances. When hauling waste materials over
more than a 240-foot vertical lift, conveyors were again found to be
more economical (Desteese et al., 1978).

Pneumatic pipelines transport finely ground coal by air pressure.
Presently, the most feasible role for the pipelines may be to replace
spur lines and operate along the short runs between mines and railroads.
Pneumatic pipelines require no water, can operate over steep inclines,
are easily relocatable, and shutdowns will not block the piping since
the particles are dry.

VIII. PUBLIC POLICY ISSUES

The transportation of coal is a difficult and often complex operation with numerous economic, environmental, and social impacts requiring
public policy decisions. Since coal use is expected to triple by 1990,
then significant public policy issues with respect to the largest modes
of transportation must be faced to allow an orderly accomplishment of
this goal.

Rail transportation is expected to absorb the greatest increase in
capacity. Since railroads have the ability to serve passenger traffic,
as well as other freight, at a very low energy cost, expansion of the
rail network is a critical element of national transportation policy.
The ancillary benefit to expanded coal hauling ability for the railroads
represents a highly significant public policy issue.

Railroads face many problems in operation, refurbishment, and new
construction. Increased rail traffic through towns can disrupt whole
communities and virtually cut towns in half. Noise, coal dust, increased
accidents, and possible cutoff of emergency services are severe problems
that must be faced. Therefore, expansion of the rail system will be very

costly and, because of past uncertainty in the demand for coal, the railroads will be reluctant to commit huge financial resources without guarantees concerning their use.

Water transportation is a cheap, energy-efficient, and relatively
problem-free mode in those areas served by the existing domestic waterway system, which is constructed and maintained by the federal government. The two major public policy issues in water transportation involve the potential for environmental damage during construction and expansion, and the question of requiring water carriers to pay user fees
that at least approximate the cost of operating the waterway network.

Pipeline transportation is also expected to grow significantly. At
present, there is one operational coal slurry with at least five new
pipelines proposed. The two major public issues regarding pipelines are
eminent domain for construction and water for operation. Although railroads have fought granting eminent domain for pipelines in order to
preserve their new monopoly on long distance coal shipment in areas without water routes, pipeline companies appear to be gaining ground. Water
availability is a major factor for pipeline operations. In the west
where there are limited supplies of water, western water rights may not
provide the water needed for coal slurry pipelines. Pipeline construction poses numerous environmental problems which may slow down or alto-
gether prevent the appearance of pipelines in some areas.

Trucks carry large quantities of coal over short hauls, frequently
damaging the roads they use. Since almost all of these roads are paid
for by taxes from federal, state or local sources, the operation of coal
trucks depends on a heavy subsidy from the public, thus raising another
public policy issue. Noise, coal dust and decreased property values
along coal truck routes are also problems that must be dealt with. Because trucking is the least energy-efficient mode of coal transportation,
it should be limited to those demands which cannot be met by other modes.

Mine-mouth power plants with energy transportation by transmission
lines present fewer significant public policy issues. Aesthetic and environmental considerations are probably the most important issues. A
related issue is whether power plants should be located, instead, near

their load centers with these areas bearing the cost of clean-up or suffering the effects of the pollution. Vacillating policies by the federal government have introduced uncertainties regarding coal demand and
transportation requirements. If the nation makes a firm commitment to
utilize coal for a large percentage of our energy needs, a national transportation policy must be implemented to expand the coal transportation
network.

IX. ECONOMIC COMPARISONS

The economics of coal transportation are based primarily on the
geographical location of supply/demand, route circuitry and length, competition among modes, and volume of throughput. The geographical distribution of coal limits the economic competitiveness of both waterway
and highway transportation as primary coal transportation modes. Waterway transportation has the lowest per ton-mile cost of all modes.
Reiber et al. (1975) reports costs of 0.25 cents per ton-mile for barge
as compared to 0.6 cents per ton-mile for long distance, high volume,
unit train shipment. However, waterway transportation is constrained by
the extent and location of navigable waterways in relation to the supply
and demand for coal. With respect to truck transportation, distance is
the constraining factor. The economics of highway transportation are
such that trucks are competitive only for very short hauls.

The greatest economic competition arises among railroads and slurry
pipelines. Both have economic factors that make them favorable, given a
certain set of circumstances. Soo et al. (1975)   state that slurry
pipelines may have a cost-advantage as great as two to one if new road-
bed must be constructed. However, they report that, when the rail line
already exists, rail costs are as little as 50% of those associated with
slurry pipelines. Davis and Associates (1976) report that in most cases,
for slurry pipelines to be economically competitive with rail, annual
volumes must exceed 5 to 6 million tons and transportation distances
must exceed 100 miles.

When making economic comparisons, it is important to consider the
system cost. Each mode receives varying subsidies from the government.

For example, waterways and interstate highways are maintained by the
federal government, while railroads must perform and pay for their own
track maintenance. The effect of increased coal transportation on shippers of other commodities may also result in larger system costs. If increased coal transportation creates bottlenecks in a transportation network, this may increase the cost of shipping other commodities. A con-
trasting view, however, is that limiting coal transportation by any particular mode to allow capacities for other commodities is in effect a
subsidy of those commodities. In addition, it may be difficult to equi-
tably assign utilization costs. For example, if railroads install heavy
duty rails to accommodate coal transportation by unit trains and raise
freight rates accordingly, in effect, other commodities will be subsi-
dizing coal transportation (Gundwaldsen et al., 1977).

X. FUTURE TRANSPORTATION ISSUES

     In addition to the broad public policy issues already identified,
other issues of a more specific nature will be surfacing in the near future. One of these is federal ownership and maintenance of rail beds.
Railroads are at a distinct disadvantage with respect to water and truck
transportation, since railbeds must be constructed and maintained by the
railroad companies themselves. Waterways and highways are constructed
and maintained by federal, state and local governments. Thus, in order
to encourage rail refurbishment and expansion, one possible approach
would be for the federal government to take over ownership of the railbed
system and allow railroad companies to run their rolling stock over the
rails. Such a plan could provide faster and more certain rail expansion
capability.

     Another issue expected to provoke wide interest is the transporta-
tion of coal waste back to original mine locations for disposal and reclamation purposes. Approximately 10% of coal is ash and another 10% is
flue gas desulfurization (scrubber) sludge. For every 100 carloads of
coal burned, 20 carloads of coal waste must be disposed of. In many areas
of the country, there are no satisfactory disposal sites near the coal use
facility. One solution would be to return the waste to the mine, using

empty shipping vehicles. Trains, barges, and trucks are possible return vehicles, but a slurry pipeline would not be amenable to this operational method. On the other hand, coal waste could be transported
to other locations if it were to be used in making such products as as-
phalt, concrete, or gypsum.

XI. CONCLUSION

At present, coal use policy for the U.S. faces great uncertainty.
Until a clear policy direction emerges, there can be no accompanying
commitment to expansion by the coal transportation network. The capability for significant expansion exists, but large capital investments
will not be made until the need for expansion is completely resolved.

Much research is needed concerning coal transportation to answer
questions regarding optimum economic, social and environmental decisions.
Modes chosen for expansion should satisfy some decision criteria for
handling the required loads while minimizing cost, social impacts, and
environmental degradation. Formulating a transportation model and an
appropriate decision function should be assigned high priority in order
to prepare for a sensible and orderly expansion plan. The transportation model should include all major modes of rail, water, pipeline,
truck, and transmission lines from mine-mouth power plants, as well as
the capability for adding futuristic modes. The decision must accurately reflect all major economic, social, and environmental factors while
being constrained by existing laws and regulations. Obviously, such a
project would require substantial time and resources to complete, but
nonetheless is a requirement, prior to any large scale funding of transportation network expansion.

XII. FURTHER REMARKS[1]

Transportation of coal by coal-oil slurries, coal-oil-water slurries and coal-methanol slurries has been discussed briefly in Section
IV as attractive alternatives to coal-water slurry systems. It is

---

[1]This section is due to F. Fardshisheh and A. E. S. Green.

noteworthy since most of the existing oil or gas burning plants cannot
be economically utilized for coal burning,that one of the most promising
techniques for increasing coal utilization for the immediate future
would be through the use of coal-oil mixtures (COM) as industrial or
utility fuels. While investigations on the use of COM go back nearly a
century, interest in this area was limited because of the lack of clear-
cut economic advantage during periods of cheap oil and gas. Now with
high oil and gas prices the advantages of COM are becoming increasingly
clear. Indeed the use of such mixtures perhaps represents the best
short-term way of increasing the utilization of coal with only minor
retrofitting of existing oil and gas burning facilities. For example, a
large fraction of the electric utilities and industrial boilers in
Florida use oil or natural gas including large facilities built within
the past decade. Accordingly, a coal-oil slurry pipeline or coal-oil-
water pipeline to Florida might not only serve as an advantageous mode
of transportation, but one which would facilitate a rapid conversion
from oil burning to COM utilization.

The literature is developing rapidly on in situ COM preparation
facilities. Much work remains to be done as to methods of preparation
of COM, as to their abrasive characteristics, pumpability, optimum percentage mixtures, burner operations, boiler characteristics, slagging
characteristics, the nature of the stack effluent and the possible use
of additives. However, data is already available which suggests that
coal-oil slurries can successfully and economically be burned in oil
burning installations with only minor alteration or modification of
equipment. Furthermore, it appears that the use of the COM might provide sufficient economies to pay for clean-up equipment.

Returning to the transportation topic, the question remains as to
whether a centralized COM preparation facility has economic advantages
with respect to COM facilities at each utility plant. Thus, economies
of scale in the transportation and use of COM must be examined. For
example, if the pipeline originates at the coal mine, then the costs of
bringing the oil to the mine and the extra pumping costs must be addressed.
Possibly the compromise of bringing both the coal and oil to a large
central port facility where the coal is cleaned and the COM prepared for

delivery through short pipelines to the utilities in a megapolis will have economic advantages over originating the pipeline at the coal mine or the oil field. All of these public policy issues should be addressed, not only in the context of the transportation problem but also in the context of how this country can reduce its use of oil most rapidly. The possible advantages of later adaptation of COM to coal-methanol mixtures, also warrants examination.

References

Banks, W. F., An energy study of pipeline transportation systems, Systems Science and Software, La Jolla, CA. Prepared for the DOE, San Francisco Operations Off., Oakland, CA, Dec., 1977.

Banks, W. F., Slurry pipelines: Economic and political issues, a review, Systems Science and Software, La Jolla, CA. Prepared for the DOE, San Francisco Op. Off., Oakland, CA, Nov., 1977.

Campbell, T. C. and S. Katell, Long distance coal transport: Unit trains or slurry pipelines, Bu Mines IC 8690, 1975.

Congressional Research Service, National energy transportation, Vol. 1, Systems and movements, Washington, D.C., May 1977.

Congressional Research Service, National energy transportation, Vol. 3, Issues and problems, Washington, D.C., March, 1978.

Davenport, C., Statement made before the subcommittee on surface transportation of the committee on public works and transportation, House of Representatives, Ninety-fifth Congress Second Session, On H.R. 1609, March 22 and April 11, 1978.

Davis and Associates, Survey of cost of rail versus new technology for long distance coal transportation, HCP/13461-01, McLean, Va., 1976.

Desai, S. A., and J. M. Whitten, Rail transportation requirements for coal movement in 1985, Input Output Services, Cambridge MA. Prepared for U.S. DOT, Office of the Secretary, Dec., 1978.

Desteese, J. G. et al., Energy material transport, now through 2000, system characteristics and potential problems, Task 2 Final Report. Coal Transportation, Pacific Northwest Laboratory, Richland, Washington, June, 1978.

Gunwaldsen, D., N. Bhagat and M. Beller, A study of potential coal utilization 1985-2000, Technology Assessment Group, National Center for Analysis of Energy System, Brookhaven National Laboratory, Upton, N.Y., Dec., 1977.

Hickman, L. S., Jr., Coal—can our railroads haul it?, proceedings of the Fourth Annual Energy Conference, Chicago, Ill., September 16-17, 1976.

Hood, R. and T. G. White, Coal transportation study, final report, U.S. DOE Rep. COO/2970-1, 9-77.

Larwood, G. M. and D. C. Benson, Coal transportation practices and equipment requirements to 1985, BU Mines IC 8706, 1976.

Lewis, L. R., Potential demand for western coal in the northeastern United States and its relationship to waterway transportation facilities, Argonne Nat'l Lab., Ill., Jan., 1978.

Reiber, M. et al., The coal future: Economic and technological analysis of initiatives and innovations to secure fuel supply independence, U. of IL. at Urbana-Champaign, May 1975.

Soo, S. L. et al., The coal future: Coal transportation - Unit trains, slurry and pneumatic pipelines, Appendix F, Univ. of Ill. at Urbana-Champaign, June 1975.

Stacy, D. M., Factors affecting future expansion of the coal transportation network: Legal and institutional constraints on accelerated coal freight. Prepared for the U.S. DOE, Morgantown Energy Technology Center, Morgantown, West Virginia, July 1978.

Stoubauh, R. and D. Yergin, Energy Future, Random House, 1979.

Szabo, M. F., Environmental assessment of coal transportation, PEDCo Environmental, Inc., Cincinnati, OH. Prepared for the U.S. Environmental Research Lab., Cincinnati, OH, May 1978.

U.S. Congress. Office of Technology Assessment, A technological assessment of coal slurry pipelines, Washington, D.C., March, 1978.

U.S. Executive Office of the President. Coal: A data book, the President's Commission on Coal, Washington, D.C., 1979.

Welty, G., Mass transportation for bulk commodities, Railway Age, V. 179, July 31, 1978.

Whitten, J. M. and S. A. Disai, Transportation requirements for coal movement in 1985, Input Output Computer Services, Cambridge, MA. Prepared for U.S. DOT, Off. of Secretary, Dec., 1978.

CHAPTER 5

COAL BURNING TECHNOLOGY

Richard T. Schneider and Michael J. Rowe

I. INTRODUCTION

Coal has been used as a combustible fuel in Western Society for
more than 1000 years. Several times in history, coal burning has been
banned due to its polluting effects. However, with today's energy
supply problems and the USA's most abundant fossil fuel being coal, it
is imperative that the proper technologies be developed for its utilization.

There are two types of coal burning technologies; namely, concentrated use, such as in electric utilities, and dilute use, as in residential coal burning. Although residential coal burning is today less
than 2 percent of the total coal used, amounting to about 12 million
tons per year, the use of coal furnaces could increase in areas where
coal is easily obtained at a low price. (Cart, Jr., 1977) At present
most of the heat produced for residential use is from oil or gas furnaces; however, areas around coal mining districts could develop serious
air quality problems (See Chapters 9 and 12).

A review of residential technologies shows that, after 1940, oil
and gas increased in popularity due to the simplicity of operation.
However, improvement in fixed-bed coal burning equipment has reduced the
difficulty of operation for commercial-residential units. Stokers for
feeding coal to the furnace can hold a week's supply of coal. Ash pans
need not be emptied for up to one week. Cyclic thermostat control,
meaning the furnace is run at full power for only 5 min. each hour

and the rest of the time at low power, eliminates the need to restart the furnace constantly. In short, a modern coal furnace can be operated with minimum periodic maintenance. Therefore, there will be an increase in the amount of coal used directly by residential consumers.

Coal technologies once popular but since forgotten, such as coal-powered cars and steam trains, may soon be back in production. The coal-powered car possessed wide popularity during World War II when there was no gasoline. Any internal combustion engine can run on "coal gas" if the displacement is larger than 2 liters. Future recurrence of gas shortages could promote the redevelopment of these older technologies. Dilute technologies do not allow economical pollutant clean-up which, on a large scale, could be a significant problem. On the other hand, concentrated techniques do allow for economical cleanups.

Concentrated technologies can be linked directly to dilute technologies through developments such as the electric car. (See Chapter 7). Using a concentrated source such as a coal fired electric utility, batteries can be charged and used for dilute, non-polluting energy sources. There are economic considerations, however. Buying coal for $30-45 a ton (Johns, 1979) and burning only a few pounds per hour for heating an average household, cannot be more expensive than buying electricity for resistive heating. Economics will dictate trends in technological utilization but, under any conditions, concentrated coal burning technology will be the major consumer of coal. (Figure 1)

II. PARAMETER OF COAL COMBUSTION

The objective of coal combustion is to convert chemical energy into heat, which is to be converted using a heat engine into mechanical (kinetic) energy and (in most cases) subsequently, into electrical energy.

This is practiced despite the fact that chemical energy is a higher form of energy than heat and could be directly converted into electricity, e.g., in a fuel cell. The reason combustion is used at all is that the direct conversion technology is not yet developed to a sufficient sophistication to allow large scale energy conversion.

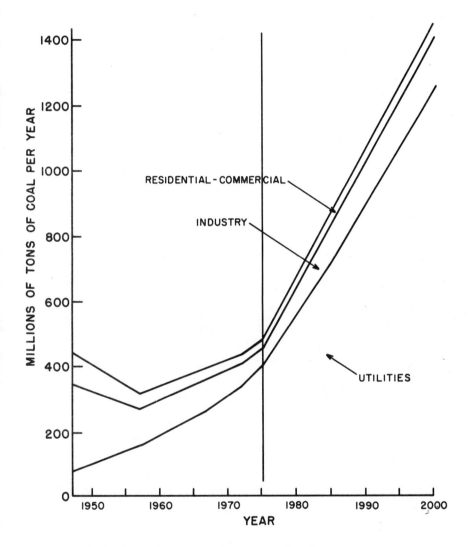

Fig. 1.   Combustion of Coal and Lignite by End Use Sector

Source: Annual Report to Congress, Energy Information Administration,
          Department of Energy, Vol. III, 1977, OTA.

If combustion of coal has to be performed, it is desirable to obtain the maximum possible amount of heat, to obtain this heat at an optimum temperature, and to avoid generating undesirable combustion products. These three needs will be the subject of the following discussion.

The combustion of coal with air involves a large number of chemical reactions. Oxidation of carbon to form carbon dioxide is the most important one, of course; but many other reactions have to be considered, some of them dealing with oxidation of impurities (especially sulfur), some of them dealing with oxidation of nitrogen and of hydrocarbons (volatile matter in coal).

In addition to chemical reactions, physical reactions also have to be considered. These are phase changes of the reaction partners prior to the chemical reaction, which require energy. These are, especially, vaporization of the moisture content of coal, sublimation of carbon prior to combustion, and release of gases and water adsorbed to the surface of coal particles.

Finally, catalytic action of impurities might be considered although this plays a minor role.

It is easy to see that a complete analytical description of all these physical and chemical processes, including their interrelations, is an almost impossible task. Therefore, all descriptions attempted have to be highly idealistic; any conclusions drawn from theoretical models should be verified by measurements. However, considering the enormous amounts of coal that will be combusted in the next 30 years, a clear understanding of all these processes is essential. Any improvement in the combustion process or any reduction of emission of undesirable combustion products, even if ever so small, will translate in enormous amounts of energy saved or harmful chemicals not discharged in the environment.

A. Temperature

The temperature (T) of any gas is related to its pressure (P) and volume (V) by the equation of state

$$PV = nRT$$ R: universal gas constant
                      n: number of moles.

Therefore, the temperature of the gas obtained from the combustion of
coal (flue gas) is related to how much gas (mass, volume and pressure)
was heated by the combustion process. The chemical reaction of oxidation, of course, can occur in principal at any temperature; however,
the reaction rate is a strong function of temperature. Therefore, a
minimum temperature (the ignition temperature) is required to ensure
that the chemical reaction proceed at a reasonable rate and be self-
sustaining. Other than this, the temperature can be manipulated by
selection of pressure, fuel to oxygen ratio and other parameters within
a--although narrow--range of achievable values. Therefore, a decision
needs to be made as to the exact desired temperature. The answer to
this question depends on the thermodynamics of the system used for
energy conversion.

If a gas (like helium CO_2 or air) is to be heated by the flue gas
for expansion in a gas turbine (closed cycle helium turbine concept) the
desired temperature is the highest temperature the heat transfer areas
of the gas heater used can tolerate (metal temperature). At present,
this temperature is around 1000°C. Due to the more complex thermodynamics of the two phase system water-steam, the metalurgical limit
dictates a different temperature. Experience of the last 10 years
indicates that operation in the supercritical region of steam does not
bring a substantial advantage. Therefore, the desirable steam temperature is around 500°C, which is far below achievable combustion temperatures. It is, of course, true that the heat transfer through the metal
walls of the boiler tubes is proportional to the temperature gradient
across the heat transfer barrier. This gradient is determined by the
combustion temperature (outside) and the desired steam temperature
(inside). First consideration would indicate that this gradient should
be as large as possible, which means a combustion temperature as high as
possible. However, all that this gradient will do is to speed up the
heat transfer, which means less heat transfer area for a given power
output--in other words a smaller boiler (and therefore less capital

investment). If, at the same time, an increased combustion temperature means increased production of NO_x which also requires capital investment for removal, obviously operation of the highest possible combustion temperature is not necessarily the optimum solution. And, as discussed above, the thermodynamic efficiency of the power plant is determined by the steam temperature, not by the combustion temperature.

However, the situation is, in reality, even more complex. Besides heat transfer by convection of the flue gas to the boiler tube surfaces and consequent heat transfer by conduction through these surfaces, there is also heat transfer by radiation and heat transfer by small particles. The latter is used extensively by the fluidized bed technology. Heat transfer by radiation can play a significant role at higher combustion temperatures.

The combustion temperature usually is computed on the basis of the equation of state and the specific heat of gases. Tables and diagrams to accomplish this can be found in the literature (Steam, Its Generation and Use and Mantell, 1968).

B. Heat of Combustion (Heating Value)

The amount of chemical energy converted into heat is called the heat of combustion, and in case of coal, the heating value is given a practical number. Different coals are ranked according to their quality and the heating value is one of the factors contributing to it.

There is, of course, only one heat of combustion for pure carbon which is determined by the binding energies of the valence electrons of participating atoms O and C and the binding energies involved in the destruction of the O_2 and formation of the CO_2 molecules. However, coal, as pointed out before, is a mixture of many different chemical compounds and the summation of all the different reaction heats form the heating value.

The size of the heating value determines the amount of heat that can be obtained from a given amount of coal, and should, therefore, determine the price of this particular brand of coal. However, the

heating value lies in between fairly narrow limits for the different
brand of coals (±15%) and there are other qualities of coal which
influence the mechanical handling required which do or do not cause
additional costs usually outweigh the difference in the heating value.
Therefore, improvement in coal burning technology has to come in the
form of better handling, transportation, and preparation of coal, as
well as in improvement of the combustion process itself.

C.   Fuel to Oxygen Ratio

The laws of combustion kinetics define a correct mixture ratio of
carbon and oxygen, the stoichiometric ratio.

However, since coal is a conglomerate of all kind of chemical
compounds, the stoichiometric correct ratio will vary for different
brands of coals. It is even questionable if it is desirable to operate
at stoichiometric conditions. Obviously, excess air will reduce combustion temperatures, since only the amount of oxygen corresponding to
the stoichiometric condition will participate in the chemical reaction
and the excess will only be in the way, being heated or, in other words,
cooling the flame. The same is, of ocurse, true for the nitrogen contained in air. Although it participates in the chemical reaction by
forming NO_x to some extent, its major contribution is lowering the
combustion speed and flame temperature. Therefore, substantial improvements in coal combustion technology could be achieved if the nitrogen
could be separated from the air by some economically feasible process
prior to combustion.

III.   COMBUSTION OF POROUS PARTICLES

In the history of the industrial revolution, a major milestone of
progress was the introduction of mechanical stokers followed in the
early 1920s by pulverized coal firing, which is the ultimate extreme in
mechanization of coal handling.

For this reason, the processes involving combustion of pulverized
coal (p.c.) are of utmost interest.

One major question is: What limits the combustion rate of p.c.--
the massfeed rate or the chemical kinetics? Obviously, one wants to
arrange the combustion processes in a way that achieves maximum speed
since the size of a combustion chamber (furnace) is proportional to the
time required for the chemical reaction to consume the p.c. particles
completely.

A. Physical Processes During P.C. Combustion

Some of the processes of interest have already occurred prior to
combustion. These involve the pulverization process itself. Not all
types of coal are equally suited for pulverization. Therefore, a
"Grindability Index" was defined for characterization of different
coals. This index (ASTM D 409) is a measure for the ease of pulverization of coal, a standard coal having the assigned value of 100 for
grindability. After injection of the p.c. particles into the combustion
chamber (furnace), a two stage process takes place:

    1. volatilization of the particle
    2. carbonaceous residue combustion.

As pointed out before, coal is a conglomerate of a large number of
chemical compounds of carbon and other chemical elements. Most of these
compounds, with only a few exceptions like SiO_2, will vaporize at a much
lower temperature than carbon requires to sublime. The triple point of
carbon occurs around 103 atm. Therefore, carbon does not melt and
vaporize at atmospheric pressure; it sublimes instead. As a result,
once the incoming p.c. particle is heated by the surrounding flue gas in
the furnace, it will volatilize in a more or less violent fashion, the
liberated gases will perform regular flame type combustion, and a particle consisting now for the most part of carbon (carbonaceous residue)
remains in suspension in the flue gas.

There is evidence that these remaining particles are more or less
sphere shaped with numerous pores typically 1 μm in diameter on the surface. The pores are probably caused as the gases being generated by
volatilization escape from the particle proper. Under extreme conditions particles in the form of hollow spheres can be formed by this

process. The fact that the particles take on spherical shape prior to
combustion, while they were irregular-shaped when injected, indicates
that, during the particle heating process, a substantial amount of
carbon sublimes from the surface of the particle before combustion takes
over. Since the sublimed material is in the gaseous or near gaseous
state, combustion of this material takes place in a flame fashion--a
mode to be preferred to the solid state chemical reaction on the surface
of the particle.

Therefore, further advances in p.c. combustion technology can be
expected along the lines of improving the velocity of this sublimation
process, e.g., preheating in precombustion chambers, etc.

B. Chemical Processes During P.C. Combustion

Two different reactions kinetics are found in p.c. combustion,
which are

1. flame combustion (gaseous fuel)

2. solid state oxidation (solid fuel)

As pointed out above, the volatile part of the material forming the p.c.
emanates from the particle in gaseous form and is combusted in the
gaseous phase. The remaining carbonaceous residue is oxidized while in
the solid state. The oxidant air has to be adsorbed in the surface of
the particle and after the oxidation takes place, the major combustion
product CO has to leave the surface to make room for fresh oxidant. For
each O_2 molecule, 2 CO molecules will be formed; therefore, the gas flow
(number of particles) away from the surface will be a factor of 2 larger
than the gas flow towards the surface of the particle. This impedes the
oxidation process in two ways: first, it limits the access of fresh
oxidant and, second, it provides effective cooling of the particle
surface, thereby keeping its surface temperature too low to ignite CO
for combustion to CO_2. This combustion eventually takes place in the
flame proper, but very little CO_2 is formed on the particle surfaces.

The described mechanism determines the combustion rate for the individual particle. The combustion of the entire particle has to be
completed before it strikes the surface of a boiler tube where it would

be cooled down, dramatically reducing the rate of combustion.   In this
case, whatever remains of the particle would leave the boiler tube
region and would constitute lost fuel.   Of course, in case of the fluid
ized bed, the particle is returned to the bed and reheated.

The description given above is very simplistic and is based on the
impervious-sphere model.   In reality, the particle is covered with
numerous pores which have two counteracting effects.   First, it drastically increases the surface area and, therefore, the area available for
oxidation; second, the oxidant has to penetrate into the pores.   If the
particles are to be considered as cylinders with diameters in the order
of 1 μm, the mean free path of the air molecules will be larger than
this at these elevated temperatures (reduced densities); therefore,
molecular rather than laminar flow will take place.

In summary, one can say that the detailed description of coal
combustion must involve very complex models.   At the present time, it is
fair to say that further research certainly is needed.   A better knowledge of the combustion process would very likely provide the stimulant
for new systems concepts for coal combustion.

IV.   COAL BURNING SYSTEMS

Coal combustion has several important parameters which were described in the last section; namely, burning temperatures, heat of
combustion, oxidant control, and burning times.   These parameters control the energy output as well as waste conditions and gaseous byproduct
production. By manipulation of these factors, desired characteristics of
coal combustion can be obtained.

Since coal is not just pure carbon, but a mixture of chemical compounds, there is a limit to the techniques in which combustion parameters can be adjusted due to physical characteristics such as size,
water content or others.   With these constraints, a suitable coal combuster can still be engineered and manufactured.

Before coal is burned by the end user, there is auxiliary equipment
necessary for most coal burning systems.   Coal is stored in bins.   To
burn the coal, it must be transported from the bin to the combuster.

Techniques can range from a shovel to a pneumatic injector. Before the coal is sent to the combustor, physical characteristics may be changed, that is, the coal may be crushed, screened, and then pulverized. Therefore, all combusters encompass a particular configuration of auxiliary equipment.

There are many types of coal combustors but there are only three main categories of size; namely, utility burners using up to 500 tons of coal per hour, industrial burners using up to 20 tons per hour, and residential burners using only a few pounds per hour.

A. Fixed Bed Burners

A fixed bed burner is a grate platform on which coal is burned. Combustion air is forced upward from underneath the grate. The coal burns proportional to the amount of air forced through the grate. The ash falls through the grate into an ash pan or flies out with the exhaust gases. Grates are one of three types; a static-grate, a traveling-grate or a vibrating-grate. A static grate utilizes gravity to force ash to fall off the coal bed. A traveling-grate is like an endless belt--one continuous loop grate driven around two rollers. A vibrating-grate shakes the bars of the grate to induce motion of the coal bed. The grate is sloped to give the motion of the coal a preferred direction.

The heat of combustion rises from the grate to an overhead heat exchanger. The heat exchanger can heat air or water. For residential use, warm air is produced and distributed through the residence by ducting. Industrial steam can be produced at rates up to 1,000,000 pounds per hour (Berkowitz, 1979). Typically, these boilers are smaller, about 100,000 pounds per hour.

A stoker is used to feed coal to the grate. The coal is replenished at the same rate it is consumed. Stokers are of three basic designs; overfeed, underfeed and crossfeed. An overfeed stoker lets coal fall down to the grate where the coal is spread by traveling or vibrating. The underfeed stoker sends coal up to the grate from below by a feed screw. Distribution of the coal on the grate can be done by travelling

or vibrating. A crossfeed stoker mechanically spreads the coal evenly
across the grate. Each stoker-grate combination operation better given
a specific type of coal. When coal supplies are predictable, a specific
system with optimum efficiency can be assembled. If coal supplies vary,
a system which can use a wider variety of coal grades can be used, but
with a loss of efficiency.

Most furnaces and boilers have a variable demand: that is, the
desired amount of hot air or steam changes. To slow the heat produced,
the combustion air is reduced. Since less coal is needed, the stoker
cycle time is changed. For example, during operation the stoker may
feed coal 20 minutes and be off 5 minutes, but during a lower demand
period, the stoker may feed coal 5 minutes and be off 20 minutes.

B. Suspension Burners

To produce more steam, the rate at which the coal is burned must be
increased. To do this, the amount of air is increased and the size of
the coal is decreased, yielding a much higher fuel surface-to air ratio.
The coal size is reduced by pulverization to about 300 μm which is the
consistency of baby powder. Coal is mixed into a stream of air and
injected into a fire box at velocities about 50 ft/sec. (Berkowitz,
1979) Since the coal is very fine and the burning time is on the order
of one second, combustion takes place "suspended" in the air of the fire
box.

The air which is mixed with the coal dust is called primary air.
Additional air is needed for complete combustion. This air is called
secondary air. The directions of the primary and secondary air determines the type of fire box: if the air and coal are injected from the 4
corners of a square fire box towards the center, this is called tangen-
tial firing. Other techniques are vertical firing (downward) and horizontal firing.

The products of combustion are hot gases, fly ash, and slag (molten
ash). Heat is removed from the hot gases in the furnace by an elaborate
set of water tubes called a water wall. By selecting the correct geometry for optimum heat transfer, superheated steam at temperatures of

1000°F and pressures up to 3200 psi can be generated. To improve combustion efficiency, the primary and secondary air is heated by the furnace to 500°F. The increased efficiency can be up to 10 percent. Suspension burners can burn 10 to 500 tons of coal per hour.

There are several problems with suspension burners. Slag can collect on the tubes in the boiler, reducing their ability to transfer heat into the water in the tubes reducing the amount of steam generated. Slag will also corrode the boiler tubes, shortening the lifetime of the boiler. Large amounts of fly ash are produced in suspension burners and the fly ash must be removed by separate systems after burning. With high flame temperatures large amounts of NOX are produced. NOX production has been a major drawback in improvements in design of suspension burners: that is, to reduce the fly ash and improve efficiency by high flame temperatures, production of NOX is greatly increased—in some cases excessively, rendering this burning technique unacceptable.

Such a design was the cyclone furnace. Combustion temperatures are greater than 3000°F which reduced the fly ash to less than 20% of the solid waste. Due to the geometry of the furnace, the slag can be removed from the base of the furnace. But NOX formation is substantial and difficult to control.

The burners described up to this point are developed and commercially available. The systems described next are very promising but are still in the pilot plant stage.

C.   Fluidized-Bed Burners

Fixed bed and suspension burners have several inherent problems, namely production of NOX, SOX and fly ash as well as poor heat transfer characteristics. NOX production increases proportional to flame temperature. It cannot be removed from flue gas by any commercially available technique. However, high temperature gradients are needed to improve heat transfer. This is a conflict which requires certain compromises. Increasing the size of the boiler so temperature can be lower increases the cost of the boiler. Capital costs are also increased due

to scrubbers and precipitators to reduce SOX and fly ash to required
levels. Scrubbers and precipitators require energy to work which will
lower the net efficiency of the system.

"Fluidized-beds" (F-b) which derive their name from liquid-like
properties of the combustion material, have several advantages over
conventional burners. F-b temperatures are maintained between 1500° to
1800°F. (Yaverbaum, 1977) Conventional burner temperatures of 3000°F
can be reduced by 1500°F, decreasing the NOX formation. At F-b temperatures, ash remains solid, not interfering with heat transfer by slag
deposits. SOX can be directly removed from the F-b by mixing limestone
or dolomite into the F-b. Heat transfer properties are increased by
immersing the boiler tubes into the F-b. The tube surface area needed
to produce steam is greatly reduced compared to conventional burners.

A F-b is generated by forcing a gas upward through a baseplate. The
fuel particles are then suspended above the baseplate. Fresh fuel is
injected at the bottom of bed. Ash and sulfate $(CaSO_4)$ are tapped from
the side of the bed. Boiler tubes are in direct contact with the F-b.
Therefore, heat conducting is exceptionally high, up to 100 BTU/hr
ft^2°F. (Berkowitz, 1979) The size of the F-b may be kept as small as
10% of conventional burners producing the same power.

Coal availability is not as critical for a F-b as for a conventional
burner. Coal size can range from 325 mesh to 1 inch and still produce a
homogeneous distribution in the bed. Physical characteristics of coal,
such as ash, moisture, and caking, do not cause difficulty in operation.
Also, limestone powder can be added to the fuel to remove sulfur dioxide

$$CaCO_3 \rightarrow CaO + CO_2$$

$$CaO + SO_2 + 1/2 \ O_2 \rightarrow CaSO_4.$$

The fly ash produced is separated out by cyclone separators. It
is claimed that the advantages of the F-b provide a 50% saving as
compared to present coal burning technologies. F-b burners look very
promising but are still an immature technology which must demonstrate a
competitive position with current technologies.

D.   Novel Approaches

1.   *Magnetohydrodynamic (MHD)*

The concept of MHD power generation consists of passing a hot electrically conductive gas at high velocity through a magnetic field. By placing electrodes on each side of the gas duct, a current is induced between the electrodes. Technological difficulties, however, have hindered the development of MHD for 100 years since it was first proposed by Faraday. Required gas temperatures as high as 5000°F cause serious material erosion and corrosion problems and, even at these temperatures, the gas has to be "seeded" with a material of low ionization energy (e.g., Cs, K, Na). Molten slag collects on the electrodes reducing the current and fly ash between the electrodes is reducing the effective conductivity of the gas.

Presently only natural gas and liquid fired systems have been constructed eliminating the ash problems inherent in coal power MHD. The best output power presently obtainable is about 10 megawatts. (Squires, 1972) However, due to high flame temperatures, formation of NOX is still a major obstacle for eventual large-scale development. Coal-fired MHD is still in the research stage.

2.   *Hybrid Systems*

With improvements in gas turbine design, hybrid gas turbine-steam turbine systems are becoming practical. An example of a hybrid system follows. Coal is first gasified in a fluidized-bed-gasifier. The products produced are called crude power gas. (See Chapter 6). Removal of SOX and particulates is done before power gas is burned. Power gas is burned and expanded through a nozzle into the gas turbine. The exhaust of the gas turbine feeds directly into a steam boiler. A conventional steam turbine system is powered by the boiler. Since the flue gas will contain no SOX, fly ash venting can be direct. Advanced NOX cleanup technologies, however, need to be utilized on flue gas.

Other combinations of techniques dependent on coal conditions such as MHD-steam turbines may be beneficial where clean coal is available.

There are certain advantages to these complex systems; however, all
are still immature or unavailable on a large scale (100 megawatt and up).

3. *Coal-Oil and Coal-Methanol Mixtures*[1]

The use of coal-oil mixtures (COM) represents perhaps the best
short-term way of increasing the use of coal while utilizing existing
oil and gas burning facilities. A typical unit designed to modify an
existing facility to fire a coal-oil mixture requires a coal pulverizing
system, a coal mixing system, COM storage, a fuel handling system to
burners, burner equipment, soot blowers, precipitators, and ash removal
systems. Economic estimates of these costs suggest that at current
day oil prices COM offers considerable advantages. The possibility of
mixing a higher sulfur fuel with a lower sulfur fuel with various
additives serving as scrubbers seems feasible.

In the longer term it appears advantageous to utilize methanol
which can be produced from coal as the liquid constituent of a coal
conveying slurry to be burned in present facilities. However, the
technology of slurries having varying proportions of fuels and additives is in its infancy and considerable research and development is
needed.

References

Berkowitz, N., *An Introduction to Coal Technology*, Academic Press, New York, 1979.
Cart, E. N *et al.*, Evaluation of the feasibility for widespread introduction of coal into the
    residential and commercial sector, Government Research Laboratories of Exxon Research and
    Engineering Company for the Council on Environmental Quality, April, 1977.
Johns, L. S. *et al.*, Direct use of coal, Office of Technology Assessment, Washington, D. C.,
    1979.
Mantell, L. C., *Carbon and Graphite Handbook*, Interscience Publishers, New York, 1968.
Squires, A. M., Clean power from dirty fuels, *Scientific American*, October 1972.
Steam: Its Generation and Use, Babcock and Wilcox, New York, 1975.
Yaverbaum, L., *Fluidized Bed Combustion of Coal and Waste Materials*, Noyes Data Corporation,
    New Jersey, 1977.

[1]This subsection has been added by A. E. S. Green.

CHAPTER 6

SYNTHETIC FUELS FROM COAL

Dennis J. Miller and Hong H. Lee

I. INTRODUCTION TO COAL CONVERSION

The world's supply of petroleum and natural gas is being depleted
at an alarming rate, while the demand for these resources remains constant or is increasing. In a few years substitutes for or alternate
ways of obtaining these fuels must be secured if our energy-consuming
and highly mobilized society is to continue to exist as we know it today. As stated elsewhere in this book, coal makes up the majority of
fossil fuel reserves in this country, far outweighing the known reserves
of petroleum and natural gas. Unfortunately, coal is of a form which
makes it difficult to distribute cleanly and efficiently. Additionally,
coal often contains high levels of pollutants which make its direct use
as a fuel environmentally damaging. If coal is to become a major fuel
source in the future, some way must be found to make it economically
and environmentally acceptable to today's standards.

Coal conversion is the process of making clean-burning and easily-
handled gaseous, liquid, and solid fuels from coal. The development of
coal conversion technology has experienced a boom in the past ten years,
even though the concepts are not new. Coal conversion takes place via
a complex series of chemical and physical reactions, usually at rather
extreme conditions of temperature and pressure. These processes can, in
general, be divided into two categories: those which produce gaseous
fuels (gasification), and those which produce liquid fuels (liquefaction). Of course, there is a good deal of overlapping of these two

111

categories; many conversion processes produce both gaseous and liquid
products.

The products of gasification and liquefaction processes cover a
broad range of chemical species and compositions. In gasification processes, products vary from a low heat value synthesis gas composed primarily of hydrogen and carbon monoxide, suitable for burning or feedstock, to a high heat value synthetic natural gas suitable for pipeline
distribution. In liquefaction processes, products range from light
gasolines through fuel oils to heavy tars which are solids at room
temperature.

The primary pollutants in raw coal are sulfur and ash. In the
process of coal conversion, most of these pollutants are removed and
disposed of, leaving the products clean-burning and pollution-free.
This separation of pollutants is quite costly, resulting in higher
prices for the synthetic fuels.

The major chemical transformations involved in coal conversion are
breakdown of the coal carbon structure and increasing the hydrogen-to-
carbon ratio of coal, together leading to the formation of gaseous and
liquid products. These transformations can be realized through the
reaction of coal with gases and liquids in gaseous or liquid media. In
gasification, synthesis gas can be produced by the reaction of coal
with steam and air or pure oxygen. Synthetic natural gas is produced
either by direct reaction of coal with hydrogen gas or by further
processing of synthesis gas. In liquefaction, liquid fuels can be obtained by dissolving coal in a solvent and then reacting it with hydrogen. Another method for obtaining liquid fuels is through rapid heating
of coal and subsequent quenching to recover the volatile products in
the coal.

At the present time there are many coal conversion processes under
development and in operation, some dating back to the early 1900s when
gas from coal was used for heating and lighting. Germany further developed coal technology during World War II, when that country's lack
of natural crude oil forced alternate methods of fuel production. With
the discovery of huge natural gas reserves in the United States and the

development of a pipeline distribution system after World War II, the
need for gas from coal ended, and coal conversion technology stagnated.
The sudden realization of the finiteness of our gas and oil reserves
led to a resurgence of interest and a resultant new wave of coal conversion technology.

Even though this resurgence of interest has led to the development
of many new processes, no processes have reached the stage of commercial
operation except those developed in World War II, namely the Lurgi,
Winkler, and Koppers-Totzek processes for producing low Btu synthesis
gas. Even though the technology is mature enough to produce commercial-
scale amounts of synthetic natural gas and liquid fuel from coal, the
economics of such processes prevents their construction at this time.
Many of these processes, however, have operated successfully on a
pilot-plant scale, proving their readiness for commercial operation.

There are essentially three types of reactor configurations used
in gas-solid coal conversion processes, and several others for liquid
state reactions. For gas-solid reactors the effect of gas flow on
solids movement is the differentiating factor in reactor type. The
three types of reactors are moving bed, fluidized bed, and entrained
flow reactors. In the moving bed, solid flow is unaffected by gas flow;
in the fluidized bed, gas flow causes a mixing effect on solids; and in
entrained flow, the solid is carried along with the gas. These three
reactor configurations all have characteristics which make them
desirable for some processes and undesirable in other situations.

Coal conversion processes are very capital-intensive, making it
difficult for private corporations to undertake construction of full-
scale commercial units. Thus, the government, through the Department
of Energy, is currently funding a great deal of the gasification and
liquefaction research, and will in all likelihood supplement the construction of early coal conversion plants.

II. COAL CONVERSION PROCESSES

Coal conversion processes are able to produce both gaseous and
liquid products by a variety of methods. Although in general the

formation of hydrocarbons is the final goal, various paths are followed
by the different processes to produce a wide variety of fuels and related products.

A summary of gasification processes which have reached the pilot
plant stage of development is given in Table 1. As can be seen from
Column 8 of Table 1, most gasifiers produce little methane (pipeline
gas) directly in the gasifier. The product gas is primarily synthesis
gas, a mixture of hydrogen and carbon monoxide. Syngas is a versatile
product, used either as a fuel gas or as the feed gas in production of
synthetic natural gas, methanol, hydrogen, ammonia, or hydrocarbon
liquids. The production of hydrocarbon liquids occurs via the Fischer-
Tropsch synthesis, which is a hybridized gasification-liquefaction
process.

The liquefaction processes which have reached the pilot plant
stage of development are summarized in Table 2. Liquefaction processes
can be divided into two categories: pyrolysis processes and dissolu-
tion-extraction systems. Pyrolysis takes place in a high-temperature
gaseous media, where volatile hydrocarbons are driven from coal and then
quenched. Pyrolysis processes produce large amounts of char and gas as
byproducts. Extraction dissolution processes operate at lower temperatures and higher pressures, and generally give higher liquid yields
and less byproducts. See Table 3 for coal conversion reactions.

There is one "liquefaction" process which actually produces a
solid as the primary product. The SRC I process generates a pollution-
free high molecular weight hydrocarbon with a melting point around
300°F. This process is somewhat similar to other extraction dissolution
processes, hence its listing in this section.

III. COMPARISON AND ECONOMICS OF CONVERSION PROCESSES

The preceding tables of coal conversion processes give little idea
as to which process or processes would be applicable to a specific gasification or liquefaction project. Many factors involving both direct
and indirect economic considerations must be considered in selecting a
process. The following section attempts to sort out the advantages and

Table 1: *Gasification Processes*

Process Name	Developer (Reference)	State of Development	Reactor Type	Temp (°F)	Pressures	Feed Gas[a]	Products %[b]	HHV, Btu/ft[3]
Bigas	Bituminous Coal Research (Grace, 1975)	120t/day pilot plant	Ent.	2400-2700	100-1500 psig	St[a], O₂	8	380
C-E	Combustion Engineering (Winterson, 1975)	5t/day pilot plant operating	Ent.	3200	1 atm	St, Air	<5	120
CO₂-Acceptor	Consolidation Coal Co. (Fink et al., 1975)	40t/day pilot plant operating	Flu.	1500-1900	150-300 psig	St	17	440
GE Gas	General Electric (Kydd, 1975)	3/4t/day pilot plant operating	Stir.	N.A.	0-300 psig	St, Air	<5	160
Hybrid	Hitachi Research, Japan (Miyadera et al., 1978)	Laboratory scale unit	Flu.	1700	70-300 psig	St, O₂	18 T	460
Hydrane	U.S. Bureau of Mines (Gray and Yovorsky, 1975)	24t/day pilot plant planned	Flu.	1500	1000-1500 psig	St	73	820
HyGas	Institute of Gas Technology (Anastasia and Bair, 1975)	75t/day pilot plant	Flu.	1750	1000 psig	St, H₂	20 C	370-550
Kiln Gas	Allis-Chalmers (Chem. and Eng. News, 1978)	10t/day pilot plant in construction	Tum.	800-1000	200-500 psig	St, Air	5	100-200
Koppers-Totzek	Heinrich Koppers (Whiteacre et al., 1975)	15 plants in commercial operation	Ent.	2750-3300	1 atm	St, O₂	0	300
Lurgi	Lurgi Mineralotecknik (Bodie and Vyas, 1975)	14 plants in commercial operation	Mov.	1150-1400	300-500 psig	St, O₂	5 TH	300
Molten Iron	Applied Technology, Inc. (LaRosa and McGarvey, 1975)	Laboratory scale unit operating	Molt.	2500	1 atm	St, Air	0	160
Molten Salt	Atomics International (Kohl et al., 1978)	1t/day pilot plant in construction	Molt.	1700	1200 psig	St, Air	0	150
Synthane	U.S. Bureau of Mines (Haynes and Forney, 1975)	70t/day pilot plant	Flu.	1800	500-1000 psig	St, O₂	15 TC	400
Texaco	(Hottell and Howard, 1971)	Pilot plant operating	Ent.	2000	1 atm	St, Air	<5	170
Tri-Gas	Bituminous Coal Research (Colaluca et al., 1979)	Laboratory unit in operation	Flu.	1000	250 psig	St, Air	<5	150
U Gas	Institute of Gas Technology (Patel and Loeding, 1975)	Demonstration unit in design	Flu.	1900	350 psig	St, Air	4	150
Westinghouse	Westinghouse (Andermann, 1978)	16t/day pilot plant operating	Flu.	1800-2000	250 psig	St, Air	0	140
Winkler	Davy Powergas, Inc. (Banchik, 1975)	16 plants in commercial operation	Flu.	1500-1850	15 psig	St, O₂	2	270

Ent. - entrained bed; Flu. - fluidized bed; Stir. - stirred bed; Tum. - tumbling bed; Mov. - moving bed; Molt. - molten bath

[a] St - steam   [b] % methane   T - tars   C - char   TH - Tar, heavy oil   TC - tar, char;   N.A. - information not available

Table 2: *Liquefaction Process*

Process Name	Developer (Reference)	State of Development	Process Type	Reactor Type	Temperature	Pressure	Liquid Product Yields	Other Products
Coalcon	Union Carbide (Martin, 1975)	3000 t/day demonstration plant in design	P	Flu.	800-1000°F	500-1000 psig	Gasoline .5 BBL/ton Fuel Oil 1.3 BBL/ton	10000 SCF SNG/ton
COED	FMC Corporation (Jones, 1975) (Hottell and Howard, 1971)	36 t/day pilot plant operating	P	Flu.	800-1600°F	6-10 psig	Fuel Oil 1 BBL/ton	8000 SCF SNG/ton 1200 lb char/ton
CSF	Consolidation Coal Co. (Phinney, 1975) (Neben, 1978)	70 t/day pilot plant operating	E-D	Stir.[a]	700-900°F	3000 psig	Naptha .5 BBL/ton Fuel Oil 1.5 BBL/ton	3400 SCF SNG/ton
Dow	Dow Chemical (Pruitt, 1978)	200 lb/day miniplant in operation	E-D	Stir.[b]	800-850°F	high, NA	Naptha 1 BBL/ton Fuel Oil 2 BBL/ton	200 lb char/ton 1000 SCF SNG/ton
Garrett	Occidental Petroleum (Green, 1975)	4 t/day pilot plant in operation	P	Ent.	1000-1600°F	0-50 psig	Light Oil 1 BBL/ton	8500 SCF SNG/ton 1200 lb char/ton
H-Coal	Hydrocarbon Research (Johnson et al., 1975) (Kunesh et al., 1978)	3 t/day pilot plant in operation	E-D	Flu.[c]	850°F	2200-2700 psig	Naptha 1.5 BBL/ton Distillate 1 BBL/ton	3000 SCF SNG/ton
SRC I	Pittsburgh & Midway Coal Mining (Bodle and Vyas, 1975)	50 t/day pilot plant in operation	S-R	Stir.	600-800°F	1000 psig	Ash and sulfur-free solid product	none
SRC	Pittsburgh & Midway Coal Mining (Anderson, 1977)	Laboratory scale unit operational	E-D	Stir.	600-800°F	1900 psig	Light and Middle Weight Oils	none
Synthoil	U.S. Bureau of Mines (Friedman et al., 1975)	1/2 t/day pilot plant operational	E-D	Fix.[c]	800-850°F	2000-4000 psig	Distillate 3 BBL/ton	none
Toscoal	Oil Shale Corporation (Carlson et al., 1975)	25 t/day pilot plant in operation	P	Tum.	800-1000°F	N.A.	Fuel Oil .5 BBL/ton	1500 SCF SYG/ton

P - Pyrolysis   E-D - Extraction Dissolution   S-R - Solids Refining
Flu. - Fluidized   Stir. - Stirred   Ent. - Entrained   Fix. - Fixed Bed   Tum. - Tumbling Bed
N.A. - not available   SNG - Synthetic Natural Gas

[a] Catalyzed, type N.A.   [b] emulsion catalyst   [c] Cobalt-moly catalyzed

Table 3: *Coal Conversion Reactions*

Name of Reaction	Reaction	$-\Delta H_R, 25°C, \frac{kcal}{mole}$	Function
Steam Gasification	$C(coal) + H_2O \rightarrow CO + H_2$	-41.9	Formation of synthesis gas (syngas)
Water-Gas Shift	$CO + H_2O \rightarrow CO_2 + H_2$	- 0.68	Increase H_2/CO ratio in syngas
Methanation	$CO + 3H_2 \rightarrow CH_4 + H_2O$	49.3	Formation of methane from syngas
Methanol Synthesis	$CO + 2H_2 \rightarrow CH_3OH$	30.6	Formation of methanol from syngas
	$CO_2 + 3H_2 \rightarrow CH_3OH + H_2O$	31.3	
Fishcher-Tropsch Synthesis	$nCO + 2nH_2 \rightarrow (CH_2)_n + nH_2O$ $2nCO + nH_2 \rightarrow (CH_2)_n + nCO_2$	>0	Formation of liquid hydrocarbon fuels from syngas
Hydrogasification	$C(coal) + 2H_2 \rightarrow CH_4$	17.9	Formation of methane directly from coal
Combustion	$C + O_2 \rightarrow CO_2$	94.0	Provide heat for gasification
General Liquefaction	$(CH_{0.5})_n (coal) + nH_2 \rightarrow (CH_{2.0})_n$	>0	Increase hydrogen content of coal to form liquid products

disadvantages of several aspects of coal conversion processes and set
up guidelines to be used when selecting a gasification or liquefaction
process for a particular application. One concept that will emerge
from the comparisons is that no one process is suitable for all applications, and a process suitable for one application may not be suitable
for another.

Somewhat unique to coal conversion processes is the fact that
indirect economic considerations may outweigh the costs directly related to construction and operation of the conversion processes. These
indirect factors include compatibility of the process with the coal to
be processed, pretreatment requirements for the coal, reliability of a
given process, environmental aspects, and usage or disposal of waste
products and byproducts. The factor that looms above these, however,
is the selection of a process that gives the desired products. In
gasification processes, the production of syngas may be only the first
step in the desired product synthesis. The gasification process chosen
should produce a synthesis gas feedstock with composition closest to
that required for synthesis of the final products. For instance, the
formation of methane in the gasifier is generally not desirable unless
synthetic natural gas is desired as the final product. For other
product syntheses, such as ammonia or methanol, methane is an impurity
which must be removed from the product stream. In liquefaction, the
wide range of liquid fuels produced by the various processes makes
proper selection of a process equally important.

Along with these general guidelines for selecting a process for a
given situation there are many other, more traditional factors contributing to the economic analysis of coal conversion. High temperatures
and pressures demand rugged equipment, resulting in higher capital
costs. Other factors contributing to capital costs are corrosion resistance, reactor type, separation processes, and purification
facilities. Some capital cost factors unique to coal conversion are
solids feeding into pressure vessels, pretreatment of coal, oxygen and
hydrogen supplies, and gas or liquid storage facilities. With regards
to operating costs, reactor type will determine pressure drop and

recycle requirements, and very high temperature processes may show
excessive heat losses. A general rule of thumb in gasification is to
have the process pressure decreasing in each successive step of the
system, since several moles of gas are usually formed for each mole of
feed gas. For all conversion processes, a simpler system usually results in lower operating costs.

Although not all of the economically-related factors for coal conversion have been listed above, those which have been listed provide a
basis upon which preliminary decisions can be made. The literature
also reports several studies that can provide insight into the appro-
priateness of a given process.

For the production of pipeline gas from raw coal, Knudsen and
Hedman (1978) have done comparative economic estimates of five processes
and found the HyGas process most economical, followed by CO_2-acceptor,
Bigas, Lurgi, and Synthane processes. Arora et al. (1977) have made
an economic comparison of U-Gas and HyGas, and found HyGas more
economical for high-Btu gas manufacture. However, the cost per Btu of
low-Btu gas produced from U-Gas was much less than the cost per Btu of
high-Btu gas produced from HyGas, showing the advantage of using U-Gas
for producing boiler fuel and HyGas for producing pipeline gas. Morel
and Kim (1977) have done comparative studies for production of methanol
and ammonia, showing that the choice of a gasifier depends heavily on
the products desired.

In liquefaction, comparative economics have been made by O'Hara
et al. (1977) between the Fischer-Tropsch and SRC II processes. The
Fischer-Tropsch process is somewhat more costly than the SRC II process
for producing gasolines and light oils. Cochran (1976) gives rough
yields and comparisons of various groups of liquefaction processes.

IV. RESEARCH IN COAL CONVERSION

The preceding pages have given an overview of the available coal
conversion technology, and have briefly compared some of the conversion
processes which have been developed to the pilot plant stage or beyond.
However, these processes by no means represent the apex of coal

conversion technology; in fact, coal conversion is one of the most
active areas of technical research in this country today. Research in
coal conversion is being carried out at all levels from fundamental
studies of reaction mechanisms to new concepts for large-scale processes. The scope of this research is too broad to be covered here,
but an attempt will be made to categorize and discuss the most important new developments.

Present day coal conversion research can be divided into two
loosely-defined categories. The first deals with the development of
new concepts in coal conversion processes; the second is centered
around the more fundamental concepts of the chemistry and physics of
the coal conversion reactions. Within both of these categories, a
great emphasis is currently being placed on catalysis. The development
of economical catalysts for the coal conversion reactions could be the
key factor that initiates large-scale construction of coal conversion
plants.

New coal gasification processes are being developed primarily
using fluidized bed and entrained flow reactors. The Hoffman process
(Hoffman, 1978) utilizes a nickel catalyst and steam feed gas in a
fluidized bed reactor to produce methane from coal. The process is
unique in that methane is formed from the intermediates, hydrogen and
carbon monoxide, directly in the fluidized bed. The nickel catalyst
promotes methane formation from the intermediates, and the product gas
is suitable for pipeline transportation without further methanation.
Another fluidized bed process has been studied by Wen (1974), in which
coal is gasified in two reactors to produce a low-Btu gas.

The entrained-solids reactor has received a great deal of attention
in recent research. Coates, Chen, and Pope (1974) have studied devolatilization of coal in a short residence time (.01-.03 sec.) reactor
and found up to 14% of the carbon in coal was gasified. Greene (1977)
discusses development of the Cities Service Short Residence Time
process, in which coal is gasified in a hydrogen stream to produce
methane, along with some tar and liquid byproduct. Hydrogen in this
process is produced by steam gasification of residual char. Another

dilute solids gasifier is under development by the U.S. Bureau of Mines
(Feldman et al., 1974) in which raw coal falls freely through a stream
of hydrogen at high temperature and high pressure. This process gives
up to 95% methane formation directly in the gasifier, with only 5% by
subsequent methanation. Also, the free-fall reactor eliminates the
need for coal pretreating, thus raising the volatiles yield.

These short residence time reactors appear practical because they
allow large throughputs of solid per volume of reactor. Research in
the catalysis of hydrogasification reactions is directed toward increasing the product yield in the reactors as well as moderating the
reaction conditions. The primary catalysts used for hydrogasification
are the alkali metal salts and oxides. The catalytic effects of these
compounds has been studied by Gardner, Samuels and Wilks (1974) on coal
chars. Using potassium carbonate as a catalyst, they report a two- to
five-fold increase in reaction rate. In a somewhat similar study,
Chauhan et al. (1977) studied calcium oxide as a catalyst on raw coal,
and reported an increase in reaction rate as well as a decrease in
agglomerating tendencies. In both of these studies the catalyst was
impregnated onto the solid coal particles. Other catalysts studied
pertaining to coal hydrogasification include nickel (Nishiyama and
Tamai, 1979) and various minerals (Tomita et al., 1977).

Also, research is being done on catalytic steam gasification of
coal, again using primarily alkali metal carbonates as catalysts.
Haynes et al. (1974) and Willson et al. (1974) both studied these
catalysts in steam gasification, and Vadovic and Eakman (1978) have
developed a kinetic model for catalytic gasification in the Exxon
Catalytic Gasification process.

Most research on catalysis focuses upon the effect of catalyst on
conversion rate, without much detailed thought about actual kinetics
or reactor design. Fundamental research in coal gasification concentrates on reaction kinetics and physical processes of the gasification
reactions. The three most studied reactions are hydrogasification of
coal chars (Walker et al., 1977; Chauhan and Longabach, 1978; Johnson,
1974), hydrogasification of raw coal (Johnson, 1977; Weil et al., 1978;

Chambers and Yavarsky, 1978; Feistel et al., 1977), and steam gasification of raw coal (Smoot et al., 1977; Miyadera et al., 1978; Schobert
et al., 1978; Feistel et al., 1977; Linares et al., 1977). In these
kinetic studies the primary results are simple models describing the
rate of coal conversion as a function of several reaction parameters.

Research in liquefaction is following many of the same directions
as gasification, but the technological level of liquefaction is somewhat below that of gasification. Conversion of fixed carbon in coal is
a thermodynamically difficult process, and yields of carbon in the form
of hydrocarbons are generally lower in liquefaction than in gasification. Thus it is not surprising that novel ways of obtaining liquid
fuels from coal are currently under investigation in all phases of
liquefaction.

Pyrolysis represents an important group of liquefaction processes,
due to its simplicity and relatively low cost. By removing volatiles
from raw coal, it essentially accomplishes the easy step in liquefaction, leaving the fixed carbon base as a char residue. The present
emphasis in new pyrolysis processes is on very short residence time
exposure to high temperature. When coal is pyrolyzed, hydrocarbons
are formed which will quickly "crack," or decompose, at the high
pyrolysis temperature. By keeping the residence time to a few seconds,
the decomposition is halted and a higher yield of volatile liquids
results. Steinberg and Fallon (1979) showed that residence times of
two to seven seconds gave a maximum yield of hydrocarbon liquids of
ten percent of raw coal weight. Willson et al. (1977) also studied
oil yields from pyrolysis.

Extraction-dissolution processes also command research attention.
Novel approaches to liquefaction include atmospheric pressure liquefaction (Mochida and Takeshita, 1978), supercritical extraction of
coal volatiles (Maddocks et al., 1979), and short residence time
extraction (Longenbach et al., 1979). The short residence time extractor uses a version of the SRC process with extractor residence
times of one to three minutes. The atmospheric pressure liquefaction
process being developed in Japan utilizes pyrenes as coal solvents,

and has extractor residence times near one hour. The supercritical extraction of volatiles from coal is based on the enhanced volatility of
hydrocarbons in coal in the presence of a gas near its critical point.
Supercritical gases can enhance the volatility of heavy molecules up to
ten thousand times, thus allowing extraction of heavy liquids below
their boiling points.

As in gasification, the use of catalysts to accelerate reaction
rates in liquefaction is currently the subject of several investigations.
The major reaction in extraction dissolution processes is the hydrogenation of carbon and hydrocarbons in the coal-solvent slurry. In this
system the catalyst is usually supported on alumina mixed with coal-
solvent slurry. The most effective catalyst is cobalt-molybdenum (Stern
and Hinden, 1978), which shows best resistance to deactivation and
greatest acceleration of reaction rate. Other catalysts for liquefaction
studied are molybdenum-iron (Morita et al., 1979), zinc chloride-
methanol (Shinn and Vermeulen, 1979; Yoshida and Bodily, 1979), nickel-
molybdenum (Veluswamy et al., 1979), stannous chloride (Loverto and
Weller, 1979) and nickel-tungsten (Veluswamy et al., 1979).

The kinetics of coal liquefaction processes are not yet well
developed; work is being undertaken to determine rate dependencies on
reaction parameters. Kinetics of coal pyrolysis have been investigated
by Suuberg et al. (1977) and Antal et al. (1977). An extensive review
of devolatilization literature is given by Anthony (1976). The kinetics
of hydrogenation processes are even less well defined, because of the
overall complexity of the extraction dissolution process. Development
of a kinetic model applicable over a wide range of operating conditions
may prove to be extremely difficult; scale-up of commercial units will
in all likelihood depend on pilot plant data.

We have active research programs here at the University of Florida
on the catalytic gasification of coal using alkali and alkaline earth
compounds as the catalyst. Many catalyst screening studies have shown
that these catalysts are effective and that optimal loading of catalyst
is 10 to 20% of the weight of coal or coal char. While there are more
effective catalysts, the high costs involved in using typical hydrogenation catalysts has led to the use of relatively cheap alkali and alkaline

earth compounds. Unlike the traditional catalytic reactions that involve gaseous reactants, the catalyst has to be continuously introduced
to the reactor with the catalyst-impregnated coal. This means that a
large amount of catalyst has to be used and that, even with efficient
catalyst recovery and reuse, the costs of catalyst become extremely high.
The other factor in the choice of alkali metal catalysts is quick deactivation of traditional catalysts by sulfur compounds in coal.

The primary goals of our research programs are twofold: a fundamental understanding of the chemical and physical rate processes of
catalytic gasification, and an understanding of the role of the catalyst
in enhancing the gasification rate. The physical and chemical rate
processes at work during gasification are the intrinsic kinetics of
gasification, mass and heat transfer, migration and chemical state of
catalyst, and deactivation of catalyst. These rate processes interact
with one another to yield the gasification rate that is actually measured
and observed. Without a good understanding of these rate processes, a
reactor design based on kinetics observed under certain conditions
would fail to perform as expected under other circumstances. In view
of the significant costs associated with the catalyst, an understanding
of catalyst deactivation figures significantly in regeneration and
reuse of the catalyst. While many laboratory studies using fresh
catalyst have shown negligible deactivational effects except in the high
conversion range, it should be recognized that to make the process
economical, the catalyst has to be regenerated and reused in plant
operation. For instance, Exxon studies on catalytic gasification of
coal (Exxon Research, 1978) on a pilot plant scale showed that the
catalyst was indeed deactivated and had to be regenerated. A fundamental understanding of these rate processes would invariably lead to an
insight into the role of the catalyst. An understanding of the dispersion and chemical state of catalyst during gasification, the effect
of pretreatment conditions on these rate processes, and the effect of
the use of various alkali and alkaline earth compounds, in particular
the anion effect and the dispersion on these rate processes would pave
the way to an understanding of the role of the catalyst.

Many studies on catalytic gasification used catalysts impregnated in fine coal particles. In order to isolate the effect of one rate process from another, use of a pellet is particularly useful, either made of impregnated coal particles or surrounded by an outer layer of catalyst. For instance, the coal pellet surrounded by an outer layer of catalyst is particularly useful for the study of catalyst migration and chemical state of catalyst during gasification. The mass transfer effect or the diffusion of gaseous reactant to the reaction sites can also be well characterized by the use of the model catalyst system of pellets. An experimental method (Lee, 1979a) has been devised to determine the change of the thickness of the outer catalyst layer of the model pellet with the extent of reaction, and to determine the diffusivity of gaseous reactant and the porosity. Rate processes for the model pellets have also been formally analyzed (Lee, 1978). The diffusional effect on the intrinsic kinetics has been studied for the model system (Lee, 1979b). While these model systems are necessary to study individual rate processes, the catalyst impregnated coal particles typically used in gasification (10 to 20% catalyst of coal weight) may or may not form a continuous outer layer of catalyst, depending on the particle size or to be more specific, its surface area. If the catalyst is finely dispersed on the particle surface, the rate process of sintering or the agglomeration of catalyst particles during the course of gasification becomes important. This sintering can significantly reduce the reactivity of the catalyst. For instance, Dalla Betta (1976) found that in an automobile catalytic converter the reactivity of the catalyst was reduced twentyfold due to the sintering. An analytical tool for sintering in this instance has been developed (Lee, 1979) which can be used for the catalytic gasification of coal. We are currently investigating these rate processes using in our laboratory a high-temperature, high-pressure reactor proper.

The coal research laboratory in the Department of Chemical Engineering is equipped with a high-temperature, high-pressure reactor designed to operate up to 1000°C and up to 1000 psi. The laboratory is equipped with a gas chromatograph, a Perkins-Elmer sorptometer for the

characterization of pellets, a pelletizer and has access to TEM, an
electron microprobe analyzer, X-ray diffraction and spectroscopy.

V.   ALTERNATE SOURCES OF SYNTHETIC FUELS

The versatility and value of high-quality hydrocarbon fuels has
been the primary motivation for development of coal gasification and
liquefaction. Because of the importance of synthetic fuels alternative
sources have been under scrutiny for economic and technological feasibility. There are three major alternatives to coal for synthetic fuels
production: heavy petroleum residuals, oil shale, and oil from tar
sands.

The production of synthesis gas from petroleum residual oil involves essentially the same processes as the production of syngas from
coal. The chemical composition of heavy oil is somewhat similar to that
of coal; steam gasification allows removal of the high levels of sulfur
and heavy metals present in the residual oils while producing the useful
syngas. In addition to gasification, heavy residuals can be cracked
and hydrogenated to form liquid hydrocarbon fuels. Processing oil shale
will be in all likelihood the first commercial operation for production
of synthetic fuels in the United States (Radding, 1977). There are
tremendous reserves of shale oil in the U.S. nearly two trillion barrels
total, with about 600 billion barrels concentrated enough to warrant
recovery. The oil in shale is present throughout the porous structure
of the inorganic rock; the removal of the oil involves pulverizing the
oil-rich rock and heating it to drive off and decompose the products.
The product is a heavy oil similar to crude oil, but with substantial
amounts of nitrogen and sulfur which must be removed before conversion
to synthetic fuels. The crude oil produced can then either be gasified
or refined in the conventional way to produce the desired synthetic
fuels. Besides the nitrogen and sulfur, other waste disposal problems
involve disposal of waste solid and large amounts of contaminated water
present in the rock. An alternative to mining and transport to a processing plant is in-situ removal of oil involving steam injection; however, in-situ oil shale processing appears very difficult, both economically and technologically.

The third major source of synthetic fuels is from the Canadian
tar sands (Radding, 1977). There are approximately one trillion
barrels of oil in the tar sands, of which about 250 billion barrels
are recoverable. As in oil shale production, there are two methods for
production of oil from tar sands: in-situ and mining-processing. In-
situ recovery appears more attractive with tar sands, due to the easier
accessibility to oil in sand than in solid rock. The processing of tar
sands involves mobilizing oil, using heat or steam as in shale oil
production. The product oil is a heavy crude, high in sulfur and heavy
metals content. Formation of synthetic fuels from this oil involves
desulfurization, then cracking or gasification to produce the desired
synthetic fuels.

In comparing synthetic fuels from coal, the technology of oil shale
and tar sands recovery processes are simpler than those in coal conversion. The oil shale technology is in use in several foreign countries, and there is a 50,000 barrel per day tar-sands oil-production
unit operating in Canada. Coal does have two important advantages over
these processes, however. The yield of oil from coal is much higher
(3 barrels per ton as compared to 0.8 barrels per ton for oil-rich shale),
and the solids disposal is much less a problem, as coal char can be
burnt to produce heat. Also, tar sands occur in Canada, and use of the
oil would require that it be imported.

VI. SUMMARY

Coal conversion is the process of making clean-burning liquid and
gaseous fuels from coal. While coal conversion goes back many years,
only in the past few years has coal conversion become a viable alternative to the use of petroleum. The basic concept of coal conversion
chemistry is to enrich raw coal with some form of hydrogen to produce
hydrocarbon products with relatively large hydrogen content as compared
to coal. This can be accomplished by reaction of coal with hydrogen or
steam, or by removing excess carbon from coal to give a hydrogen-rich
product and a carbonaceous byproduct. Coal conversion processes can be
divided into two categories: those which produce gaseous fuels

(gasification), and those which produce liquid fuels ranging from light
gasolines to heavy tars (liquefaction). These two types of processes
produce much overlapping, and many systems produce both gaseous and
liquid products.

There are currently several commercially-operating gasification
processes, and many more in the pilot plant stage of development. These
gasification processes can be categorized in terms of reactor type and
products formed in gasification. The raw synthesis gas produced in
most gasifiers can be used either as a fuel or as a feedstock to make
synthetic natural gas, methanol, hydrogen, ammonia, or liquid hydrocarbon fuels via the Fischer-Tropsch synthesis.

Liquefaction processes have not yet reached the commercially
feasible stage of development because the high cost of the resulting
clean-burning fuels cannot compete with the lower cost of petroleum-
derived fuels. Liquefaction processes can be classified as either
pyrolysis or extraction types; the products formed range from natural
gas byproducts to gasolines to high molecular weight fuel oils and tars.

Research in coal conversion is a very active field, as science
strives to develop ways of making coal conversion competitive with crude
oil for fuels production. Primary emphasis in the research has been
placed on catalysis and reaction kinetics of coal conversion reactions.
Effective catalysis of the conversion reactions will lower the extreme
pressures and temperatures necessary to drive the reactions, thus
lowering capital and operating costs for the processes operating under
these extreme conditions.

Finally, coal conversion to synthetic fuels has the potential to
supply an important part of our need for clean energy in the next several decades. Technology is available to produce both gaseous and
liquid fuels from coal; only economic considerations of the processes
prevent large-scale production at present. Ever-increasing cost of
crude oil, however, almost assures early, widespread, large-scale production of coal-derived fuels.

VII. CONCLUDING REMARKS

The technology for producing gaseous and liquid fuels from coal has been developed to the point where large-scale production is possible. But, at the present time, hydrocarbon products derived from coal cost about one-and-a-half times as much as those products derived from crude oil. However, the ever-increasing price of crude oil and more stringent environmental controls will in all likelihood close this price gap in the near future. In addition, research is being carried out related to catalysis and new coal conversion processes, with the ultimate goal being a reduction in product prices.

Making fuels and related products from coal will soon be attractive for other than purely economical reasons. Most importantly, the products found in coal conversion are essentially pollution free, having had most of the pollutants removed during the conversion process. The recent realization that further damage to our environment must be prevented makes these clean-burning fuels attractive for future use. Secondly, the coal reserves in this country far outweigh the petroleum and natural gas reserves, assuring a supply of synthetic fuels far past the time when petroleum reserves are exhausted. Our society's dependence on fossil-fuel transportation demands that we secure adequate fuel supplies for the years ahead.

In retrospect, it should be mentioned that coal conversion will be neither an easy nor a final answer to the energy problem. Producing synthetic fuels from coal in quantities similar to those produced presently from crude oil would require a huge national monetary commitment. Coal conversion processes are very capital-intensive; converting to coal derived fuels would incur a cost similar to that of rebuilding the nation's petroleum refining system. Besides the cost, the effect on the environment of waste byproduct (primarily sulfur) disposal and large-scale mining must be studied before commercial processes can operate.

Finally, our coal reserves are finite, as are all our fossil fuel reserves. Thus, coal conversion should be seen as only an intermediate solution to the problem of supplying fuel to our transportation-oriented society. In effect, we are buying time while yet undeveloped and somewhat more permanent sources of energy to satisfy our needs can be brought into use.

Heavy petroleum residuals, oil shales and tar sands can serve as
alternatives to coal as feedstock for synthetic fuel. Shale oil processing will become a major source of fuel if the equipment for handling
vast amounts of solids can be financed and constructed, while the development of tar sands processing is being carried out in Canada. Coal
will be used on perhaps a more limited scale for the production of high
quality fuels because of its greater value as a power producer by
direct combustion.

Appendix 1: Conversion Efficiencies

In this appendix we will take a brief look at the thermal efficiencies associated with conversion of coal to synthesis gas and methane.
Synthesis gas is produced by the reaction of steam with raw coal

$$C + H_2O \rightarrow CO + H_2 .$$

This reaction is highly endothermic, requiring about 40 kcal for
each mole of carbon gasified. In addition, steam must be supplied at
the reaction temperature. For a gasifier at 1000°C, an extra 20 kcal
must be supplied for steam generation per mole of carbon gasified.
This heat is supplied by coal combustion in the gasifier,

$$C + O_2 \rightarrow CO_2; \qquad \Delta H_R = - 94 \text{ kcal/mole C.}$$

Thus, for the gasification of one mole of carbon, an extra .64 moles of
carbon must be burnt to provide the necessary heat.

The gasification produces two moles of synthesis gas, a 50-50 mixture of hydrogen and carbon monoxide. The heat obtainable from the
combustion of this synthesis gas is shown in Table 4.

$$68.3 \text{ kcal/mole } H_2 + 67.6 \text{ kcal/mole CO} = 135.9 \text{ kcal/mole}$$

TABLE 4:	*Heats of Combustion*
Species	$-\Delta H_C$, 25 °C, $\frac{kcal}{mole}$
C	94.0
CO	67.6
H_2	68.3
CH_4	212.8

If the coal had been burnt instead of gasified, the heat obtained would have been

$$1.64 \text{ moles} \times 94 \text{ kcal/mole} = 154 \text{ kcal.}$$

Thus, the thermal efficiency for the production of synthesis gas could be calculated as

$$\frac{\text{heat supplied through gasification}}{\text{heat supplied from direct burning}} = \frac{135.9}{154} = 88\%$$

This calculation is a very rough one, and does not take into account such factors as incomplete coal conversion, sensible heat losses, coal preheating, etc. It does, however, give a rough estimate of the maximum efficiency that could be expected in syngas formation.

In the production of methane, the thermal efficiency is much lower. This is because the methanation reaction occurs at lower temperatures than gasification, and the heat produced during methanation cannot be recycled to the higher temperature gasifier, but instead goes out as waste heat.

The methanation process yields .5 mole of methane for every mole of carbon gasified by steam. The other .5 mole of carbon is converted to carbon dioxide in the shift reaction. The estimated thermal efficiency is

$$(.5 \text{ mole } CH_4)(212.8 \text{ kcal/mole})/(154 \text{ kcal}) = 69\%.$$

Once again the actual thermal efficiency will be much lower, usually closer to 50%.

References

Anastasia, L. J. and W. G. Bair, The HYGAS process, Papers from Clean Fuels from Coal Symposium II, IGT, p. 177, Chicago, 1975.

Andermann, R. E. and G. B. Haldipur, Development of an advanced fluidized bed coal gasification process, Preprints from Div. of Fuel Chem., ACS, 23(3), 142, 1978.

Anderson, R. P., The SRC II process, Preprints from Div. of Fuel Chem., ACS, 22(7), 132, 1977.

Antal, M. J., E. G. Plett and T. P. Chung, Recent progress in kinetic models for coal pyrolysis, Preprints from Div. of Fuel Chem., ACS, 22(1), 137, 1977.

Anthony, D. and J. B. Howard, Coal hydrogasification and devolatilization - A journal review, Amer. Inst. of Chem. Eng. Journal, 22(4), 625, 1976.

Arora, J. L., K. B. Burnham, and C. L. Tsaros, High- and low Btu gas from Montana sub-bituminous coal, Preprints from Div. of Fuel Chem., ACS, 22(7), 72, 1977.

Banchik, I. N., The Winkler process for the production of Low-Btu gas from coal, Papers from Clean Fuels from Coal Symposium II, IGT, p. 359, Chicago, 1975.

Batchelder, R. F., and Y. C. Fu, Evaluation of use of syngas for coal liquefaction, Preprints from Div. of Fuel Chem., ACS, 23(1), 30, 1978.

Bertalacini, R. J., L. C. Gutberlet, O. K. Kim, and K. K. Robinson, Relation of coal liquefaction catalyst properties to performance, Preprints from Div. of Fuel Chem., ACS, 23(1), 1, 1978.

Bodle, W. W. and K. C. Vyas, Clean fuels from coal-introduction to modern processes, Papers from Clean Fuels from Coal Symposium II, IGT, p. 14, Chicago, 1975.

Bodle, W. W., K. C. Vyas, and A. T. Talwalkar, Clean fuels from coal - technical historical background and principles of modern technology, Papers from Clean Fuels from Coal Symposium II, IGT, p. 53, Chicago, 1975.

Carlson, F. B., L. H. Yardumian, and M. T. Atwood, The TOSCOAL process-coal liquefaction and char production, Papers from Clean Fuels from Coal Symposium II, IGT, p. 495, Chicago, 1975.

Chambers, H. F. Jr., and P. M. Yavorsky, Production of SNG by free fall, dilute-phase hydrogasification of coal, Preprints from Div. of Fuel Chem., ACS, 23(3), 150, 1978.

Chauhan, S. P., and J. R. Longanbach, Determintion of the kinetics of hydrogasification of char using a thermobalance, Preprints from Div. of Fuel Chem., ACS, 23(3), 73, 1978.

Chauhan, S. P., H. F. Feldman, E. P. Stambaugh, and J. H. Oxley, A novel approach to gasification of coal using chemically incorporated catalysts, Preprints from Div. of Fuel Chem., ACS, 22(1), 38, 1977.

Low-Btu gasification process nears market, Chemical and Engineering News, p. 29, May 8, 1978.

Coats, R. L., C. L. Chen, and B. J. Pope, Coal devolatilization in a low pressure low residence time entrained flow reactor, Coal Gasification, ACS Advances in Chemistry Series 131, p. 9, Washington, D.C., 1974.

Cochran, N. P., Oil and gas from coal, Scientific American, 234(5), 24, May, 1976.

Colaluca, M. A., M. A. Paisley, and K. Mahajan, The tri-gas gasification process, Chem. Eng. Prog., 75(6), 33, June, 1979.

Dalla Betta, R. A., R. C. McCune, and J. W. Sprys, Ind. and Eng. Chem. Prod. Res. Develop., 15, p. 169, 1976.

Exxon Research and Engineering Company, Exxon catalytic coal gasification process: Predevelopment program, Annual Report to Dept. of Energy, FE-2369-20, 1978.

Feistel, P. P., K. H. van Heek, H. Jungten, Gasification of a German bituminous coal with H_2, H_2O, and H_2-H_2O mixtures, Preprints from Div. of Fuel Chem., ACS, 22(1), 53, 1977.

Feldman, H. F., J. A. Mina, and P. M. Yavorsky, Pressurized hydrogasification of raw coal in a dilute-phase reactor, Coal Gasification, ACS Advances in Chemistry Series 131, Washington, 1974.

Fink, C., G. Curran, and J. Sudbury, CO_2-acceptor process pilot plant-1974, Papers from Clean Fuels from Coal Symposium II, IGT, p. 243, Chicago, 1975.

Friedman, S., P. M. Yavorsky, and S. Akhtar, The SYNTHOIL process, Papers from Clean Fuels from Coal Symposium II, IGT, p. 481, Chicago, 1975.

Gardner, N., E. Samuels, and K. Wilks, Catalyzed hydrogasification of coal chars, Coal Gasification, ACS Advances in Chemistry Series 131, p. 209, Washington, 1974.

Grace, R. J., Development of the BiGas process, Papers from Clean Fuels from Coal Symposium II, IGT, p. 207, Chicago, 1975.

Gray, J. A., and P. M. Yavorsky, The hydrane process, Papers from Clean Fuels from Coal Symposium II, IGT, p. 159, Chicago, 1975.

Green, N. W., Synthetic fuels from coal - the Garrett process, Papers from Clean Fuels from Coal Symposium II, IGT, p. 299, Chicago, 1975.

Greene, M., Engineering development of the cities service short residence time (CS-SRT) process, Preprints from Div. of Fuel Chem., Acs, 22(7), 133, 1977.

Han, K. W., V. B. Dixit, and C. Y. Wen, Analysis and scale-up considerations of bituminous coal liquefaction rate processes, Ind. Eng. and Chem. Process Des. & Dev., 17(1), 16, 1978.

Haynes, W. P., and A. J. Forney, The SYNTHANE process, Papers from Clean Fuels from Coal Symposium II, IGT, p. 149, Chicago, 1975.

Haynes, W. P., S. J. Gasior, and A. J. Forney, Catalysis of coal gasification at elevated pressures, Coal Gasification, ACS Advances in Chemistry Series 131, p. 179, Washington, 1974.
New process may make eastern coals easier to gasify, Heating, Piping and Air Conditioning, p. 44, September 1978.
Hoffman, E. J., Coal Conversion, Modern Printing, Laramie, Wyoming, 1978.
Hoogendoorn, J. C., New applications of the Fischer Tropsch process, Papers from Clean Fuels Coal Symposium II, IGT, p. 343, Chicago, 1975.
Hottell, H. C. and J. B. Howard, New Energy Technology - Some Facts and Assessments, pp. 103-227, MIT Press, Cambridge, Mass., 1971.
GE's pilot coal gasification plant, Iron and Steel Engineer, p. 120, September, 1977.
Johnson, J. L., Kinetics of initial coal hydrogasification stages, Preprints from Div. of Fuel Chem., ACS, 22(1), 17, 1977.
Johnson, J. L., Kinetics of bituminous coal char gasification with gases containing steam and hydrogen, Coal Gasification, ACS Advances in Chemistry Series 131, 145, Washington, 1974.
Johnson, C. A., M. C. Chervenak, E. S. Johanson, H. H. Stotler, O. Winter and R. H. Walk, Present status of the H-COAL process, Papers from Clean Fuels from Coal Symposium II, IGT, p. 525, Chicago, 1975.
Jones, J. F., Project COED (Char-Oil-Energy Development), Papers from Clean Fuels from Coal Symposium II, IGT, p. 323, Chicago 1975.
Knudsen, C. W. and P. O. Hedman, Fossil fuel economics, Preprints from Div. of Fuel Chem., ACS, 23(3), 1, 1978.
Kohl, A. L., R. B. Harty, and J. G. Johanson, The molten salt gasification process, Chem. Eng. Prog., 74(8), 73, August 1978.
Kunesh, J. G., M. Calderon, G. A. Popper, and M. S. Rakow, Economics of the H-COAL process, Preprints from Div. of Fuel Chem., ACS, 23(3), 25, 1978.
Kydd, P. H., The Gegas process, Papers from Clean Fuels from Coal Symposium II, IGT, Chicago, 1975.
LaDulfa, C. J. and M. T. Greene, Comparative economics of the cities service CS-SRT process with the Lurgi process, Preprints from Div. of Fuel Chem., ACS, 22(7), 94, 1977.
LaRosa, P. and R. J. McGarvey, Fuel gas from molten iron gasification, Papers from Clean Fuels from Coal Symposium II, IGT, p. 227, Chicago, 1975.
Lee, H. H., Determination of diffusion properties of double layer catalyst pellets, to be published in Amer. Inst. of Chem. Ing. Journal, 1979a.
Lee, H. H., Rate processes of catalytic gasification of coal, Paper presented at 71st National Meeting of Amer. Inst. of Chem. Eng., Miami, Florida, November 1978.
Lee, H. H., The effectiveness factors for solid catalyzed gas-solid reactions, Chemical Engineering Science, 34, p. 5, 1979b.
Lee, H. H. Kinetics of sintering of supported metal catalysts - the mechanism of atmo migration, submitted to J. Catalysis, 1979.
Lee, J. M., A. R. Jarrer, J. A. Guin, and J. W. Prather, The selectivity of coal minerals as catalysts in coal liquefaction and hydrodesulfurization, Preprints from Div. of Fuel Chem., ACS, 11(6), 120, 1977.
Linares, A., O. P. Mahajan, and P. L. Walker, Jr., Reactivities of heat treated coals in steam, Preprints from Div. of Fuel Chem., ACS, 22(1), 1, 1977.
Loganbach, J. R., J. W. Droege, and S. P. Chauhan, Short residence time coal liquefaction, Preprints from Div. of Fuel Chem., ACS, 24(2), 52, 1979.
Loverto, D. C. and S. W. Weller, Stannous chloride and cobalt molybdenum alumina catalysts in hydrogenolysis of solvent refined coal, Preprints from Div. of Fuel Chem., ACS, 23(1), 71, 1978.
Maddocks, R. R., J. Gibson, and D. F. Williams, Supercritical extraction of coal, Chem. Eng. Prog., 75(6), 49, June 1979.
Martin, J. R., Union carbide's "COALCON" process, Papers from Clean Fuels from Coal Symposium II, IGT, p. 869, Chicago, 1975.
Michaels, H. J., and H. F. Leonard, Hydrogen production via K-T gasification, Chem. Eng. Prog., 74(8), 85, August 1978.
Miyadera, H., M. Hirato, S. Koyama, and S. Gomi, Effects of reaction conditions on gasification of coal-residual oil slurry, Preprints from Dkv. of Fuel Chem., ACS, 23(3), 160, 1978.
Mochida, I., and K. Takeshita, Coal liquefaction under atmospheric pressure, Preprints from Div. of Fuel Chem., ACS, 24(2), 98, 1979.
Morel, W. C. and Y. J. Yim, Economics of producing methanol by entrained- and fludized-bed gasifiers, Preprints from Div. of Fuel Chem., ACS, 22(7), 94, 1977.
Morita, M., S. Sato, and T. Hashimoto, Kinetics of direct liquefaction of coal in the presence of Mo-Fe catalyst, Preprints from Div. of Fuel Chem., ACS, 24(2), 62, 1979.
Morita, M., S. Sato, and T. Hashimoto, Effect of hydrogen pressure on rate of direct coal liquefaction, Preprints from Div. of Fuel Chem., ACS, 24(2), 270, 1979.
Neben, E. W., The economics of advanced coal liquefaction, Chem. Eng. Prog., 74(8), 95, 8/1978.
Netzer, D. and J. Moe, Ammonia from coa, Chemical Eng., 84(23, 129, Oct., 1977.
Nishiyama, Y. and Y. Tamai, Hydrogasification of carbon catalyzed by nickel, Preprints from Div. of Fuel Chem., ACS, 24(2), 219, 1979.

O'Hara, J. B., N. E. Mentz, and R. V. Teeple, Conversion of coal to liquids by Fischer Tropsch and Oil/Gas Technologies, Preprints from Div. of Fuel Chem., ACS, 22(7), 20, 1977.

Patel, J. G., and J. W. Loeding, IGT U-gas process, Papers from Clean Fuels from Coal Symposium, II, IGT, p. 193, Chicago, 1975.

Phinney, J. A., Clean fuels via the CSF process, Papers from Clean Fuels from Coal Symposium II, IGT, p. 467, Chicago, 1975.

Pollaert, T. J., Hydrogen from coal, Chem. Eng. Prog., 74(8), 95, August, 1978.

Pruitt, M. E., Dow details new coal liquefaction process, Chem. and Eng. News, p. 43, Sept. 1978.

Radding, S. B., Director of Publications, Symposium on Oil Sand and Oil Shale, Preprints from Div. of Fuel Chem., ACS, 22(3), 1977.

Rudolph, P. F. H., and H. H. Bierbach, Fuel gas from coal, Papers from Clean Fuels from Coal Symposium II, IGT, p. 85, Chicago, 1975.

Schobert, H. H., B. C. Johnson, and M. M. Fegley, Carbonization reactions in the Grand Forks fixed bed slagging gasifier, Preprints from Div. of Fuel Chem., ACS, 23(3), 136, 1978.

Shinn, J. H. and T. Vermeulen, High yield coal conversion in a zinc chloride-methanol melt under moderate conditions, Preprints from Div. of Fuel Chem., ACS, 24(2), 80, 1979.

Smoot, L. D., F. D. Skinner, and R. W. Hanks, Mixing and reaction of pulverized coal in an entrained gasifier, Preprints from Div. of Fuel Chem., ACS, 22(1), 77, 1977

Stanalonis, J. J., B. C. Gates, and H. H. Olson, Catalyst aging in a process for liquefaction and hydrodesulfurization of coal, Amer. Inst. of Chem. Eng. Journal, 22(3), 576, 1976.

Steinberg, M., and P. Fallon, Flash hydropyrolysis of coal, Chem. Eng. Prog., 75(6), 63, June 1979.

Stern, E. W. and S. G. Hinden, Catalyst development for the hydroliquefaction of coal, Preprints from Div. of Fuel Chem., ACS, 23(1), 21, 1978.

Suuberg, E. M., W. A. Peters, and J. B. Howard, Product composition and kinetics of lignite pyrolysis, Preprints from Div. of Fuel Chem., ACS, 22(1), 112, 1977.

Thomas, M. G., B. Granoff, G. T. Noles, and P. M. Baca, Hydrogen consumption in non-catalyzed coal liquefaction, Preprints from Div. of Fuel Chem., ACS, 23(1), 42, 1978.

Tomita, A., O. P. Mahajan, and P. L. Walker, Jr., Catalysis of coal gasification at elevated pressures, Coal Gasification ACS Advances in Chemistry Series 131, p. 179, Washington, D. C., 1974.

Vadovic, C. J. and J. M. Eakman, Kinetics of potassium catalyzed gasification, Preprints from Div. of Fuel Chem., ACS, 23(3), 89, 1978.

Veluswamy, L. R., J. Shabtai, and A. G. Oblad, Hydrogenation of coal liquids in the presence of sulfided N_1-Mo/Al$_2$O$_3$, Preprints from Div. of Fuel Chem., ACS, 23(2), 280, 1979.

Vernon, L. W., Free radical chemistry of coal liquefaction, Preprints from Div. of Fuel Chem., ACS, 24(2), 143, 1979.

Walker, P. L., O. P. Mahajan, and R. Yarzab, Unification of coal char gasification reactions, Preprints from Div. of Fuel Chem., ACS, 22(1), 7, 1977.

Weil, S. A., M. Onischack, and D. V. Punwami, Peat hydrogasification, Preprints from Div. of Fuel Chem., ACS, 23(3), 149, 1978.

Wen, C. Y., R. C. Bailie, C. Y. Lin, and W. S. O'Brien, Production of low-Btu gas involving coal pyrolysis and gasification, Coal Gasification, ACS Advances in Chemistry Series 131, p. 9, Washington, 1974.

Whiteacre, R. C., J. F. Fransworth, and D. M. Mitzak, The Koppers-Totzek process and its applications to industrial needs, Papers from Clean fuels from Coal Symposium II, IGT, Chicago, 1975.

Willson, W. G., S. A. Qader, and E. W. Knell, Flash pyrolysis coal tar, Preprints from Div. of Fuel Chem., ACS, 22(6), 131, 1977.

Willson, W. G., L. J. Sealock, Jr., F. C. Hoodmaker, R. W. Hoffman, D. C. Stinson, and J. L. Cox, Alkali carbonate and nickel catalysis of coal-steam gasification, Coal Gasification, ACS Advances in Chemistry Series 131, p. 203, Washington, 1974.

Winterson, H. M., Low-Btu coal gasification test facility, Mechanical Engineering, p. 58, 12/77.

Yoshida, R. and D. M. Bodily, The initial stage of coal hydrogenation in the presence of catalysts, Preprints from Div. of Fuel Chem., ACS, 24(2), 371, 1979.

CHAPTER 7

TECHNOLOGICAL INNOVATIONS

Harold P. Hanson et al.

I. INTRODUCTION

The American public has faith that technological advances will
ameliorate, postpone, or even eliminate the energy shortage. To a certain extent this is true, and the search for new approaches with greater
efficiency and effectiveness continues, with many ideas showing substantial promise. This chapter will be devoted to a consideration of
various technological innovations or approaches that supplement the
arsenal for waging the energy war through increased coal utilization.
It should be noted that many of the ideas broached here are not new.
Rather they are ideas whose time has returned.

It should also be noted that the discussions of various ideas
incorporated in this chapter have been prepared in large part by the
collaborators indicated with the title of each section, rather than by
the author of record alone.

II. INTEGRATED UTILITY SYSTEMS

In the effort to respond to the urgent need for more energy production, it is natural to (1) search for new energy sources or (2) attempt
to improve the extant technology. It is less natural to search for new
combinations of old technologies. Yet this approach has attractive
features. If well-established devices and processes can merely be combined so as to increase over-all efficiency and effectiveness, the result

can be as beneficial as developing wholly new energy sources or significantly improving an extant technology.

A case in point is the Integrated Utility System. This system, as
the name implies, integrates various techniques of energy production
into one central production unit. A few of these are in operation at
various places in the nation and they have demonstrated their economic
feasibility. There has not been general adoption of the concept because
of concomitant technical complexities. Further, as will be pointed out,
some previous options are disappearing. Nevertheless, the concept will
become more generally accepted, with coal being the prime consumable
fuel in such systems.

A. Overview

From the time of the Romans, at least, public utilities have been
established for the benefit of the members of the community. The great
aqueducts and public baths of Rome were the antecedents of the power
plants and sewage disposal systems of today.

It is characteristic of most municipalities that utilities are
provided through specialization of function. In general there will be
a "gas company," an "electric company," etc. Even when the facilities
are municipally owned, or at least under single ownership, the utility
operations are separate and distinct.

The concept of the Integrated Utility System is to combine as many
functions as is economically feasible so that economies of size and
place can be effective. In the limit it is conceivable that all utilities could be furnished from a single plant. This could run the gamut
from furnishing potable water to distributing coolness. There is no one
single configuration which represents the optimum Integrated Utility
System. Each one has to be engineered for the locale and condition in
which it will operate.

However, most of the Integrated Utility Systems which have been
built or planned are limited in function to providing energy by burning
various types of fuels. We shall, for the most part, consider only such
units.

Further, while the concept is broad enough to encompass the provision of power to a city or even a region, in actual practice it appears
that it would be most effective for a specialized entity such as a university campus, a hospital, or a correctional institution. The federal
government has been particularly interested in exploring possibilities
for educational institutions.

The general plan that appears to be most promising is one in which
the Using Unit produces all of its own heat and generates most of its
own electricity. Rather than produce air conditioning with electrically
driven centrifugal chillers, the heat would be utilized to drive absorption chillers. Electric power would be generated by backpressure turbines operating from coal-fired spreader stoker-equipped boilers.

Two cautionary comments should be made. Even though the Using Unit
may generate most or even all of its own electrical power, it must have
the capability of accepting electric power from outside sources. Clearly
there is an attendant cost. Secondly, any built-in versatility has its
attendant cost, primarily in capital construction, but also in operation.
Thus one should not look for a panacea in the Integrated Utility System;
rather it provides a possibility for marginally more economic operation.

B. The Role of Coal

The idea of Integrated Utility Systems has been around for decades.
As originally conceived, the plan called for tactical shifts from one
type of fuel to another as the availability and price dictated.

However, with the growing realization of the limited supplies of
gas and oil, it has become clear that the main dependence has to be on
coal. In fact, with government regulations as they are on Major Fuel
Burning Installations (MFBI), Integrated Utility Systems must be
basically versatile coal-burning units which have the capacity to be
augmented by burning other selected fuels, principally solid waste
material.

Thus, in point of fact, the problems and opportunities of Integrated
Utility Systems are basically the problems and opportunities of coal-
fired plant operation.

C. Should Waste be Wasted?

The most attractive feature of the Integrated Utility System is
that it incorporates solid waste management. It is a depressing fact
of modern civilization that vast amounts of garbage will be produced
as a by-product of normal living. In America, the garbage is produced
more copiously than in any other industrial nation. The occasional
garbage-collector's strikes in large cities give graphic proof of the
problem as mountains of rubbish accumulate in short order in our
cities' streets.

Survival requires that garbage be disposed of with reasonable sani-
tation at a reasonable cost. Normally there is a rather high cost
associated with hauling away refuse, which is only fractionally offset
by the use of garbage as landfill material.

The economics of processing waste is fraught with anomalies. For
an Integrated Utility System, it is frequently possible to decrease
hauling costs by assembling the garbage at a conveniently located pro-
cessor rather than bringing the material to a remote land-fill site.
Thus there is a lesser cost associated with processing a material which
is now going to be useful. However, very careful engineering studies
must be made regarding the cost of the garbage processing per se.

D. Garbage In, dRDF Out

In general, the processing of garbage for energy production in
Integrated Utility Systems takes two forms. They relate to the source
of the supply of garbage. There will be garbage derived from the Using
Unit itself. In the case where the Using Unit is a university campus
there will be a particularly high fraction of paper. On the other hand,
there will be a larger supply of lower quality garbage that is available
from the surrounding community. These two garbages, differing in
inherent fuel quality, are best processed differently.

Thus, if the supply of waste material is not great, and if the
waste is relatively dry, the most economic use is obtained through a
solid waste gasification system. In this process the waste is introduced into a high temperature refractory vessel which operates in a

condition of oxygen starvation. The resultant incomplete combustion
produces a gaseous effluent which contains carbon monoxide, soot, and
various combustible gases, all of which serve as energy sources. The
effluent is much like "producer gas" which was used in the last century
in many American towns. It is this effluent gas which is ducted to a
burner which, in turn, is an auxiliary heater for the coal boiler.

We now consider the second situation. When the supply of garbage
is great, but perhaps of lower heating value, a suitable process of
mixing the garbage directly with the coal is found to be most economical.
To be specific, a boiler of growing popularity is one which employs a
spreader stoker-equipped unit for burning coal. A pelletized fuel can
be prepared from the solid waste which can then be transported, handled,
and used just as is done for coal, without facility or site modification.
In a recent series of experiments, it has been shown that this pelletized fuel is a suitable stoker coal substitute (Rigo, et al., 1978).
This fuel is called "densified refuse derived fuel" or dRDF.

The actual pelletizing is carried out as follows: Solid waste is
loaded onto a conveyor which feeds a primary trommel screen. As is
indicated in Figure 1, all pieces smaller than 5 inches fall through the
holes of the trommel and are removed. This includes most of the glass
bottles and beverage cans. Large materials are then passed through a
flail mill where the large objects are reduced to less than the 5-inch
size. This material is screened and added to the original sub-5-inch
stream.

By a process of "lofting," an air classifier separates out about 55%
of the material as being the lighter component. This lofted fraction
is principally paper and plastic, and it constitutes the fuel material.

After cleaning by fine-screening, the cleaned material is processed
and compressed in a hammer mill, and extruded by a pelletizer as cylin-
drical plugs approximately one-half inch in diameter and one inch long.
The material in this form has about half the fuel value of an equivalent
mass of coal.

The dRDF is then trucked to the Integrated Utility System where it
is blended and stored with the coal for eventual firing of the boiler.
Figure 2 shows schematically the overall operation of the plant.

The heavy residue which was rejected by the air classifier is passed over a drum magnet for recovery of the magnetic scrap. The residue may be further processed or used as landfill as economics justify.

E. Environmental Impact

Air Pollution: Since the Integrated Utility System of the future will be basically a coal-fired unit, comparison need only be more between a coal operation and a coal:dRDF operation. The gasification process provides only a small fraction of the energy, and could add almost no increment to power plant pollution simply on a quantitative basis. Further, the burn is clean relative to coal itself.

The case of coal:dRDF bears study because of the sulfite processing of paper. However, studies indicate that a coal:dRDF mixture for a typical Integrated Utility System will have only 85% of the sulfur content of a coal burner that complies with federal standards.

There is a dust problem due to the dRDF both in the processing plant and in the Integrated Utility System. This must be dealt with.

Water Pollution: No different than coal-only operation.

Noise Pollution: Except for the dRDF processing plant, little different from coal-only operation.

Land Pollution: Bottom and fly ash residues essentially the same as for coal-only operation.

F. Economic Analysis

As has been stressed, each Integrated Utility System would be unique. The economic benefits, if any, would vary from installation to installation. To give rough figures, however, it has been estimated that for a fairly typical large university, the Integrated Utility System would (1) make possible a three percent saving on the cost of coal by using the Solid Waste Gasification System and (2) make possible a twenty-two percent saving through the use of regionally accumulated waste to produce dRDF.

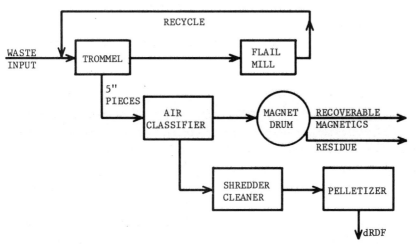

Fig. 1. Densified Refuse Derived Fuel Plant

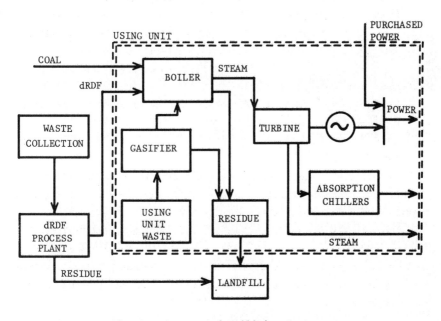

Fig. 2. Integrated Utilities System

These savings are sufficient that·with respect to any greater
capital costs or greater operating costs of the waste processing, the
Integrated Utility System would begin producing a net pay-back at 1/4 to
1/3 of the life of the plant.

III. CO-GENERATION SYSTEMS

As was noted in the previous section, it has been common practice
to have each utility provided by individually dedicated units. This
has been even more true in recent years than it was in the early part
of the century, and even up until World War II. After that time, huge
specialized electric power stations became the vogue, and the heat they
generate is simply discarded.

The idea of co-generation is, of course, to combine the production
of electricity and steam. District heating would then replace the
individual furnaces that rely on expensive oil or natural gas. Further,
the steam can be used to power central chillers for air conditioning.
Co-generation does not require any technological breakthrough. It would
mean resuming an approach that was all but abandoned several decades ago.

It has been estimated that waste heat could satisfy a third of the
nation's total energy needs. About half as much fuel would be used to
produce electricity with steam as is needed to produce the two separately. Basically this is because fossil-fuel-fired central stations
now convert about one-third of their energy input into electricity,
allowing two-thirds to escape unused. By using reject heat, co-generation plants can achieve a thermal efficiency of 70-80%. Thus, government sources speak confidently of a saving through district heating of
the equivalent of six million barrels of oil per day.

Some cautionary comments should be made. It would seem that a
one-third reduction in our energy needs would provide an immediate
major remedy for the energy problem. However, the technology was
abandoned because it was not economical. The economics are changing,
but the old distribution systems are inefficient and even dangerous.
Distributing centrally produced steam and hot water to homes and fac-
tories would require major and expensive modifications. In principle,

large cities could be important benefactors of this technique, but putting a distribution system on top of or underneath an already built-up city represents a problem of vast difficulty requiring appropriately vast resources.

There are other difficulties as well, principally bureaucratic and regulatory in nature. Presumably, however, the regulatory thicket could be thinned and various incentives could be instituted to encourage co-generation. Thus, the use of waste heat represents one idea whose time has returned. Clearly, it offers promise for the future, but it will be phased in as the need to follow this route becomes more obvious.

IV. ELECTRIC AUTOMOBILE (M. J. Rowe)

The conversion of central power stations to coal firing, away from oil, gas and nuclear, which is under way was mandated in order to reduce foreign oil imports. Ironically, it is not electrical power generation but cars which are the major consumer of foreign oil. Therefore, logically, after conversion of central power stations to coal, every effort should be made to develop the electric car, which can charge its batteries at night, when electricity is not in demand, yet the production facilities exist and are idle.

The major problem which keeps the electric car from the market place is not technical, it is psychological. The public, as well as industry and government, expect the electric car to perform in the same way as a gasoline fueled automobile. Such a vehicle is still to be developed, and there is no evidence which points towards the possibility of realization of such a goal in the near future.

However, a prolonged gasoline crisis will change the psychology drastically, and soon enough the public will be satisfied with an electric car costing twice as much as the gasoline fueled family car and having a range of 50 miles at 35 mph maximum speed (seating only one or two persons). This is a car which the public needs to go to work and back, but does not want as yet. Such a car is technologically feasible at present. (It always was, since cars have existed). If this type of commuter car would be deployed in large numbers, the

gasoline consumption would be reduced drastically, since the majority
of miles driven by private individuals are performed exactly within
the limits quoted above.

Operation of such a car, due to the requirement of frequent battery
replacement, may not be even cheaper than the gasoline fueled car (and
there is no reason why it should be) but electricity will be available
and gasoline will not be, at least not on an unlimited basis.

Materials required for the batteries can be recycled; therefore,
battery manufacture is mainly a question of labor cost and availability,
a situation much healthier for the national economy than the costs and
availability of gasoline.

As a matter of fact, the availability of such an electric car
against the background of electricity generation from coal, would have
an enormous dampening effect on gasoline prices, since the kind of
private driving absolutely requiring gasoline is more easily avoidable
than is going to work.

V. AIR POLLUTION CONTROL TECHNOLOGY (M. A. Kenney)

As will be discussed in Chapters 9-14 of this book, the combustion
of coal produces major emissions of sulfur and nitrogen oxides, particulate matter, and volatile trace hazardous and toxic substances. The
types and relative amounts of these compounds emitted from a furnace
depend mainly on the quantity and the composition of the coal burned,
on furnace design, and on furnace operation. The local micrometeorology
and plume dispersion patterns are important in determining how much of
the pollutants will be found at ground level under normal as well as
worst-case conditions. Compliance with the ambient air quality standards, which prescribe health and welfare endangering levels when surpassed, often requires the design and incorporation of air pollution
control technology into the combustion process. The location and the
degree of emissions control required will be determined by both the
quality of the area's ambient air and by the potential impact of any
environmentally protected areas in the vicinity.

The types of techniques employed to reduce emissions of particulates and gases in the coal combustion process vary considerably and
the technology is proceeding very rapidly. A detailed description and
evaluation of each is beyond the scope of this chapter. Rather, only
a brief mention of those procedures most commonly practiced will be made.

A. Particulates

Flue-gas cleaning is the most common technique used for control
of particulate emissions from coal combustion sources (NAPCA, 1969a).
Among those gas-cleaning methods employed are settling chambers, single
and multiple stage cyclones, wet scrubbers, fabric filters, and electrostatic precipitators.

Electrostatic precipitation has played a major role in controlling
particulates from coal-fired boilers and should continue to do so in the
future. It offers the highest efficiency of particle control, but at
enormous costs. Electrostatic precipitators are often used in combination with cyclone collectors, which usually precede the electrostatic
precipitators. Scrubbers, likewise, are used in combination with precipitators. However, unlike cyclones, they also aid in reducing gaseous
pollutants in the flue gas.

The principle of fabric filtration has only recently been accepted
as a practical and economical means of controlling particulates from
the burning of coal (Frederick,1977).

An essential part of any particulate control system is the necessary equipment and methods to remove the collected fly ash and transport
it for safe and proper disposal. The development of useful applications
for this waste material, in addition to landfill is an important
problem.

B. Sulfur Oxides

Sulfur oxide is the most abundant gaseous air pollutant emitted
during coal combustion. There are three general methods of reducing
the amount of SO_2 released into the atmosphere: desulfurization of
fuels, process modification, and flue gas desulfurization. This section

will deal only with the present status of techniques for the removal
of sulfur oxides from flue gas. This is an area that is receiving
considerable attention at the present time, particularly from the
power industry.

Progress in developing efficient desulfurization processes for flue
gases has been extremely slow because of the complexity and magnitude of
the problem. The majority of processes are expensive and only a few
have been tried on large plants.

Sulfur dioxide removal processes generally use either an absorption
or adsorption technique followed by some means of separating the sorbent
from the flue gas for regeneration. There also exist wet or dry techniques that are differentiated simply by whether or not the active
removal agent is contained in a liquid solution. The removal system
typically involves the use of absorption, adsorption, or catalytic
processes.

Limestone scrubbing is presently favored by many power companies
as the method for controlling sulfur dioxide emissions (Nannen et al.,
1974). The principal advantages of the lime/limestone scrubbing
technique are the relative ease by which it can be incorporated into
existing facilities and the ability to remove particulates as well as
SO_2.

A brief comparison of some of the major processes under development
for SO_2 removal from stack gases appears in Table 1. In some processes
a high-efficiency particulate collector, such as an electrostatic precipitator, must precede the SO_2 removal equipment because particulates
are not acceptable in its operation.

C. Nitrogen Oxide

Generally speaking, emission of nitrogen oxide during coal combustion is largely determined by the design and operation of the equipment used. Therefore, the most effective and applicable means of NO_x
control is through combustion system modification; for a detailed
discussion see EPA, 1979. In contrast to SO_2 control, very little
effort to date has gone into research and development of techniques

Table 1: *Descriptions of Sulfur Dioxide Removal Processes*

Process	Method of Removal	Active Material
A. Nonregenerative scrubbing processes		
1. Lime or limestone	Slurry scrubbing	CaO, $CaCO_3$
2. Sodium	Na_2SO_3 solution	Na_2CO_3
3. Double alkali	Na_2SO_3 solution, regenerated by CaO or $CaCO_3$	$CaCO_3/Na_2SO_3$ or CaO/NaOH
B. Regenerative scrubbing processes		
1. Magnesium oxide	$MgSO_3$ slurry	MgO
2. Electrolytic	NaOH solution, acid decomposition, electrolytic regeneration	NaOH
3. Citrate	Sodium citrate solution, reaction with H_2S to form sulfur	H_2S
4. Sulfoxel	Na_2CO_3 solution, reaction with CO to form sulfur	Na_2CO_3/CO
5. Formate	KCOOH solution, reaction with CO to form H_2S	KCOOH/CO
C. Dry Processes		
1. Catalytic oxidation	Oxidation at 725°K, scrubbing with H_2SO_4	V_2O_5 catalyst
2. Carbon adsorption	Adsorption at 400°K, reaction with H_2S to S, reaction with H_2 to H_2S	Activated carbon/H_2
3. Dry limestone injection	Powdered limestone in combustion chamber	CaO, $CaCO_3$

Ref. Adapted from Wark, 1972.

for removing nitrogen oxides from flue gas. Among those methods
currently being considered are catalytic decomposition, catalytic
reduction, adsorption, and absorption (NAPCA, 1970a).

D. Carbon Monoxide and Hydrocarbons

For minimum CO emissions, combustion equipment is designed for
rapid reaction rates and long reaction time which tend to facilitate
complete fuel combustion. However, some of these same conditions promote formation of nitrogen oxides. As with the control of nitrogen
oxides, the most practical techniques for reduction of CO emissions from
coal combustion are good design and firing practices (NAPCA, 1970b).

Hydrocarbons are also products of incomplete combustion. However,
when properly operated and designed, coal combustion units are not a

source of organic emissions, and control equipment is not usually
required (NAPCA, 1970c).

Carbon dioxide is the major gaseous emission from all fossil
fuel plants. While it has no direct health impact at current levels,
and hence is not normally regarded as a pollutant, it has potentially
adverse impacts upon the atmosphere (see Chapter 11). Relatively little
effort has been made to trap CO_2 emissions from power plants to this
point. Presumably, when the problem becomes sufficiently urgent
scientists will address the problem. Generally such techniques are
costly. However, the possibilities that useful by-products can be
formed out of these waste products which will off-set these costs
represents one of the challenging issues of coal utilization.

VI. COAL CLEANING (M. J. Rowe)

The ideal combustion of coal would produce CO_2 as the only by-
product. Since coal is far from ideal, there is a serious problem of
unwanted by-products or pollutants. Present methods for handling this
problem are to clean up the gaseous by-products and store the solid by-
products. There is an alternative method for dealing with problems,
namely coal cleaning or removing the unwanted products before the coal
is burned. As in all fuel quality improvements, there is a price in
dollars which must be paid. But coal cleaning offers several cost
benefits which until now have not counteracted the expenditure.

Cleaned coal has a higher energy density than raw coal. The com-
bustibles are not improved but the incombustible ashes are removed. Also
pyritic sulfur is removed. Organic sulfur, however, is very expensive
to remove but since pyritic sulfur can be up to 50% of the total sulfur,
raw coal with higher sulfur content can be cleaned and burned with
developed technologies meeting Federal and State regulations. Larger
usable coal reserves are, therefore, available for clean energy production. High water content in coal increases the energy lost during combustion. Some water can be removed from coal easily. Removal of these
three unwanted products before burning decreases the cost of waste
clean up. In some cases, such as residential heating, no clean up

is attempted. Clean coal, in this case, would decrease the undesirable effects while not increasing the cost of the furnace.

The most popular technique for cleaning coal is mass separation. Coal ash has a higher specific gravity than coal. By mixing raw coal in a working fluid, the ash will settle out to the bottom of the fluid. The difficulty of the technique is for some raw coals the density difference between coal and ash may be small. This will reduce the efficiency of the separation technique wasting some usable coal. Improvement in the separation of close specific gravity coals can be done with a working fluid which can improve the buoyancy of coal and not the ash. Such a technique is to put raw coal into an oil bath in which the oil will stick only to coal and not to the coal ash. Mixing the oil-coal into a tank of water with air bubbles (which adhere only to the oil) will reduce the specific gravity of the coal, forcing the coal to the surface. The coal ash will settle to the bottom. The oil can then be recovered for recycling.

Pyritic sulfurs and coal ash can be separated out by the above described techniques, but this leaves the coal wet. Thus, the final step in coal cleaning is to dry the coal. There are three major techniques for drying coal. The first technique is to shake coal on a screen letting the water drip off. The second technique is to use a cyclone dryer, that is, to spin off the water in a high wind "cyclone." The third technique is to use a thermal heater. Unless the coal is dried, freezing or clumping can occur during transportation.

There are other techniques for cleaning coal which are less efficient or under development. Such techniques are: air fluid mass separators, magnetic separators or electrostatic separators. Whichever technique is used does improve the quality of the raw coal reducing the difficulties of transportation and burning. Also, the by-product production of ash and SO_x are directly reduced. Each of thse advantages reduce the cost of coal combustion which must balance the cost of coal cleaning.

VII. OFFSHORE COAL-FIRED POWER PLANTS (A. E. S. Green and J. M. Schwartz)

The concept of offshore power generating plants was originally formulated with nuclear reactors furnishing the heat for steam production. With

the present slow-down in nuclear plant licensing and construction, the offshore approach using coal-burning technology has features which warrants investigations, particularly since certain conditions favor offshore
electric utilities. Roughly 42% of today's total U.S. demand for electrical
energy exists within a 200-mile strip along the Atlantic, Gulf, and Pacific
coasts; and that percentage is expected to increase. There are thousands of
suitable sites for power plants off the 5,700-mile coastline of the United
States, and sites need only be selected and developed. The only requisites:
a geologically stable floor and an ocean depth of 20 to 70 feet. For land-
based generation plants, suitable sites are becoming more scarce, more expensive, and more distant from load centers. For offshore sites, only unused ocean floor is needed, compared to 300 acres or more for a land-based
plant. Ocean sites would be nearer to coastal load centers than is now possible, with no major investment in real estate needed years in advance of
actual construction (Ashworth, 1973 and Westinghouse Tenneco, 1975)

Construction time for land plants has lengthened to the point where, in
many areas of the country, some utilities may be unable to meet load demands
especially with oil or gas shortages. For offshore coal plants, construction time would be shorter because a standardized 500 MW plant could be
"assembly line" produced on a single floating 400 foot x 400 foot platform,
towed into place and permanently moved to an offshore location. The coal-
fired boiler, turbine generator and auxiliary systems are all based on existing technology for coal burning utilities. At the utility's site, a breakwater would surround and protect the plant and form a basin around the
power plant. The breakwater is founded on the seabed in shallow water.
The water depth would be on the order of 40 feet so that the base of a
short smoke stack could be imbedded in the seabed.

The standardized plant is constructed in a permanent shipyard-like
facility which has access to the sea. The basic platform is constructed
in a dry dock and floated, equipment installed, and the complete plant
towed to the offshore utility's site.

The plant can be designed to stay afloat for its lifetime or permanently settled in the basin formed by the breakwater. The water in the
basin is in communication with the sea outside. The off-site power con-

nection to the plant is by two sets of 350KV underwater cables. Because
the cables are submerged, they are generally unaffected by bad weather
conditions and should prove to be reliable once the basic undersea - high
voltage transmission technology is developed.

Coal loaded barges could be directly unloaded at the plant with coal
ash residue taken aboard for return as possible landfill to halt land erosion along the coastlines. Water for conversion to steam in the boiler
could be obtained from processed municipal waste water piped to the plant
before dumpage into the ocean, with sea water used for cooling purposes
where feasible.

The breakwater would prevent collision of any ocean vehicles with the
plant. It would also diminish any force of large waves, with the plant
remaining in service even in a "once in a 100 years" storm wind and waves.
Coal storage capacity for at least two weeks, living facilities and a heli-
copter landing port would be part of the facility. These aspects are
similar to the oil-drilling rigs now dotting the oceans throughout the
world.

Only minor adverse effects are expected on the ecosystem which would
be partially offset by the artificial reef (breakwater) and mildly heated
water for marine life. While usually low quality coal could be used with
low smoke stacks, low sulfur or preprocessed coal will be used to keep
the SO_2 concentrations at reasonable levels during sea breezes. Normally
over 50% of the output effluents would blow into the ocean, a sink that
has far greater capacity than do our lakes.

Perhaps a better technique would be to use a high capacity pump to
inject stack effluents directly into the ocean. The ocean would probably
absorb some of the CO_2 which might promote the growth of seaweed and algae.
Fish and marine life might welcome the extra heat. The ocean could probably accommodate the SO_2 and NO_x effluents without significant adverse
ecological impact. The cost effectiveness of this technology and the question of whether the ocean has this much capacity must, of course, be examined.

The possibility of alleviating electric power deficiencies through
transfer of the innovated technologies developed for nuclear offshore
plants warrants further consideration.

VIII. COAL PLANT SITING (M. J. Ohanian)

As has been noted, the major user of coal in the forseeable future
will be the electric utility sector. Moreover, this application is expected to grow significantly as the use of oil and gas for electricity
generation is curtailed by government regulation, and nuclear energy is
faced with an indeterminate period of uncertainty. This increasing dependence on coal for central station steam plants for electricity production will bring into sharper focus the important technical and socio-
economic issues related to the siting of these plants. While the issues
are not as difficult as those related to the siting of nuclear power generating facilities, to the extent that coal-fired plants also compete for
resources such as land and water and affect regional and local air quality,
their siting also creates problems. For example, a 1,000 MW coal-fired
plant requires approximately 100 acres of land if once-through cooling is
used and approximately 150 acres when cooling towers are employed; in addition 400-500 acres are required for coal and waste storage. In the
case of cooling towers, 12-15 thousand gallons per minute of water is consumed in evaporative and blowdown losses for a typical 25-30°F rise in
condenser temperature. On an annual basis such a plant consumes $2\frac{1}{2}$ million tons of coal and produces 25,000 tons of ash, as well as 9 million
tons of gaseous waste products (Loftness, 1978).

A number of techniques have been developed over the past few years
for the selection of central station power generating sites (DuBois,
et al., 1973). These widely and successfully used methods employ quantitative and qualitative measures to distinguish among sites based upon
the key criteria of land and water availability, environmental considerations (such as air and water quality), access to coal transportation modes
and electrical transmission considerations. The objective of these siting
methodologies is to select the "best" sites in a region from the point
of view of resource availability, economic considerations and minimizing
environmental and related impacts.

For the siting of nuclear power generating facilities, concentrated
vs. dispersed siting is a major current issue (see, for example, Burwell,
Ohanian and Weinberg, 1978), and compelling reasons (such as safety, pro-
fessionalism of the nuclear operating cadre, security, minimizing radiation impacts) exist for concentrated siting of nuclear plants. However,
the same is not necessarily true for fossil-fired stations and the long-
standing trend for dispersed siting of coal-fired steam electric plants
is expected to continue. There have been a number of unpublished studies
of "energy parks" based on coal-fired electric generating plants (most
notably for Pennsylvania), but none of these studies found compelling
reasons for adopting concentrated siting of such facilities. In fact,
it is likely that the stresses on a given locale will be less with dispersed siting (i.e., 5 GW or less); therefore, barring compelling institutional reasons such siting will likely continue in a dispersed mode.
It should be noted, however, that a recent study by General Electric-
Tempo and Mathtech (Males, 1979) arrived at a different conclusion.
They find a large economic advantage to the construction of a single
2200 MW unit with respect to four dispersed 550MW units.

Within the framework of dispersed siting there are a number of potential innovations which can have longer term (mid-1990's and beyond)
implications for improved utilization of coal.

The first refers back to the discussion on co-generation (Tourin,
1978; see also Section III). While co-generation technology has received
significant acceptance in Europe (e.g., Sweden, East and West Germany,
Poland, U.S.S.R. where it is used for district heating), in the United
States it is only moderately applied and then mainly in the industrial
sector. The primary requirement for co-generation is that the power
plant be located close to population centers -- this would become increasingly feasible with improved coal-burning and pollution control
technologies.

A second potential innovation is the platform mounted, off-shore
coal-fired steam-electric plants discussed earlier. The primary advantage of this concept is that, the facility could be located close to

load centers without utilizing valuable land; it would also minimize environmental problems due to the waste heat which can now be directly discharged into deeper sections of the sea. Such a plant could utilize a
slurry pipeline for bringing the coal to the platform; this pipeline
could also be used to remove the ashes from the plant to an on-shore
(remote) storage area which would have the common function of being the
coal storage facility.

A third innovation and one which is appropriate to more concentrated siting relates to the development of synfuel "energy parks" whose
output would be hydrogen or other synthetic fuels and not electricity
per se. These types of energy parks seem ideally suited to mines remote
from load centers--specifically the Western coal mining regions, which
are a major repository of the U.S. coal resource base.

References

Ashworth, J. A., Offshore nuclear power plants "Atlantic Generating Station," presentation to
American Nuclear Society, Palm Beach Shores, Florida, April 25-27, 1973.
Burwell, C. C., M. J. Ohanian, and W. M. Weinberg, A siting policy for an acceptable nuclear
future, Science, 204, June 1978.
DuBois, O. B., B. E. Holmes, J. A. Hancock and M. J. Ohanian, Current methodologies in power
plant site selection, Trans. Am. Nic. Soc. 16:2, 1973.
Environmental Protection Agency, U.A., Industrial Environment Research Laboratory, Applica-
bility of the thermal DeNo process to coal-fired utility boiler, EPA-600/7-79-079,
Research Triangle Park, N.C., 1979.
Frederick, E., Electric utility applications of fabric filters, J. of the Air Poll. Contr.
Assoc., 27:11, 1977.
Loftness, R. L., Energy Handbook, Van Nostrand Reinhold Co., New York, 1978.
Males, R., R & D status report - Energy Analysis and Environment Division, EPRI Journal,
p. 56, Oct. 1979.
Nannen, L. W., R. E. West, and F. Kreith, Removal of SO_2 from low sulfur coal combustion gases
by limestone scrubbing, J. of Air Poll. Contr. Assoc.; 24:1, 1974.
National Air Pollution Control Administration (NAPCA), Control techniques for particulate air
pollutants, U.S. Department of Health, Education, and Welfare, Washington, 1969a.
National Air Pollution Control Administration (NAPCA), Control techniques for nitrogen oxides
from stationary sources, U.S. Department of Health, Education and Welfare, Washington,
D.C., 1970a.
National Air Pollution Control Administration (NAPCA), Control techniques for carbon monoxide
emissions from stationary sources, U.S. Department of Health, Education, and Welfare,
Washington, D.C., 1970b.
National Air Pollution Control Administration, Control techniques for hydrocarbon and organic
solvent emissions from stationary sources, U.S. Department of Health, Education and
Welfare, 1970c.
Rigo, H. G. et al., A field test using coal: dRDF blends in spreader stoker-fired boilers,
EPA Contract 68-03-2426, May, 1978.
Tourin, R. N., District meeting with combined heat and electric power generation, in Advances
in Energy Systems and Technology, Vol. I, Peter Auer (ed.), Academic Press, N.Y., 1978.
Wark, K., and C. F. Warner, Air Pollution: Its Origin and Control, Harper and Row, New York,
1976.
Westinghouse Tenneco, Offshore Power Systems, Company Brochure, 1975.

CHAPTER 8

WATER RESOURCES

Joel G. Melville and William E. Bolch, Jr.

I. INTRODUCTION

To attain the goal of national energy self-sufficiency, the world's recoverable coal reserves of 650 billion tons must be exploited efficiently without unreasonable environmental stress. The United States produces 650 millions tons of coal per year and production is likely to double before the year 2000. If a state is not a coal producer, it will certainly be a consumer. Thus, the switch to coal for power will have national implications. One of the primary constraints on coal energy is water availability and the pollution potential for water directed to other industries, public water supplies or agricultural use.

II. WATER AVAILABILITY AND POTENTIALS FOR CONTAMINATION

Current world and national water balance statistics are shown in Table 1. Observing that 94% of the water volume is saline, it is apparent that groundwater accounts for a large portion of our supply of freshwater.

Lvovitch (1970) estimates the most active groundwater regimes to be at 4×10^6 km^3 rather than 60×10^6 km^3 shown in Table 1. A detailed analysis of the available freshwater reveals a breakdown of 95% groundwater; 3.5% lakes, swamps, reservoirs, and river channels; and 1.5% soil moisture (Freeze and Cherry, 1979). While it is recognized that surface water demands and protection must be considered, it is generally accepted that surface water and air pollution standards have been reasonably suc-

Table 1: *Estimate of the Water Balance of the World*

Parameter	Volume (km^3) 10^6	Volume (%)	Residence Time
Oceans and seas	1370	94	4000 years
Lakes and reservoirs	0.13	0.01	10 years
Swamps	0.01	0.01	1-10 years
River channels	0.01	0.01	2 weeks
Soil moisture	0.07	0.01	2 weeks 1 year
Groundwater	60	4	2 weeks 10,000 years
Icecaps and glaciers	30	2	10-1000 years
Atmospheric water	0.01	0.01	10 days
Biospheric water	0.01	0.01	1 week

Source: Nace, 1971

cessful at an expense to groundwater resources. In the remainder of this chapter, the groundwater stress will be emphasized and shown to be the most threatening. For many reasons, groundwater is more desirable than surface water for many reasons. Groundwater usually needs no treatment for removal of pathogenic organisms, turbidity or discoloration, chemicals, or biological contamination. Also, the long residence time of groundwater in the hydrologic cycles (Figure 1) results in a storage supply not subject to short term droughts or seasonal variations in precipition. The underground reservoirs also supply water at uniform temperature. About 71% of the precipitation that falls on the continental United States is returned as evapotranspiration (Linsley and Franzini, 1979). Some 34% of streamflow is directed for use in either irrigation, public water supply, or industry. About 23% of that volume diverted is consumptively used. Infiltration into the groundwater supplies continues to diminish with land development.

Individual use of water increases greatly with improvement in the quality of life. For example man can survive easily with 20 ℓ/day for drinking, cooking, and washing. However, with modern (water consuming) conveniences the rate may be 600 ℓ/day. Some typical water costs related to specific aspects of modern life are: 10 liters of water per kg of meat, 100 liter of water per kg of paper, and 200 liter of water per kg of steel.

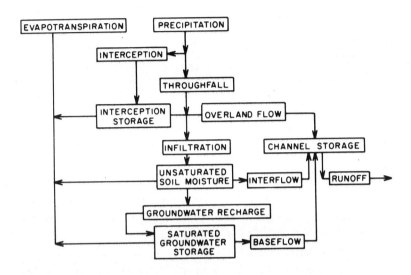

Fig. 1. Schematic of Hydrologic Cycle (Freeze and Cherry, 1979)

From the energy point of view, individual water costs can also be estimated. For example, Morgan, 1975, states that individual energy use has grown rapidly over the past 4 generations, as shown below:

Our Grandparents	110 kwh/day/per
Our Parents	150
We	250
Our Children	350

Electric power stations require water for steam condensation. Plants may use either once-through cooling or some form of cooling tower. Natural and forced-air evaporative cooling towers are the most common.

For once-through cooling systems, water demands are (Hampton, 1978) 45.4 gal/kwh - 11.99 ℓ/kwh. Based on the projected needs of an indivi-

dual, the water demand per capita per day for our children is given by
the expression

$$350 \ \frac{kwh}{d} \ x \ \frac{11.99 \ \ell}{kwh} = 4196 \ \ell/day$$

The consumptive use of water for cooling towers is 0.83 gal/kwh = 0.22
ℓ/kwh. This rate gives an individual water demand of

$$350 \ kwh/d \ x \ 0.22 \ \ell/kwh = 77 \ \ell/d$$

Thus, water demands for individual energy use is high.

Water use in the United States is summarized in Table 2 and Figure
2 (Freeze and Cherry, 1979).

Table 2: *Water Use in the United States*

	1950–1970			
	Cubic Meters/day 10^6			
	1950	1960	1970	Percent of 1970 Use
Total water withdrawals	758	1023	1400	100
Use				
Public supplies	53	80	102	7
Rural supplies	14	14	17	1
Irrigation	420	420	495	35
Industrial	292	560	822	57
Source				
Groundwater	130	190	262	19
Surface water	644	838	1150	81

Source: Murray, 1973

The total water resource potential of the nation is estimated at 4550 x
10^6 m^3/day (Freeze and Cherry, 1979) and projected demands for 1980 and
2000 will be 1700 x 10^6 m^3/day. As groundwater use also grows, the
aquifer residence time will decrease, water table will fall, and poten-
tiometric surfaces will decline. These effects will promote more rapid
recharge of aquifers and reduce the aquifer storage capability to meet
fluxating demands. Thus, the interaction of surface and groundwater

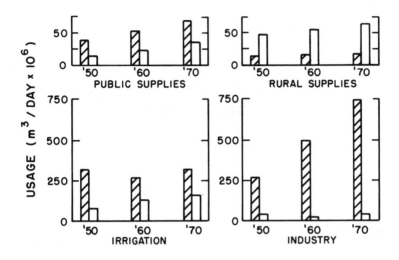

*Fig. 2.   Surface Water (hatched) and Groundwater (Stippled)
Use in the United States, 1950-1970 (after Murray, 1973).*

will become progressively more important and segregated views of the two
water supply systems will be impossible.

The demand on the groundwater will be double edged: the demand for
greater quantities (up to 1120 x 10^6 m^3/day by 2000), and the pollution
threat to groundwater supplies by solid waste disposal.

Surface water pollution is easily detectable. Effluent plumes are
observable from aircraft. Fish kills are immediate. Although more subtle changes in ecology due to small thermal loadings are possible, the
difficulty of recognition and correction of surface water pollution is
small compared to those of groundwater pollution.

The more subtle component of groundwater pollution is that it occurs
very slowly; often many years pass before the effects of severe pollution
is detected. The most menacing characteristic of groundwater pollution
is that damage is probably irreversible or, at least, corrective time

scales can be measured only in terms of years or even generations.

As shown in Figure 3, groundwater pollution can come from many sources. Increased use of coal will enhance some pathways and generate new potentials for contamination. In Figure 4, the tortuous path of water is shown moving slowly (1 m/day for dine sands) through the aquifer skeleton. Obviously, flushing rates would be equally slow and in fact blind pore spaces many never flush. Chemical attractions between the pollutant and skeleton may further retard flushing. A list of groundwater pollutants is shown in Table 3 (Fried, 1975). In Table 4, water quality norms for drinking water as specified by the World Health Organization are specified.

Table 3: *List of Main Possible Groundwater Pollutants and Pollution Indicators*

Total dissolved solids	Free CO_2	Phosphate (HPO_4^{2})
COD (Chemical Oxygen Demand)	Biocarbonates (HCO_3)	Zinc
BOD (Biological Oxygen Demand)	Iron (Fe)	Lead
Carbon (organically linked)	Total iron (Fe & Fe)	Copper
Hydrogen (organically linked)	Manganese	Arsenic
Nitrogen	Sodium	SiO_2
Detergents	Potassium	Temperature
Phenols	Calcium	pH
Oxygen	Mangesium	Conductability
Sulfates	Total hardness	Redox potential
H_2S	Chloride	
Nitrates (NO_3)	Fluoride	
Nitrates (NO_2)		
Ammonium (NH_4)		

III. ENERGY DEVELOPMENT VERSUS WATER RESOURCES

The report Water Related Constraints on Energy Production (Hampton and Ryan, 1978) presents a comprehensive review of water resource needs and problems associated with the increased energy production, with coal as a growing contributor. In addition to energy scenarios and projections, the report addresses in detail the topics of water resource demands on energy production, competition for resources, availability of

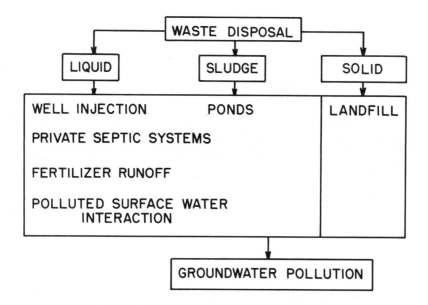

Fig. 3. Waste disposal and relation to groundwater

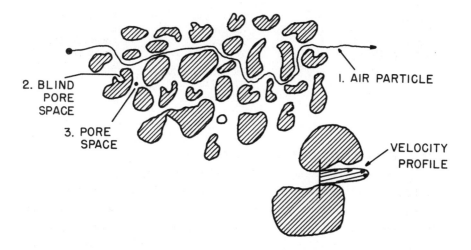

Fig. 4. Particle pathlines

Table 4a: *Water-Acceptability Characteristics (from W.H.O., 1972)*

Compound or Property	Colour (platine-cobalt Colour scale)	Odour	Turbidity	Total Solids (mg/1)	pH	Total Hardness (mg/1CaCO$_2$)
Maximum concentration proposed	5 units	No limit	5 units	500	7-8.5	100
Maximum concentration admissible	50 units	No limit	25 units	1500	6.5-9.2	500

Anionic detergents (mg/1)	Mineral oil (mg/1)	Phenols (mg/1)	Calcium (mg/1)	Chloride (mg/1 of Cl)	Copper (mg/1)	Iron (mg/1)
0.2	0.01	0.001	75	200	0.05	0.1
1.0	0.3	0.002	200	600	1.5	1.0

Magnesium (mg/1)	Manganese (mg/1)	Sulfates (mg/1 of SO$_4$)	Zinc (mg/1)
30 if sulfate concentration 250 otherwise 150	0.05	200	5.0
150	0.5	400	15

Table 4b: *Toxic Compounds (from W.H.O., 1972)*

Compound Expressed As	Arsenic	Cadmium	Cyanides	Total Mercury	Lead	Selenium
Maximum Concentration (mg/1)	0.05	0.01	0.05	0.001	0.1	0.01

water in the United States, characteristics and control of wastewater
produced on site, effluent guidelines and impacts on ambient water quality, applicable legislation and regulations and policy, and the issues
of availability constraints, competitive demands and potential solutions.

The conclusions of the Hampton and Ryan (1978) report are based on
general precipitation and runoff hydrologic data. While general discussions are presented of special problems beyond water availability (treatment costs, effluent guidelines, competing water uses), the conclusions
are that water demands will be severe in arid regions west of the Mississippi River.

A. Water Constraints

Suggested actions for reduction of water constraints includes:

(1) Water importation to a constrained area by interbasin
transfer. Examples are:

     a. Snake River basin in Colorado River basin;

     b. Suwanee River "pipeline" to Tampa Bay area; and

     c. Such groundwater "mines" as those in Arizona, where
it is assumed that transfer will become economically
feasible.

(2) Increased storage facilities to take advantage of seasonal variation of surface water supply. Large dams, however, have a
number of serious environmental impacts that must be considered.

(3) Improved agricultural methods, for example,

     a. Control of the 22% of irrigation water lost through
seepage from conveyance systems;

     b. Remove lands of marginal agricultural production; and

     c. Grow less water-intensive crops.

(4) Groundwater exploitation. This may be an irreversible
commitment of resources in many areas.

(5) Desalination. This becomes more viable as cost of energy increases. A similar solution is the use of brackish water in cooling systems.

(6) Cloud seeding.

(7) Improved water conservation within energy facilities.

B. Competive Water Demands

One of the key issues in the increased use of coal will be allocation of water resources to competitive users. The competing water demands include:

1. municipal,

2. industrial,

3. agricultural,

4. navigational,

5. recreational, and

6. general desire and necessity to maintain our water
    quality for future generations.

It is relatively easy to assign benefits (dollars) to the first 4
items above, but benefits associated with the in-stream water requirements (5 & 6) are difficult to assign. Improvement of water quality is
not only of increasing importance economically but also desirable for
the environment. Actually, it is very difficult to separate economical
and environmental importance. For example, to improve flood control and
drain wetlands for agricultural use, Florida's Kissimmee River channeli-
zation was completed in 1970. However, transformation of a shallow,
fast-moving, meandering river into a narrow, deep, and sluggish canal
has resulted in water quality decay in the river and in Lake Okeechobee.
The costs of proposed river restoration projects are astronomical. Regardless of the chosen restoration plan, the economic value of points 5
and 6, above, is clearly demonstrated. This example shows the short-
coming of non-site specific water constraint studies. There was a sur-
plus of water in the Kissimmee Basin (flooding was the problem); however,
the use of water (to provide rapid runoff during storms) turned out to
be an expensive decision. Similarly, if increased water demands (rapid
runoff, leachate pollution, groundwater withdrawal, surface water diver-
sion, etc.) are imposed by power plants without specific information on
impacts, the consequences may be undesirable and expensive.

C. Point Source versus Non-point Source Pollution

Point source pollution is defined as that wastewater source whose location, plow, degree of treatment, quality, and point is generally well defined. The general conclusion is that point-source pollution at facilities is manageable and that increased energy production will not have detrimental impacts on ambient water quality in the United States. Point-source pollution at coal-burning facilities is manageable; the existing technology is available; and treatment costs are site-specific and not prohibitive.

Trade-off at each site must, however, be seriously considered. For example, the elimination of once-through cooling may either require a demand for a large area of land for cooling ponds or may produce a waste treatment and residue disposal problem associated with cooling tower blowdown.

Non-point source pollution includes runoff, overflow storage pile leachates, groundwater movement through landfill, etc. The location, flow, degree of treatment, quality and endpoint may all be ill-defined. The prevention of pollution from these sources is much more difficult, but still not an insurmountable task. Above-ground solutions include items such as exclusion barriers (diversions, storage pile forms, liners, and catch basins) and the availability of on-demand treatment facilities. Most of these solutions are based upon design with provision for a statistically predictable storm of predicted magnitude. But design storms are always exceeded. Site-specific analyses should consider the consequences of sudden pollutant discharge under the stress of a possible 100-year, 5-day storm.

The landfill versus groundwater contamination question is difficult. New regulations call for extensive leach phases for waste materials. Sophisticated installations that include impermeable liners and monitoring programs are now generally accepted.

D. Instream Flow Requirements

Many studies ignore the requirements of instream flow. Thus, a project that is economically feasible can be meaningless. When costs

are passed on to consumers, or when alternative schemes are considered
within limited domains (i.e., when all types of cooling systems are
cost-evaluated), corrective measures for instream requirements may completely overshadow previous efforts to find the system efficient as to
water use. Instream flow requirements means water necessary to provide
for:

> fish and wildlife habitat
>
> water quality maintenance
>
> satisfaction of treaties
>
> allowance for navigation

E. Coal Burning By-products

Large coal stockpiles will become more prevalent with the switch to
coal. Coalpile leachate forms when moisture comes in contact with stock-
piled coal. The leachate is highly acidic and may contain serious levels of certain pollutants. See Table 5.

Table 5: *Coalpile Leachate Quality (EPA survey, Nichols, 1974)*

Parameter	Mean concentration (milligrams/liter)	Range of concentration (milligrams/liter)
pH	2.7	2.1 – 3.0
Iron	19.540	0.11 – 93.000
Sulfate	9,006	525 – 21,920
Zinc	3.64	1.6 – 23
Copper	2.10	1.6 – 3.4
Chromium	3.27	0 – 15.7
TDS	16,440	720 – 44,050

The quantity and quality of CPL are site-dependent. During periods
of no precipitation, moisture within the coalpile dissolves minerals
which are flushed furing rainfall. Minor concentrations still exist in
late stages of CPL flow (with sufficient precipitation). Acidity, however, is rather uniform throughout the flow.

Solutions to the problem include the development of flow and con-

taminant hydrographs. Also essential is design of attenuation systems
such as basins and liners, control of water table, and lateral flow of
groundwater.

IV. WATER FROM COAL[1]   .

Many of the lignite coals found in the northcentral regions of the
United States have substantial amounts of water trapped within the lignite. This trapped water can cause problems in combustion and other operations, and often must be used before lignite can be utilized. Since
most lignite is mined in dry regions where water is at a premium, recovery and use of the water in lignite could provide a significant contribution to the water required in many coal processes. In this way the
problem of water supply and the problem of water removal and disposal
can be integrated to provide an economically promising solution.

The water content in lignite can range up to fifty percent by weight,
corresponding to about two hundred and forty gallons of water per ton of
dried lignite produced. If the lignite were gasified to produce synthesis gas, this water would provide about two-thirds of the steam required
in the gasification process.

Other possible applications of entrapped water should be considered
in attempts to make an asset of what is currently regarded as a liability.

The economic feasibility of water recovery from lignite depends on
the cost of recovery compared to the cost of obtaining water from another
source. An advantage of recovering water from lignite is that the water
obtained will be of a high purity, since it is evaporated from lignite
and then condensed by cooling.

---

[1] This section is due to D. J. Miller and A. E. S. Green

References

Anderson, W. E. and M. P. Youngstrom, Coal pile leachate - quantity and quality characteristics, J. of Environmental Eng. Div., ASCE, 1234-1253, December 1976.

Bechtel, Florida Gas Coal Slurry Pipeline: Feasibility Study, confidential report, Nov. 1978.

Brandi, R., P. Chan et al., Impacts of future coal use in California, Interim Regional Report for Coal Utilization Assessment, LBL-8402 UC-90j, Lawrence Berkeley Laboratory.

Cole, J.A. (ed.), Groundwater pollution in Europe, Proceedings of Water Research Assoc. Conference, England, p. 547, Sept. 1972.

Dewiest, R. J. M., Geohydrology, John Wiley and Sons, 1965.

Freeze, A. R. and J. A. Cherry, Groundwater, Prentice Hall, 1979.

Fried, J. J., Groundwater pollution, theory, methodology modelling and practical rules, Elsevier Scientific Publishing Co., 1975.

Gainesville/Alachua County Regional Utilities, Site Certification Document (2 volumes) for Deerhaven II, 1978.

Hampton, N. F. and B. Y. Ryan, Jr., Water related constraints on energy production, The Aerospace Corporation, Germantown, Maryland, Report No. ATR-78 (9409)-1.

Leach, S. D., Source, use and disposition of water in Florida, 1975, USGS, Water Resources Investigations 78-17, April 1978.

Linsley, R. K. and J. B. Franzini, Water Resources Engineering, McGraw-Hill 3rd Ed., 1979.

Lowenbach, W., F. Ellerbusch and J. A. King, Leachate testing techniques surveyed, Water and Sewage Works, pp. 39-46, December 1977.

McMullan, J. T., R. Morgan and R. B. Murray, Energy Resources and Supply, John Wiley, p.508, 1977.

Morgan, M. G. (ed.), Energy and Man: Technical and Social Aspects of Energy, IEEE Press, 1975.

Nace, R. L. (ed.), Scientific framework of world water balance, UNESCO Tech. Papers, Hydrol., 7, p. 27.

Nichols, C. R., Development document for effluent limitations, guidelines and new source performance standards for the steam electric power generating point source category, U.S. Environmental Protection Agency, Washington, D.C., 1974.

Rittenhouse, R. C., Fuel transportation: Meeting the growing demand, Power Engineering, pp. 48-56, July 1977.

Schlottmann, A. M., Environmental Regulation and the Allocation of Coal: A Regional Analysis, Prager Publishers, N.Y., p. 143, 1977.

Todd, D. K. and D. E. O. McNulty, Polluted Groundwater, A Review of the Significant Literature, (based on report EPA-600/4-74-001, 1974) Water Information Center, Huntington, N.Y., 1976.

U.S. Department of Energy, Executive summary, an assessment of national consequences of increased coal utilization, prepared for U.S. DOE, TID-29425 (Vol. 2), UC-88.89.90 THRU 90J.

World Health Organization, Normes internationales pour l'eau de EEINOS Gen, 1972.

CHAPTER 9

ATMOSPHERIC POLLUTION

Paul Urone and Michael A. Kenney

I.   EMISSIONS AND AMBIENT AIR QUALITY

The increased use of coal as a source of energy involves a wide
range of technical, economic, social, and environmental problems.   In
this chapter we will outline those problems associated with actual as
well as potential impacts on the atmosphere resulting from a shift to
the increased use of coal.

In addition to the heat that is used for energy, the combustion
of coal produces large quantities of waste products that must be
disposed of in one manner or another.   Historically, little thought was
given to controlling the atmospheric emissions of burning fuels.   Smoky
stacks and dirty neighborhoods were considered unpleasant but gratifying
signs of industrial activity, jobs, and prosperity.   However, death-
dealing air pollution episodes and various health, plant, and material
damage studies have shown that the air of a community or region can be
harmfully overloaded with the emissions from man's activities (Stern,
1977).

Within the last decade, federal, state, and local governments have
been passing increasingly restrictive regulations for controlling
polluting emissions from industrial, commercial, automotive, and general
everyday activities.   Air quality standards have been developed which
protect the health (called primary standards) and the welfare (called

169

secondary standards) of the general population. In addition, lists of
toxic substances and priority pollutants have been developed for intensive study to determine the possible need for developing regulations for
future control requirements (Clean Air Act, 1977).

Ambient air quality standards have been set for particulate matter,
sulfur oxides, nitrogen oxides, carbon monoxide, hydrocarbons, oxidants,
radioactive substances, and a number of toxic trace substances (40 (CFR)
50). Fossil fuel combustion contributes heavily to the emissions of
most of the controlled pollutants and is a major emitter of particulate
matter, sulfur oxides, and nitrogen oxides (U.S. EPA, 1977). Efficient
combustion in industrial and power plant boilers reduces the emissions
of hydrocarbons and carbon monoxide to a minimum except under breakdown
conditions. Residential and small commercial users of coal as a fuel
generally operate in a less efficient manner.

To estimate the air quality impact of increased coal use is not a
simple matter of taking ratios of smoke stack emissions or air measurements to the amount of increased use. Among the factors that must be
considered are the types and properties of the coal to be used. Heating
value, sulfur, nitrogen, ash, and trace element content must be considered. The local micrometeorology and plume dispersion patterns are
important in determining how much of the pollutants will be found at
ground level under normal as well as worst-case conditions. Finally,
the location and the degree of emissions control required will be
determined by both the quality of the area's ambient air and by the
potential impact on any environmentally protected areas in the vicinity.
Proper proportioning of all the above factors requires an indepth
knowledge of each of the mechanisms involved, as well as a detailed.

II. COMBUSTION PRODUCTS EMITTED TO THE ATMOSPHERE:
     TYPES AND RELATIVE AMOUNTS

Air pollutants are emitted into the atmosphere from a wide range of
natural and man-made sources. Emissions of pollutants from natural
sources are difficult to estimate and for the most part cannot be

controlled. In general, it is estimated that natural sources of emissions are greater than man-caused sources taken on a worldwide basis (Robison and Robbins, 1972). However, emissions from natural sources are often surpassed in urban areas by the local anthropogenic sources.

Man's energy requirements contribute a major share to the anthropogenic sources of pollution. Electric power generation is the greatest stationary source of sulfur oxides. The industrial and electric utility sectors contribute a large fraction of the particulate matter in the atmosphere. The residential and industrial sectors and the transportation networks account for much of the organic, carbon monoxide, and carbon dioxide pollutants. Table 1, adapted from U.S. EPA/DOE (1978), indicates gross quantities of air pollutants released nationwide by the major sources for 1975. Table 2, also adapted from U.S. EPA/DOE (1978), separates stationary sources of pollution into four subtypes and shows the relative proportion of air pollutants from each.

TABLE 1: *Estimated Nationwide Emissions - 1975**
(10^6 tons/year)

Source	Particles	Sulfur Oxides SO_x	Nitrogen Oxides NO_x	Hydrocarbons HC	Carbon Monoxide CO
Nature	Unknown	4.2	Unknown	30.7	Unknown
Stationary Combustion	7.1	22.1	11.0	0.4	1.1
Transportation	0.8	1.1	11.2	19.8	111.5
Industrial Processes	14.4	7.5	0.2	5.5	12.0
Miscellaneous	12.8	0.4	2.4	11.2	26.1
Total	35.1	35.3	Unknown	67.6	Unknown

*Reference: U.S. EPA/DOE (1978)

In discussing the combustion products emitted to the atmosphere from coal, it should be noted that the amounts of the pollutants depend

TABLE 2: *Air Emissions from Stationary Combustion Systems in the USA** *(Expressed as Percentages)*

| | Particles | SO_x | NO_x | HC | CO | Organic Compounds | | |
						BSO	PPOM	BaP
Electric Generation	63.8	72.5	64.8	34.0	33.6	8.8	0.3	0.2
Industrial	28.3	14.5	24.7	22.3	14.9	20.0	0.5	1.3
Commercial/ Institutional	4.9	6.7	7.3	12.2	7.7	16.0	0.2	0.4
Residential	3.0	6.3	3.2	31.5	44.7	55.2	99.0	98.1
Total 10^3 ton/year	7,060	22,100	10,950	353	1,070	125	4.14	0.40

SO_x = Sulfur oxides   BSO = Benzene soluble organics

NO_x = Nitrogen oxides   PPOM = Particulate polycyclic organic material

HC = Hydrocarbons

CO = Carbon monoxide   BaP = Benzo (a) pyrene

*Reference: U.S. EPA/DOE (1978)

on several factors, including the following:

1. Major and minor elemental concentration in the coal
2. Boiler configuration and firing conditions
3. Flue gas emission control devices
4. Properties of the element and its compounds.

The chemical composition of coal ash and coal derived gases depends largely on the geologic and geographic factors related to the coal deposit. However, one must use caution in attempting to characterize and quantify the effluents from coal combustion based on the average elemental concentrations from the raw coal. Many elements exhibit a great variability within this regard. One may be certain, however, that the quantities of potentially hazardous pollutants entering the environment as a result of coal combustion will increase with current increases in the amount of coal being used. Table 3 shows some typical values of major elemental composition of coal.

TABLE 3: *Major Elemental Analysis of Coal**

	C	H	O	N	S	Ash	H/C
			Weight Percent				
Coal (moisture-free)							
Subbituminous (Big Horn)	69.2	4.7	17.8	1.2	0.7	6.5	0.81
Bituminous (Pittsburgh)	78.7	5.0	6.3	1.6	1.7	6.9	0.76

*Reference: U.S. EPA (1979)

Coal has been shown to contain many trace elements in addition to carbon, hydrogen, nitrogen and sulfur (see Chapter 12 of this monograph by Bolch). All of these elements are potential emissions when the coal is burned either as part of the fly ash or in a volatilized form.

A.   Gaseous Emissions

1.   *Sulfur Oxides*

The sulfur in coal is present in both organic and inorganic forms. Sulfur is emitted to the atmosphere largely (greater than 95 percent) as SO_2 with smaller amounts of H_2S, H_2SO_4 (approximately three percent) and particulate sulfates.

Table 4 depicts SO_2 emissions from burning coals of differing heat capacity and sulfur content.

Over the past two decades, the shift to cleaner (low-sulfur) fuels such as oil and natural gas has resulted in a decline in the level of SO_2 in the atmosphere. This trend has leveled off recently and there have been slight increases in parts of the northeast. A shift to coal from oil and natural gas would create nationwide elevated SO_2 trends if strict emission control measures are not enforced.

2.   *Nitric Oxide and Nitrogen Dioxide*

Nitrogen contaminants in coal are present primarily as organic compounds. Emissions of oxides of nitrogen (NO_x) result not only from

TABLE 4: *SO_2 Emissions from Burning Different Coals**

% Sulfur in Coal	Lbs $SO_2/10^6$ BTU	
	Western Coal at 9000 BTU/lb	Eastern Coal at 13000 BTU/lb
.2	.4	.3
.6	1.3	.9
1.0	2.1	1.5
3.0	6.3	4.4
5.0	10.6	7.3
7.0	14.8	10.2

*Reference: U.S. EPA/DOE (1978)

the combustion of the nitrogenous compounds contained in the fuel but also from high temperature reactions of atmospheric nitrogen and oxygen in the combustion zone. The important factors that affect NO_x production are: flame and furnace temperature, residence time of combustion gases at the flame temperature, rate of cooling of the gases, and amount of excess air present in the flame. Nitric oxide (NO) is the major nitrogenous pollutant of combustion processes. However, nitric oxide is rapidly oxidized to NO_2 in a sunlit irradiated smoke plume. As NO_2 it may either initiate the photochemical formation of ozone (O_3) or it may be further oxidized to nitrate aerosol.

Though mobile sources contribute over half of the anthropogenic output of NO_x, fossil fuel combustion also contributes heavily (see U.S. EPA/DOE, 1978, Table 1).

3. *Carbon Monoxide and Carbon Dioxide*

Carbon monoxide (CO) has long been considered an important toxic atmospheric pollutant. It is prevalent in automobile exhaust and in the effluents from poor combustion. The quantity of CO emitted from a combustion process is dependent on the combustion efficiency of the system. Many of the factors that affect the quality of combustion are better controlled on larger boiler installations. Major CO emissions from coal come from power development, industrial operations, residential

and local uses, and coke and gas plants. Although emissions of CO from
pollutant combustion sources are second only to CO_2 in total mass, the
quantity of CO emitted from stationary sources is small when compared
with that of mobile sources (see U.S. EPA/DOE, 1978, Table 1).

The most commonly emitted combustion gas is carbon dioxide (CO_2).
It is such an integral part of all biological activities that CO_2
emissions are generally not considered as pollutants. In spite of this
fact, however, it is an air pollutant of global importance to man's
environment. Some monitoring studies indicate a gradual global increase
in CO_2 concentration from 275 ppm in 1850 to 330 ppm at present (Urone
in Stern, 1977). This increase in CO_2 concentration in the atmosphere
may serve to trap heat and cause an increase in global temperatures due
to carbon dioxide's ability to cause a greenhouse effect (see Chapter 11
of this monograph).

4.   *Hydrocarbons*

Hydrocarbon emissions obtained from coal-burning units vary widely,
depending on the quality of combustion achieved. A comparison of hydrocarbon emissions with other products of incomplete combustion indicates
that organic emission rates are generally high when carbon monoxide
emission rates are high. However, it has also been observed that hydrocarbon emission rates from larger coal-fired units are less, relative to
smaller units, and much less, relative to automobile exhaust.

A trace organic pollutant category produced during combustion of
coal is that of polynuclear aromatic hydrocarbons (PAH). Recent data on
the composition of emissions from coal combustion show that between
eight and fifteen PAH compounds have been identified in power plant
effluents (U.S. EPA, 1978). These same researchers state that compounds
formed during combustion will, upon cooling, condense out as discrete
particles or condense onto the surface of existing particles. Hence,
control technology which is effective in reducing particulate emissions
will also significantly reduce PAH emissions. Other observations
indicate that most organic compounds are in the vapor phase within
stacks and do not adsorb onto particles until the stack plume has cooled

several tens of meters from the stack mouth (Torrey, 1978). It is
generally thought that the atmospheric burden of these trace substances
will not be significantly increased by coal-fired units, but this
conclusion must be considered provisional until additional data are
obtained.

5. *Water Vapor*

Moisture content of most coals is about 10%. Vapor emissions from
coal-fired units are clearly visible as white billowy plumes. Water
vapor offers no direct pollutant hazard but may facilitate other chemical
reactions within the plume or cause visibility problems.

B.   Volatile Toxic Substances

1. *Trace Elements and Heavy Metals*

Trace element composition of raw coal varies greatly. Most trace
elements found in coal occur in concentrations which approximate the
earth's continental crustal material. As the coal is burned, some of
these elements concentrate in the residue, while others are emitted into
the atmosphere in gaseous or particulate forms. More important, some
elements are not emitted during combustion in concentration ratios
identical to those in the original coal or coal ash. Selective volatilization for these elements results in an enrichment in the fly ash,
while other elements are depleted. Enrichment tends to be greatest in
the smallest particles emitted from a coal-fired unit. These elements
include lead, antimony, selenium, and arsenic. Most of the mercury,
some selenium and most of the chlorine and bromine is discharged into
the atmosphere in the gaseous form (see Chapter 12 of this monograph by
Bolch).

2. *Radioactive Isotopes*

Coal contains small quantities of radioactive isotopes. Studies
have been conducted to investigate the release of natural radionuclides
from the combustion of coal and have attempted to compare radioactive
effluents from fossil fuel plants to those of nuclear plants (McBride et
al., 1978). Again, it is thought that many radioactive elements undergo

preferential volatilization and partitioning in the flue gas much like
the nonradioactive trace elements already described. Studies have shown
that releases of radioactive materials from coal-fired plants were well
within the limits imposed on nuclear plants. It is concluded that the
public health significance of these emissions is relatively minor
(McBride et al., 1978). Health effects associated with airborne releases
of nonradioactive material from coal-fired plants (particles, NO_x, SO_2,
etc.) would appear to be many times more significant. However, increased
coal utilization, larger coal-burning plants, higher radionuclide content
and ash releases would result in increased doses from the coal plant
(see Chapter 12 of this monograph by Bolch).

C.   Particulate Matter

Atmospheric particulate matter associated with the production and
use of coal is emitted from the following sources and processes: coal
storage piles, coal crushing, coal transportation, coal receiving, and
fly ash from combustion.

The amount of fly ash emitted from combustion sources is related to
many factors, the most important of which are: amount of ash in the
coal, method of burning the coal, rate at which the coal is burned, and
efficiency of control equipment. As noted above, the chemical composition of this fly ash is very important. However, the most important
characteristic of particulate matter emitted from coal combustion is its
size. The public is more likely to be exposed to these fine particles
(<10 μm) because of their larger atmospheric residence times and because
they can enter the respiratory system and constitute a hazard for human
inhalation (see Chapter 14 of this monograph by Jaeger and Schlenker).

Estimated coal-fired emission particle size distribution for four
broad classifications of combustion equipment are listed in Table 5.
All distributions represent the size of the particles leaving the system
in advance of any control equipment. Size distribution may vary widely
and these shown are considered "typical."

TABLE 5: *Particle Size Distribution of Typical Combustion Fly Ash**
(Weight Percent Less than Stated Size)

Particle Size (μ)	Pulverized Fuel-Fired Furnace	Cyclone Furnace	Spreader Stoker-Fired Furnace	Stoker-Fired (Other than Spreader)
10	30	76	10	7
20	50	83	20	15
40	70	90	37	26
60	80	92	47	36
80	85	94	54	43
100	90	95	60	50
200	96	97	--	66

*Reference: NAPCA (1969a)

D.   Emission Factors

To help evaluate the potential environmental effects of coal-fired combustion, tables of emission factors have been developed. Emission factors are weighted/average factors used to estimate the amounts of emissions to be expected from a given process or activity. Table 6 shows the emission factors for various coal combustion processes. The letters indicate that the percent ash (A) or sulfur (S) should be multiplied by the value given.

III. AIR POLLUTION CHEMISTRY

Polluted atmosphere is a highly complex mixture of completely or partially-consumed fossil fuels, fugitive dusts, and gases and vapors from either natural phenomena or human activities. Chemical reactions of the pollutants are affected by highly variable concentrations, temperatures, fluctuating wind patterns, and sunlight.

A.   Types of Chemical Reactions

The types of chemical reactions that can and probably do occur in a polluted atmosphere include most of those known to man. A complete

TABLE 6: *Emission Factors for Selected Categories of*
*Uncontrolled Pollutants from Coal Combustion**
(Lb/ton coal burned)

Furnace size (10^6 Btu/hr heat input)	Partic-ulates	Sulfur Oxides	Carbon Monoxide	Hydro-carbons	Nitrogen Oxides
>100 (utility and large industrial boilers):					
Pulverized:					
General	16A	38S	1	.3	18
Wet bottom	13A	38S	1	.3	30
Dry bottom	17A	38S	1	.3	18
Cyclone:	2A	38S	1	.3	55
10 - 100 (large commercial & general industrial):					
Spreader Stoker	13A	38S	2	1	15
All others	5A				
<10 (commercial & domestic furnaces):					
Under-Feed Stoker	2A	38S	10	3	6
Hand-Fired Units	20	38S	90	20	3

*Reference: U.S. EPA (1977)

understanding of all the reactions is an unrealistic goal as of this
time and may be so into the indefinite future. The science of
atmospheric chemistry, consequently, becomes one of attempting to
understand those reactions that proceed rapidly enough to dominate rate-
determining mechanisms or that have an important environmental impact
or stress.

Table 7 lists in broad general terms the types of reactions that
may occur in air. They cover the gamut of general, physical, organic,
and biochemical reactions. Indeed, the interaction of pollutant gases
and particulate matter with human, animal, and plant systems is one of
the principal driving forces for studying and reducing the concentrations
of the offending pollutant substances. As of this time, some of the

TABLE 7: *Types of Chemical Reactions in the Atmosphere*

Thermal:	Homogeneous Gaseous Aqueous Catalytic	Photochemical:	Primary Excitation Free radical formation (dissociation)
	Heterogenous Adsorption Absorption Catalytic		Secondary Free radical reactions Chain reactions - polymerization Thermal reactions
	Inorganic		
	Organic		
	Biological/physiological		

general inorganic, organic, and photochemistry is understood, but many
of the organic and physiological reactions are not known or understood.

Because oxygen is by far the dominant reactive gas in air, most
reactions proceed directly or indirectly toward higher states of oxidation. The stratosphere, with its ozone layer and intense sunlight,
forms the ultimate oxidative sink for even the most resistant of chemicals. Many of the gaseous substances emitted by industry and the
automobile are either acid anhydrides or the precursors of acid anhydrides (CO, CO_2, SO_2, NO, NO_2). Ammonia (NH_3) as a decomposition product
of urea and other biological processes is the principal alkaline gas
distributed widely throughout the troposphere. Consequently, it is not
surprising that a large fraction of the air pollution aerosols are
composed of oxidized, acidic residues and ammonium compounds (HNO_3,
NH_4NO_3, NH_4HSO_4, etc.). Although the worldwide natural production of
atmospheric ammonia is of the order of one billion tons per year, ammonia
has a short residence time due largely to the presence of acidic
pollutants. The background concentration consequently is low (0.01
ppm), and in polluted areas, the background ammonia is often readily
depleted.

B.   Scavenging Processes

There are several processes which remove pollutants from the atmosphere (Table 8).   The residence time for the more reactive pollutants is measured in hours or days.   The residence times for the less reactive substances (methane, nitrous oxide, freons) are measured in years and decades.   It is fortunate that most major pollutants are chemically and/or physically active.   They or their reaction products are removed from the atmosphere by rainout, washout, or dry deposition.

TABLE 8:   *Scavenging Processes in the Atmosphere*

Rainout (snowout)

Washout

Dry deposition

    Sedimentation
    Impaction
    Adsorption
    Absorption

Transport into stratosphere

    Photochemical conversion and eventual return

Rainout occurs in clouds where water vapor condenses on pollutant particles or condensation nuclei.   Water-soluble or hydrophilic substances may also be absorbed during the rain formation process.   The pollutants are subsequently carried to earth as part of the rain. Washout occurs while the rain (or snow) falls through a polluted atmosphere, resulting in a scrubbing or "washout" action.   Washout phenomena largely account for the preponderance of sulfate in acid rain.

Dry deposition includes those processes other than rainout or washout which remove pollutants from air.   These include sedimentation by gravitational forces, impaction on solid or liquid surfaces, and absorption or adsorption processes.   Particles larger than 10 μm in diameter settle rapidly while smaller particles take days to weeks to settle, depending on their size and the altitude at which they exist.

Very small particles (<0.1 μm) will coagulate to form larger particles
much more rapidly than particles in the size range from 0.1 to 1 μm.
This latter size range is too large to diffuse rapidly through Brownian
motion and too small to settle or impact rapidly. Unfortunately, the
0.1 to 1 μm size is also the size that scatters visible light (λ = 0.4
to 0.7 μm) most effectively, contributing in a large measure to the haze
effect of polluted atmospheres. Particles which intrude or are formed
in the stratosphere have residence times ranging from two to twenty
years--again depending upon particle size as well as distance from the
earth's surface.

Many gaseous substances are removed from the atmosphere by absorption or adsorption processes. The various water bodies, solid surfaces,
trees and other vegetative forms play important roles in this respect.
In Europe and Japan, "green belt" methods of pollution control have been
used to reduce the impact of heavy industrial areas upon residential
areas.

Inert gases such as Freon 11 and 12 and nitrous oxide (N_2O) accumulate in the atmosphere and gradually diffuse into the stratosphere. In
the intermediate to the upper regions of the stratosphere, such
substances interact with the unfiltered, highly energetic ultraviolet
rays of sunlight to form various breakdown products, including some
reactive free radicals. The free radicals enter into and disrupt the
normal stratospheric reaction cycles to an extent that depends upon the
amount and persistance of the intruding substances. Destruction of
ozone and its important ultraviolet filtering capacity is a potentially
serious long-term effect of inert or persistant pollutants.

C.   Atmospheric Chemistry of Coal Smoke Plumes

The chemistry of the gases and particulate matter in smoke plumes
has not been subjected to as much study as the urban photochemical smog
chemistry. A limited number of studies have used instrumented aircraft
to measure some of the gases and the particulate matter in power plant
plumes (Flyger et al., 1978). The rate of disappearance of sulfur

dioxide, as well as the ratio of sulfur dioxide to either particulate matter or sulfate, has been measured (Burner, 1977). Some recent studies have evaluated the long range transport of sulfur dioxide and have shown trajectories of thousands of kilometers in distance (Altshuller, 1976). Sulfur dioxide and particulate matter emitted as far west as Kentucky or Arkansas contribute to the atmospheric load of New York and the New England states (U.S. EPA/DOE, 1978).

The reaction of sulfur dioxide in a stabilized plume is slow: no more than one to two percent per hour (Meagher et al., 1978). Its residence time in air has been estimated to be six days (Kellogg et al., 1972). It will rapidly undergo catalyzed heterogeneous oxidation to sulfuric acid on particulate matter until the particle surface or water droplet attains a pH of about 3. At this point sulfur dioxide is repelled by the particle or surface and the reaction slows considerably.

As the plume moves downwind under sunlight irradiation, the nitric oxide (NO) becomes oxidized to nitrogen dioxide (NO_2) and possibly nitric acid (HNO_3). The sulfur dioxide reacts photochemically with the nitrogen dioxide to become sulfuric acid. Both the nitric acid and the sulfuric acid, or their anhydrides, will react rapidly with ammonia or alkaline particulate matter (such as lime or limestone fly ash).

The scavenging processes described above operate on the plume at all times. Dry deposition removes particulate matter and the pollutant gases as they reach ground level. Vegetative growth and topographic structures help remove considerable fractions of the pollutants. If the concentrations are high enough, plant (including tree) damage is observed. Corrosion and soiling of buildings and art structures by coal burning power plant plumes has been documented frequently (NAPCA, 1969b).

Rainout and washout processes also operate. Rainout, or in-cloud, processes remove water-soluble gases and both water-soluble and water-insoluble particles that form raindrop nucleating centers at cloud level. Washout, or below-cloud, processes will tend to remove more of the sulfur dioxide than the nitrogen oxides, as the latter are not very soluble in water. Both processes help remove gaseous and particulate

matter from the atmosphere. Unfortunately, they result in making the
rain more acidic, giving rise to the phenomenon of "acid rain" which is
of grave concern to environmentalists (see Chapter 17).

Frequently, coal smoke plumes become part of, or are intermixed
with, the emissions of an urban area. The relatively higher amounts of
sulfur dioxide in the smoke plume enter into the photochemical smog
reactions of the urban atmosphere. Since sulfur dioxide is a reducing
agent, there is a tendency for a slowing down of the photochemical
oxidant formation processes with a corresponding increase in the rate of
formation of sulfate particulate matter, acid rain, light scatter, and
harmful synergistic respiratory health effects (U.S. EPA/DOE, 1978).

D.   Atmospheric Chemistry of Sulfur Dioxide

The atmospheric chemistry of sulfur dioxide, a major coal combustion
pollutant, has been the subject of a large number of applied and theoretical studies. One of the more recent reviews was written by Urone
and Schroeder (1978).

Sulfur dioxide in air undergoes a wide range of reactions. In the
laboratory the photochemical reaction of sulfur dioxide in clean air is
quite slow. When other pollutants such as nitrogen oxides and hydrocarbons are present, the reaction is much more rapid. Some of the
mechanisms are shown in Table 9. The presence of particulate matter and
high relative humidity add additional impetus to the rate of reaction.
A more detailed discussion is beyond the scope of this monograph.

The ultimate product of the reactions of sulfur dioxide is a
sulfate salt such as ammonium sulfate or calcium sulfate. A number of
intermediate compounds are known or are suspected to be present in a
polluted atmosphere. In addition to sulfuric acid, these would include
the bisulfate salts and possibly salts of zinc and cadmium near smelters.
Organic sulfur compounds are also hypothesized to be present.

Sulfur dioxide and its reaction products have been present in
relatively large amounts (0.1 - 1 ppm) in all major air pollution
episodes. Because sulfur dioxide in itself is not highly toxic at those

TABLE 9:   *Reactions of SO_2 in Presence of NO_2 and Hydrocarbons*[*]

(1)	$SO_2 + h\nu \rightarrow {}^1SO_2$
(2)	${}^1SO_2 \rightarrow {}^3SO_2$
(3)	${}^3SO_2 + O_2 \rightarrow SO_4$
(4)	$SO_4 + SO_3 \rightarrow 2SO_3$
(5)	$NO_2 + h\nu \rightarrow NO + O$
(6)	$O + O_2 + M \rightarrow O_3 + M$
(7)	$O_3 + NO \rightarrow NO_2 + O_2$
(8)	$O + NO + M \rightarrow NO_2 + M$
(9)	$O + NO_2 + M \rightarrow NO_3 + M$
(10)	$O_3 + NO_2 \rightarrow NO_3 + O_2$
(11)	$O + O_2 + \text{olefin} \rightarrow RO_2\cdot, \text{ etc.}$
(12)	$RO_2\cdot + NO_2 \rightarrow PAN, \text{ etc.}$
(13)	$RO_2\cdot + SO_2 \rightarrow RO_2SO_2, SO_3 + RO\cdot, \text{ etc.}$

$$-\frac{d(SO_2)}{dt} = \frac{k_{11}k_{13}k_5(\text{olefin})(NO_2)(SO_2)}{[k_{12}(NO_2) + k_{13}(SO_2)] \; k_{11}(\text{olefin}) + M[k_9(NO_2) + k_8(NO) + k_6(O_1)]}$$

*Reference: Smith and Urone (1974)

concentrations and because many other pollutants were present at the
same time, it is felt that synergistic effects from two or more of these
substances resulted in the severe respiratory difficulties that caused
increased numbers of deaths for the young and the elderly.

References

Altshuller, A. P., Regional transport and transformation of sulfur dioxide to sulfates in the U.S., *J. of the Air Pollution Control Association*, 26:318, 1976.

Burner, C. A., *Concentration and Size Distribution of Particulate Sulfate and Ammonium Ions in Florida*, M.S. Thesis, University of Florida, Gainesville, 1977.

The Clean Air Act, U.S. Government Printing Office, Washington, D.C., 1977.

Department of Energy and Environmental Protection Agency, U.S. (DOE/EPA), *Energy/Environmental Fact Book - Decision Series*, Research Triangle Park, N.C., 1978.

Environmental Protection Agency, U.S. (EPA), *Compilation of Air Pollutant Emission Factors*, Third Edition, Research Triangle Park, N.C., 1977.

Environmental Protection Agency, U.S. (EPA), *Fuel Contaminants: Control of Coal-Related Pollutants*, Vol. 3, Research Triangle Park, N.C., 1979.

Federal Regulations, Code of, 40 (CFR) 50, U.S. Government Printing Office, Washington, D.C., revised July 1, 1977.

Flyger, H., E. Lewin, E. L. Thompsen, J. Fenger, E. Lyck and S.E. Gryning, Airborne investigations of SO_2 oxidation in the plumes from power stations, *Atmospheric Environment*, 12: 295, 1978.

Kellogg, W. W., R. D. Codle, E. R. Allen, A. L. Lazrus and E. A. Martell, The Sulfur Cycle, *Science*, 175:4022, 11 February 1972.

McBride, J. P., R. E. Moore, J. P. Witherspoon, R. S. Blanco, Radiological impact of airborne effluents of coal and nuclear plants, *Science*, 202:1045, 8 December 1978.

Meagher, J. F., L. Stockburger, E. M. Bailey and O. Huff, The oxidation of sulfur dioxide to sulfate aerosols in the plume of a coal-fired power plant, *Atmospheric Environment*, 12: 2197, 1978.

National Air Pollution Control Administration (NAPCA), *Air Quality Criteria for Particulate Matter*, U.S. Department of Health, Education, and Welfare, Washington, D.C., 1969a.

National Air Pollution Control Administration (NAPCA), *Air Quality Criteria for Sulfur Oxides*, U.S. Department of Health, Education, and Welfare, Washington, D.C., 1969b.

Robinson, E., and R. C. Robbins, in W. Strauss, Ed., *Air Pollution Control*, Wiley-Interscience, New York, Part II, 1972.

Smith, J. P. and P. Urone, Static Studies of Sulfur Dioxide Reactions, *Environmental Science and Technology*, 8:742, 1974.

Stern, A. C., Ed., *Air Pollution*, Vol. II, Third Edition, New York Academic Press, N.Y., 1977.

Torrey, S., Ed., *Trace Contaminants from Coal*, Noyes Data Corporation, Park Ridge, N.J., 1978.

Urone, P., in A. Stern, Ed., *Air Pollution*, Third Edition, Vol. 1, Chpt. 2, New York Academic Press, Inc., N.Y., 1977.

Urone, P. and W. H. Schroeder, in J. O. Nriagu, Ed., *Sulfur in the Environment: Part 1 - The Atmospheric Cycle*, Chpt. 8, John Wiley & Son, Inc., N.Y., 1978.

CHAPTER 10

AIR POLLUTANT DISPERSION MODELING

Raymond W. Fahien

I. INTRODUCTION

A. Purpose of Dispersion Modeling

   The quantitative prediction of the impact of increased coal burning
on air quality requires not only a knowledge of the emissions to be expected from coals of certain chemical compositions but also a quantitative method to relate emission rates to air quality. To do the latter
requires the use of a dispersion model. The purpose of an air pollutant
dispersion model is to predict the concentration of a given pollutant at
a specified location in an urban area for which source locations, source
strengths (emission rates), and meteorological conditions (wind velocity
and direction; atmospheric stability) are known. Since these data change
with time, concentrations averaged over a specified period of time are
usually obtained. These results can then be compared with federal and
state standards in order to predict probable violations. Thus the impact of a new power plant or of a change in the nature of the emissions
(e.g., those resulting from a change from oil to coal burning) can be
predicted *a priori* and, if necessary, steps can be taken (e.g., by reducing emission levels) to ensure a desired air quality level. Since
federal and state standards exist for the maximum allowable concentrations of pollutants, the effect of coal burning consequently may not only manifest itself in terms of increased concentrations, but also in
terms of increased stack height or greater control of emissions. Through
the use of a dispersion model the effect of such changes can then be

quantitatively predicted. The model is thus an important link between
input conditions and dose-response or other health effect studies.

B. Selection of Models

The accuracy of a model is usually evaluated by applying it to a
previous situation for which emission rates, meteorological data, and
measured concentrations are known. The latter are then compared to the
predicted concentrations. Unfortunately, however, the state-of-the-art
of dispersion modeling is such that more is unknown than known and, as
a result, discrepancies of several hundred percent between observed and
predicted concentrations are not uncommon. Furthermore, the results depend upon the dispersion model that is used and the approximations or
assumptions that are made. Inasmuch as the Clean Air Act Amendments of
1977 (CAAA) require permits and preconstruction reviews for new and modified stationary sources (in order to prevent "significant deterioration
of the environment"), the EPA was obliged to "specify with reasonable
particularity each air quality model or models to be used under specified sets of conditions". In compliance, the EPA has made available
computer programs for various "off-the-shelf" models. In addition, in
April 1979, the Agency published a Modeling Guideline. For areas that
do not meet federal standards (as shown by monitoring data or calculated
air quality modeling) the EPA Guideline indicates that their "recommended" models should not be considered rigid requirements". However, Mirabella (1979) indicates that the EPA regional offices usually adhere
quite closely to the Guidelines. Egan (1979) and Sklarew (1979) are
critical of the EPA as being too rigid in its acceptance of new models,
which they state are subjected to validation tests more stringent than
those previously required of the currently recommended models. Turner
(1979) of the EPA acknowledges that the EPA models have many limitations,
such as their capacity to predict long range transport and dispersion
with photochemical transformation.

II. GAUSSIAN MODELS

A. Gaussian Distribution

The fundamental problem with the EPA-recommended models is that they are extensions of a Gaussian dispersion model which assumes that the concentration C of a pollutant emanating from a point source (as a stack is assumed to be) obeys a bivariate Gaussian distribution (i.e., it follows the standard probability or error curve) in the vertical and lateral cross wind directions (See Figure 1) modified by a correction term that applies after the plume reaches the ground. The standard deviations of these curves (σ_y for the lateral and σ_z for the vertical direction) are called "dispersion coefficients". They are a measure of the size of the plume and therefore increase with distance from the source.

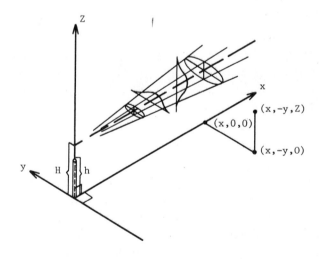

Fig. 1. Coordinate system showing Gaussian distributions in the horizontal and vertical.

Source: Perkins, 1978

If the source is emitting a pollutant at the rate Q (mass/time) and the wind velocity is u, the concentration (mass/volume) is

$$C = \frac{Q}{2\pi\,\sigma_y\,\sigma_z\,u}\,\exp\left(-\tfrac{1}{2}\left(\frac{y}{\sigma_y}\right)^2\right)\left\{\exp\left(-\tfrac{1}{2}\left(\frac{z-H}{\sigma_z}\right)^2\right)+\exp\left(-\tfrac{1}{2}\left(\frac{z+H}{\sigma_z}\right)^2\right)\right\}$$

In the above equation, H is the effective height of the stack. Since the hot gases emanating from a stack are bouyant, they will rise a considerable distance. This plume rise, when added to the actual stack height, h, gives an effective stack height, H, which must be used in the Gaussian model. The most widely used equations for calculating plume rise are those of Holland (1953) and Briggs (1969), but numerous other equations have been proposed. A comparison of these has been carried out by Stumke (1963) who found errors of several hundred percent. Thus the inaccuracy of the Gaussian model is compounded by an uncertainty in estimating the plume rise.

B. Atmospheric Stability

The dispersion coefficients vary with meteorological conditions (i.e., they depend upon the stability of the atmosphere). An unstable atmosphere is one which is bouyant to the extent that a particle of air in it has a tendency to rise; a stable atmosphere is one in which a particle of air tends to fall; and a neutral atmosphere is one in which a particle will have neither tendency. The so-called Pasquill-Gifford curves are plots of σ_y and σ_z vs. x (the downward distance) with six stability classifications (A through F) as parameters. The most unstable classification is A and the most stable is F with D being neutral. These discrete classifications are quite arbitrary since stability is actually a continuous function of the temperature gradient in the atmosphere and of the wind velocity. Since the temperature gradient is not always readily available, the stability classification is usually obtained from a table based on the estimated solar radiation (day) or fractional cloud cover (night), and the wind velocity. The assignment of a stability class is therefore somewhat subjective and non-quantitative and, since the dispersion coefficients can vary by several hundred-fold, depending on class, the procedure can lead to significant errors.

C. Pasquill Gifford Charts

The Pasquill-Gifford charts are only weakly based upon G. I. Taylor's (1921) theory of diffusion and they are supported by only a paucity of data - mostly for ground level or near ground level sources for smooth terrain, and for short distances from the source (Hay and Pasquill, 1957, 1959; Pasquill, 1951; Meade, 1960; Gifford, 1961; Bryant, 1969; Turner, 1970). Yet these charts are often used for almost all conditions of source height, source distance, and surface roughness.

D. Limitations of the Gaussian Model

The Gaussian dispersion model itself assumes (1) that the wind is always blowing in one direction only and that it is independent of distance from the ground, (2) that the dispersion coefficients are independent of distance from the ground, (3) that there is no chemical reaction or deposition of the pollutant, (4) that the terrain is level and smooth, (5) that meteorological or emissions conditions are independent of time (i.e., that a steady state exists), and (6) that the plume emanates from a point source. The model is nevertheless widely used even when these assumptions are not valid.

In view of the complexity of the problem and the simplicity of the Gaussian model, it is perhaps surprising that it gives results as good as it does. Since the model does not take surface and other conditions into consideration, the common practice was to "calibrate" it, i.e., to determine a regression equation between calculated concentrations at various locations and those actually measured. The assumption was then made that the same equation applies to all calculated concentrations under all conditions. Calibration procedures are not recommended by EPA [Turner (1979)].

E. Line, Area and Multiple Sources

The Gaussian distribution previously discussed applies strictly to

dispersion from a single point source but, in an integrated form, it is
also used for line or area sources. In order to use it for multiple
sources, as in an urban area, contributions from each source are calculated and added together linearly. However, this assumption has been
criticized by Sklarew (1979) as being fundamentally incorrect.

F. Climatological Models

    Several climatological dispersion models for estimating annual
averaged pollutant concentrations use a joint frequency distribution
of wind direction, wind speed, and stability class ("wind rose") as
meteorological input data. The average is obtained by summing the contributions due to each direction, speed, and stability class. An analytical representation of the joint frequency distribution of wind direction and wind speed has recently been developed by Green et al.
(1979). In addition the Pasquill-Gifford curves have been accurately fit
to empirical equations and the Gaussian equation has been expressed in
convenient polar coordinate form. These extensions of the Gaussian
model will greatly reduce the cost and complexity of computer calculations.

G. "Off-the-Shelf" Models

    For certain recommended "off-the-shelf" models, the EPA has made
available through NTIS a system of computer programs called UNAMAP
(User's Network for Applied Modeling of Air Pollution). These models
are called APRAC, CDM, HIWAY, CDM QC, CRSTER, PAL, Valley, RAM, PTMAX,
PTDIS, and PTMTP. Each has a special purpose but all are based on the
Gaussian plume concept. A summary of the EPA models (Turner, 1979) follows:

    APRAC: Stanford Research Institute's urban carbon monoxide model
computes hourly averages for any urban location. It requires an extensive traffic inventory for the city of interest.

    CDM: The Climatological Dispersion Model determines long-term
(seasonal or annual) quasi-stable pollutant concentrations at any ground
level receptor using average emission rates from point and area sources

and a joint frequency distribution of wind direction, wind speed, and
stability for the same period.

HIWAY: Computes the hourly concentrations of nonreactive pollutants downwind of roadways. It is applicable for uniform wind conditions
and level terrain. Although best suited for at-grade highways, it can
also be applied to depressed highways.

CDMQC: The Climatological Dispersion Model (CDM) is modified to
provide implementation of calibration, of individual point and area
source contribution lists, and of averaging time transformations. The
basic algorithms to calculate pollutant concentrations used in the CDM
have not been modified and results obtained using CDM may be reproduced
using the CDMQC.

CRSTER: Estimates ground-level concentrations resulting from up to
19 colocated elevated stack emissions for an entire year and prints out
the highest and second-highest 1-hr., 3-hr., and 24-hr. concentrations,
as well as the annual mean concentrations at a set of 180 receptors (5
distances by 36 azimuths). The algorithm is based on a modified form
of the steady-state Gaussian plume equation which uses empirical dispersion coefficients and includes adjustments for plume rise and limited
mixing. Terrain adjustments are made as long as the surrounding terrain
is physically lower than the lowest stack height. Pollutant concentrations for each averaging time are computed for discrete, non-overlapping
time periods (no running averages are computed) using measured hourly
values of wind speed and direction and estimated hourly values of atmospheric stability and mixing height.

PAL: Point, Area, Line source algorithm. This short-term Gaussian
steady-state algorithm estimates concentrations of stable pollutants
from point, area, and line sources. Computations from area sources include effects of the edge of the source. Line source computations can
include effects from a variable emission rate along the source. The algorithm is not intended for application to entire urban areas but for
smaller scale analysis of such sources as shopping centers, airports,
and single plants. Hourly concentrations are estimated and average
concentrations from 1 hour to 24 hours can be obtained.

Valley: This algorithm is a steady-state, univariate Gaussian
plume dispersion algorithm designed for estimating either 24-hr. or
annual concentrations resulting from emissions from up to 50 (total)
point and area sources. Calculations of ground-level pollutant concentrations are made for each frequency designed in an array defined by
six stabilities, 16 wind directions, and six wind speeds for 112 pro-
gram-designed receptor sites on a radial grid of variable scale. Empirical dispersion coefficients are used and the model includes adjustments for plume rise and limited mixing. Plume height is adjusted according to terrain elevations and stability classes.

RAM: Gaussian-Plume Multiple-Source Air Quality Algorithm. This
short-term Gaussian steady-state algorithm estimates concentrations of
stable pollutants from urban point and area sources. Hourly meteorological data are used. Hourly concentrations and averages over a number of
hours can be estimated. Briggs plume rise is used. Pasquill-Gifford
dispersion equations with dispersion parameters thought to be valid for
urban areas are used. Concentrations from area sources are determined
using the method of Hanna; that is, sources directly upwind are considered representative of area source emissions affecting the receptor.
Special features include determination of receptor locations downwind
of locations of uniformly spaced receptors to ensure good area coverage
with a minimum number of receptors.

The following point source models use Briggs plume rise methods and
Pasquill-Gifford dispersion methods, as given in EPA's AP-26, "Workbook
of Atmospheric Dispersion Estimates", (Turner, 1970) to estimate hourly
concentrations for stable pollutants.

PTMAX: Performs an analysis of the maximum short-term concentrations from a single point source as a function of stability and wind
speed. The final plume height is used for each computation.

PTDIS: Estimates short-term concentrations directly downwind of a
point source at distances specified by the user. The effect of limiting
vertical dispersion by a mixing height can be included and gradual plume
rise to the point of final rise is also considered. An option allows
the calculation of isopleth half-widths for specific concentrations at

each downwind distance.

PTMTP: Estimates, for a number of arbitrarily located receptor points at or above ground-level, the concentration from a number of point sources. Plume rise is determined for each source. Downwind and crosswind distances are determined for each source-receptor pair. Concentrations at a receptor from various sources are assumed additive. Hourly meteorological data are used; both hourly concentrations and averages over any averaging time from one to 24 hours can be obtained.

III. TRANSPORT MODELS

A. Analytical

A transport model is one in which the law of conservation of mass is applied for a pollutant species at an arbitrary location in the atmosphere. The resulting differential equation is quite general with wide applicability. The conservation equation is originally expressed in terms of the turbulent diffusive fluxes (i.e., the rates of diffusion per unit area in each of the three directions in space). These fluxes are then usually related to the concentration gradients by means of eddy diffusivities or "K" values -- it usually being assumed that only K_y and K_z (for lateral and vertical diffusion) are needed. Such a model is called a "K theory" model . If the K values are related to the dispersion coefficients through Taylor's (1921) theory and if the assumptions described earlier are made, the Gaussian equation results. Seinfeld (1975) discusses other analytical solutions of the conservation equation. Seinfeld et al. (1972), Lamb and Seinfeld (1973) and Seinfeld (1976) discuss the limitations of transport models.

B. Numerical

In most cases an analytical solution is not possible and a numerical solution using a computer is required. This method has been used with good results by Eschenroeder and Martinez (1972) and by Reynolds et al. (1973a and b; 1974) for photochemical models in the Los Angeles area. Sklarew (1979) discusses the general applicability of numerical methods as compared to Gaussian models and argues for their acceptance by EPA.

C. Other Transport Models

Other transport models are listed in Table 1 (Seinfeld, 1975).

D. Urban Model Studies

Numerous urban modeling studies have been reported in the literature. Some of these are shown in Table 1 (Seinfeld, 1975).

Table 1: *Urban Modeling Studies*

Model Used	Investigator	Region Applied to	Pollutants
	Steady-State Models		
Box model	Reiquam (1970 a, b)	Willamette Valley	SO_2
		Northern Europe	SO_2
Gaussian Plume Model	Pooler (1961)	Nashville, TN	SO_2
	Turner (1964)	Nashville, TN	SO_2
	Clarke (1964)	Cincinnati, OH	SO_2, NO_x
	Fortak (1966)	Breman, Germany	SO_2
	Hilst and Bowne (1966)	Ft. Wayne, IN	Particles
	Pooler (1966)	St. Louis, MO	Particles
	Miller-Holzworth (1966)	Washington, DC	NO_x
		Los Angeles, CA	NO_x
		Nashville, TN	SO_2
	Davidson (1967)	New York, NY	SO_2
	Koogler et al. (1967)	Jacksonville, FL	SO_2
	Panofsky/Prasad (1967)	Johnstown, PA	Particles
	Hilst et al. (1967)	Connecticut	Hydrocarbons, CO, NO_x, SO_2, particles
	Martin (1971)	St. Louis	SO_2
	Johnson et al. (1972;3)	Chicago, San Jose	CO
	Dynamic Models		
Lagrangian puff model	Croke et al. (1968 a,b,c) Roberts et al. (1971)	Chicago; New York	SO_2
	Shich et al. (1970)	New York	SO_2
	Lamb-Neiburger (1971)	Los Angeles	CO
K theory	Randerson (1970)	Nashville	SO_2
	Eschenroeder and Martinez (1972)	Los Angeles	CO, NO, NO_2, O_3, hydrocarbons
	Reynolds et al. (1973 a,b; 1974) Roth et al. (1974)	Los Angeles	CO, NO, NO_2, O_3 hydrocarbons
	Shir and Shieh (1973)	St. Louis	SO_2
	MacCracken et al. (1972)	San Francisco	CO

Source: Seinfeld, 1975

IV. STOCHASTIC MODELS

A stochastic model is one in which the turbulent transport is treated as a random process. The location of a particle of the diffusing substance at a given time is treated as a random variable, as are the fluctuations of the velocity and concentration about their mean values.

Suppose one wants to determine the concentration distribution of particles diffusing in a turbulent field. The diffusing path of each particle will be random and will be distinct from the path of every other particle. If a large number of particles are followed and the average displacement is measured, this will be a measure of the diffusion. More formally, the average of the displacements of a large number of particles leads to a certain probability distribution. The probability distribution is proportional to the concentration distribution (Csanady, 1973).

Such a model has been developed by Fahien et al.(1976). The predicted concentrations agreed well with the experimental values. This model can be used for any condition of wind velocity, stability, surface roughness, stack height, and terrain evenness or unevenness.

Using an equation for the velocity profile as a function of elevation, surface roughness, and stability and using various statistical parameters for the turbulence as a function of stability and elevation, the authors were able to use this model to calculate dispersion coefficients and eddy diffusivities. These were then correlated with downwind distance, surface roughness, stack height, and stability-making possible the use of either a Gaussian model with more generally correlated dispersion coefficients or a transport model with more generally correlated eddy diffusivities.

V. CALIFORNIA STUDY ON IMPACT OF COAL BURNING

An example of the use of a dispersion model to estimate the effect of a new coal burning power plant is A Study on the Impact of Future Coal Use in California (Brandi et al., 1978). The Climatological Dis-

persion Model (as modified to include the effects of chemical reaction
and pollutant deposition) was used to calculate the concentration of various pollutants (SO_2, SO_4, NO_x, and particulates) that might be expected as a result of a proposed 800 M We coal-burning power plant in Tehama
County whose completion in 1985 is expected. It was assumed that control
devices would remove 90% of the SO_2 from coal having a sulfur content of
1%. The SO_2 emissions would then be below the equivalent output of a
plant burning $\leq$ 0.5% sulfur fuel, in accordance with the State Implementation Plan for Tehama County. It was also assumed that the plant had a
precipitator for the control of particulates with a removal efficiency
ranging from 72% for particles < 5 μm to 100% for particles > 49 μm. Annual averaged climatological data (wind rose) for Sacramento were used
together with a constant inversion layer height of 305 m.

The location and magnitude of the predicted peak annual average
ground concentrations for each pollutant (SO_2, SO_4, NO_x, and TSP) was
calculated and these values were compared with federal and state standards.

Results indicated that, with the controls, federal and state air
quality standards are not exceeded. However, if there were no control
on sulfur, the SO_2 and SO_4 concentrations would increase by a factor of
10. Even then, the SO_2 concentration would be below the standard. But,
if a linear relationship between maximum SO_2 concentration and energy
output were assumed, the maximum concentration would be above the standard if another plant with no control were to be located on the same
site with a capacity of 3600 MWe.

With sulfur controls and assuming no background, the primary NO_x
standard would be reached with a plant 12.5 times the modeled capacity
and, again assuming no background, the standard for TSP would be reached
with a capacity 13 times the modeled capacity.

However, data on the background TSP concentration at Red Bluff indicated that the primary TSP standard has been exceeded by almost a factor of two in Red Bluff, Tehama County. Similar results are found for
both Chico and Redding in nearby counties. The NO_x concentration presents a similar picture for both Chico and Red Bluff, with the mean concentration in Chico about one-third that of the standard.

Hence it was concluded that any precise determination of air quality impacts in Tehama or surrounding counties due to a coal-fired power plant would depend upon a more detailed examination of ambient air quality levels in the surrounding areas.

The authors also used a method developed by General Electric Company and described in a recent report by Argonne National Laboratory to estimate short-term concentrations of pollutants. For the one- and three-hour averaging times, they assumed the plant was operating at 100% capacity.

For the 24-hour average, they assumed the plant would operate at the average capacity factor of 75%. When results were compared with the most stringent federal or state standards, a short-term violation appeared in the 24-hour average for total suspended particulates. They noted, however, that an increase in plant size by a little more than a factor of two would result in violation of the one-hour California NO_x standard.

VI. THE ICAAS-FSOS STUDY

In the Florida Sulfur Oxide Study (ICAAS, 1978) calculations were made of the dispersion of emissions from coal and oil burning power plants in the Tampa Bay region of Florida. These dispersion calculations played a critical role in an interdisciplinary benefit/cost analysis. Table 2 gives the 1975 base case which used actual emissions from the Hooker (H), Gannon (G) and Big Bend (BB) electric generators. It should be clear that the SO_2 and TSP emissions from the coal plants were quite substantial.

Table 2: *1975 Base Scenario Emissions*

Generator	Fuel	BTU/lb	% SO_2	SO_2(ton/yr)	TSP(ton/yr)
H	oil	24,000	0.94	3,267	176
G	coal	11,127	3.2	92,465	5,779
BB	coal	11,195	3.1	104,425	2,334

Average yearly emissions corresponding to the base case and to various pollution control policies were determined based on actual 1975 emissions after adjustment for the difference in heat content of the alternative fuels. Ambient levels for these three scenarios were then determined using the Climatological Dispersion Model (CDM) developed by EPA.

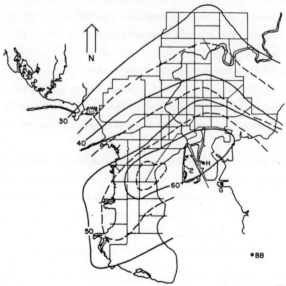

Fig. 2.   Ambient level sulfur dioxide contours in μg/m³ for the control scenario are shown relative to the Tampa census tracts used in the study.

Source: ICAAS, 1978

Meteorological information was obtained for the Tampa Airport Station from the National Climatic Center (NCC) in Asheville, N.C. The day-night version of the NCC program called STAR gives the proper form of the joint frequency function. A rectangular 5 km grid array of uniform-sized squares was used to overlay the region of interest and to catalogue the emission inventories by area sources. The geographical centers for selected census tracts were used as receptor locations in the model. Calibration was obtained from regression analysis of observed pollution data (air quality measuring stations or Hillsborough County) and the computed concentrations produced by the model. The contours in Figure 2 show ambient levels for the baseline scenario obtained from the model.

The availability of a spatial map of pollutant concentrations sets the stage for consideration of the distributional aspects of benefits and costs, one of the timely problems in B/C analyses in general. These aspects of the FSOS analysis will be discussed in Chapter 15.

VII. CONCLUSIONS

If the concentrations of various pollutants that may occur due to a new coal burning plant are estimated through the use of a dispersion model and if these are found to exceed federal or state standards, steps would have to be taken to ensure future compliance. These might include (a) reduction of emissions through more effective control devices, (b) an increase in the stack height to provide greater dispersions, or (c) plant relocation. Since any of these steps can be quite expensive, it is extremely important that an accurate modeling study be carried out in order to assess the effect of their adoption on predicted concentrations. There is, therefore, great need to develop reliable yet practical methods for the quantitative prediction of the dispersion of air pollutants emitted in coal burning.

References

Brandi, R., P. Chan, D. Ermak, M. Horovitz, J. Kooser, R. Nyholm, R. Ritschard, R. Ruderman, J. Sathaye, R. Sectro, and W. Siri, Impact of future coal use in California in Interim Regional Report for National Coal Utilization Assessment, Lawrence Berkeley Laboratory, LBL-8402, UC-90j, Nov. 1978.

Briggs, G. A., Plume rise, USAEC Critical Review Series, TID-25675, NTIS, Springfield, VA, 1969.

Bryant, P. M., Methods of estimation of the dispersion of windborne material and data to assist in their application, United Kingdom, AERE, H264/2719, Harwell, England, 1969.

Clarke, J. F., A simple diffusion model for calculating point concentrations from multiple sources, J. Air Pollut. Control Assoc., 14, 347, 1964.

Croke, E. et al., Chicago air pollution model, Argonne N. L., 1st Q. Prog. Rept., ANL/ESCC-001, 1968a.

Croke, E., J. Carson, D. F. Gatz, H. Moses, F. L. Clark, A. S. Kennedy, J. A. Gregory, J. J. Roberts, R. P. Carter, and D. B. Turner, Chicago air pollution system model, Argonne N. L., 2nd Q. Prog. Rept., ANL/LES-CC-002, 1968.

Croke, E. J., J. E. Carson, D. F. Gatz, H. Moses, A. S. Kennedy, J. A. Gregory, J. J. Roberts, K. Croke, J. Anderson, D. Parsons, J. Ash, J. Norso, and R. P. Carter, Chicago air pollution system model, Argonne N. L., 3rd Q. Prog. Rept., ANL/LES-CC-003, 1968.

Csanady, G. T., Turbulent Diffusion in the Environment, D. Reidel, Dordrecht, Holland, 1973.

Davidson, B., A summary of the New York urban air pollution dynamics research program, J.A.P.C.A., 17, 154, 1967.

Egan, B. A., Comments on review paper by Turner, J.A.P.C.A., 29, 927, 1974.

Eschenroeder, A. Q. and J. R. Martinez, Concepts of photochemical smog models, Adv. Chem, 113,1972.

Fahien, R. W., D. W. Kirmse, and S. H. Pahwa, Statistical modeling of atmospheric transport, AIChE Symp. Series No. 156, 72, 389, 1976.

Fortak, H. Rechnerische Ermittlung der SO_2-Grundbelastung aus Emissiondaten, Anwendung auf die Verhaltnisse des Stadtgebietes von Bremen mit Abbildungsteil, Institut fur theoretische Meteorologie der freien Universitat Berlin, 1966.

Gifford, F. A., Use of routine meteorological observation for estimating atmospheric diffusion, Nuclear Safety, 2, 47, 1961.

Green, A. E. S., R. P. Singhal, and R. Venkateswar, Analytical extensions of the Gaussian plume model, submitted to J. A. P. C. A., 1979.

Hay, J. S., and F. Pasquill, Diffusion from a fixed source at a height of a few hundred feet in the atmosphere, J. Fluid Mech., 2, 299, 1957.

Hilst, G. R., and N. E. Bowne, A study of diffusion of aerosols released from aerial line sources upwind of an urban complex, a final report to the U.S. Army Dugway Proving Ground, vol. I, The Travelers Research Center, Hartford, Conn., 1966.

Hilst, G. R., F. I. Badgley, J. B. Yocum, and N. E. Bowne, The development of a simulation model for air pollution over Connecticut, a final report to the Connecticut Research Commission, vols. I and II, The Travelers Research Center, Hartford, Conn., 1967.

Holland, J. Z., A meteorological survey of the Oak Ridge area, U. S. Atomic Energy Commission
    Report ORO-99, 1953.
ICAAS: E. J. Loehman, S. V. Berg, A. A. Arroyo, R. A. Hedinger, J. M. Schwartz, M. E. Shaw,
    R. W. Fahien, V. H. De, R. P. Fische, D. E. Rio, W. F. Rossley, and A. E. S. Green, Distribu-
    tional analysis of regional benefits and cost of air quality control, J. Environmental
    Economics and Management, 1978.
Johnson, W. B., F. L. Ludwig, W. F. Dabberdt, and R. J. Allen, An urban diffusion simulation
    model for carbon monoxide, J.A.P.C.A. 23, 490, 1973.
Koogler, J. B., R. S. Sholtes, A. L. Davis, and C. I. Harding, A multivariable model to atmos-
    pheric dispersion prediction, J.A.P.C.A., 17, 211, 1967.
Lamb, R. G., and M. Neiburger, An interim version of a generalized urban air pollution model,
    Atmos. Environ., 5, 239, 1971.
Lamb, R. G., and J. H. Seinfeld, Mathematical modeling of urban air pollution, Environ. Sci.
    Rwxh., 7, 253, 1973.
Macracken, M. C., T. V. Crawford, K. R. Peterson, and J. B. Knox, Initial application of a
    multi-box air pollution model to the San Francisco Bay area, 1972 Joint Automatic Control
    Conference, Stanford Univ., Stanford, CA, 1972.
Martin, D. O., An urban diffusion model for estimating long term average values of air quality,
    J.A.P.C.A., 21, 16, 1971.
Meade, P. J., Meteorological Aspects of the Peaceful Uses of Atomic Energy, WMO, Geneva,
    Switzerland, T.N. 33, 1960.
Miller, M. E., and G. C. Holzworth, An atmospheric model for metropolitan areas, J.A.P.C.A.,
    17, 46, 1967.
Mirabella, V. A., Comments on review paper by Turner, J.A.P.C.A., 29, 931, 1974.
Panofsky, H. A., and B. Prasad, The effect of meteorological factors on air pollution in a
    narrow valley, J. Appl. Meteorol., 6, 493, 1967.
Pasquill, F., The estimation of the dispersion of windborne material, Meteorol. Mag., 90, 1961.
Perkins, H. C., Air Pollution, McGraw Hill, New York, 1974.
Pooler, F., A prediction model of mean urban pollution for use with standard wind roses,
    Int. J. Air Water Pollut., 4, 199, 1961.
Pooler, F., A tracer study of dispersion over a city, J.A.P.C.A., 16, 677, 1966.
Randerson, D., A numerical experiment in simulating the transport of sulfur dioxide through
    the atmosphere, Atmos. Environ., 4, 615, 1970.
Reiquam, H., An atmospheric transport and accumulation model for airsheds, Atmos. Environ.,
    4, 233, 1970a.
Requam, H., Sulfur: Simulated long-range transport in the atmosphere, Science, 170, 3955, 1970b.
Reynolds, S. D., P. M. Roth, and J. H. Seinfeld, Mathematical modeling of photochemical air
    pollution. 1., Formulation of the model, Atmos. Environ., 7, 1033, 1973a.
Reynolds, S. C., M. Liu, T. A. Hecht, P. M. Roth, and J. H. Seinfeld, Further development and
    validation of a simulation model for estimating ground level concentrations of photochemical
    pollutants, Systems Applications, Inc., Rept., R73-19, San Rafael, CA, 1973b.
Reynolds, S. C., M. Liu, T. A. Hecht, P. M. Roth, and J. M. Seinfeld, Mathematical modeling of
    photochemical air pollution III. Evaluation of the model, Atmos. Environ., 8, 1974.
Roberts, J. J., E. S. Croke, and A. S. Kennedy, An urban atmospheric dispersion model, Proc. of
    Symp. on Multiple Source Urban Diffusion Models, A. C. Stern, Ed., U.S.E.P.A. Pub. No AP-86,
    1970, Chapter 6.
Roth, P. M., P. J. W. Roberts, M. Liu, S. D. Reynolds, and J. H. Seinfeld, Mathematical modeling
    of photochemical air pollution, II. A model and inventory of pollutant emissions, Atmos.
    Environ., 8, 97, 1974.
Seinfeld, J. H., Air Pollution, Physical and Chemical Fundamentals, McGraw Hill, New York, 1975.
Seinfeld, J. H., S. D. Reynolds, and P. M. Roth, Simulation of Urban Air Pollution, Adv. Chem.
    Series 113, American Chemical Society, Wash. D.C., 1972.
Seinfeld, J. H., Accuracy of prediction of urban air pollutant concentrations by diffusion
    models, assessing transportation-related impacts. Special report 167, Trans. Resrch Bd.,
    Natl. Academy of Sciences, Washington D.C., 1976.
Shieh, L. J., B. Davidson, and J. P. Friend, A model of diffusion in urban atmospheres: SO_2
    in greater N.Y., Proc. of Symp. on Multiple Source Urban Diffusion Models, U.S. Environ.
    Prot. Agency Publ. AP-86, 1970.
Shir, C. C., and L. J. Shieh, A generalized urban air pollution model and its application to
    the study of SO_2 distributions in the St. Louis metropolitan area, IBM Res. Lab. Rept.
    RJ 1227, San Jose, Calif., 1973.
Sklarew, R. C., Comments on review paper by Turner, J.A.P.C.A., 29, 935, 1974.
Stumke, H., Vorschlag einer empirischen Formel fur die Schornstein uberhohung, Staug23, 549, 1963.
Taylor, G. I., Diffusion by continuous movements, Proc. Math. Soc. London, 20, 196, 1921.
Turner, D. B., A diffusion model for an urban area, J. Appl. Meteorol., 3, 85, 1964.
Turner, D. B., Workbook of atmospheric dispersion estimates, Office of Air Programs Publication
    AP-26, U.S.E.P.A., Research Triangle Park, NC, 1970.
Turner, D. B., Atmospheric dispersion modeling, a critical review, J.A.P.C.A., 29, 502, 1979.

CHAPTER 11

ATMOSPHERIC MODIFICATIONS

Karl E. Taylor, William L. Chameides, and Alex E. S. Green

I. POTENTIAL ATMOSPHERIC AND CLIMATIC IMPACTS

In addition to leading to local and regional degradation of air
quality, the release of pollutants from combustion can also cause
global perturbations by changing the average composition of the
earth's lower atmosphere. Whether such perturbations will occur
depends upon the magnitudes of the combustion sources as compared to
other natural sources and the ability of the atmosphere to shed these
pollutants before they are dispersed throughout the atmosphere. Of
prime concern in this regard is the release of CO_2 in fossil fuel
burning, which will likely lead to an increase in surface temperature,
but in addition it is possible that in the coming decades chemical
perturbations may result from other by-products of combustion such
as NO_x ($NO + NO_2$) and SO_2. Furthermore, because of the generally
lower efficiencies of coal burning relative to oil and gas, it is
likely that, in the absence of stringent pollution controls, a major
shift to coal utilization will accelerate the rate of release of
CO_2, NO_x, and SO_2 into the atmosphere.

The primary question is, of course, how changes in atmospheric
chemical composition or aerosols may affect climate. It should become
clear from our discussion below that the actual impact of coal burning
on climate cannot yet be accurately predicted; in fact, we shall only
discuss the most direct possible climatic consequences. Still, it would

be foolish to forget that climatic change might be indirectly triggered
by any of the potential atmospheric modifications that will be discussed
here.

Because of the complex nature of climatic change and due to the
transport properties of the atmosphere and oceans, perturbations in
one region can have global ramifications. Similarly, changes in the
abundance of one atmospheric constituent can affect the concentrations
of several others via chemical reactions. In light of these apparent
complexities it might be structurally useful to classify the climatic
implications of coal mining and coal burning according to the time
scales over which they could manifest themselves. A particular
climatic impact might also be characterized by how _directly_ it is
related to coal utilization and by whether it results uniquely from
using coal instead of some other energy source.

On a regional scale, for example, climates may be influenced
during open-pit mining operations through the drastic disruption of
local flora. Removal of plants and the resulting decrease in transpiration can quickly lead to a dryer and warmer local climate.
This rather direct and immediate regional influence of coal mining,
which is unique to this energy source, can be contrasted with an
indirect climatic consequence of coal burning such as the one described below.

During coal combustion sulfur oxides are emitted to the atmosphere
and within a few days these gases dissolve in cloud drops and fall out
as "acid rain." This acidic rain can strongly influence the health
of plants and possibly alter, over several years, the composition of
the plant canopy itself. Any change in the earth's surficial characteristics (the biosphere, in particular) alters to some degree the exchange
of energy, water vapor, and trace gases (CO_2, hydrocarbons, etc.)
between the surface and the atmosphere. Since the chemical composition
of air affects the way it absorbs and radiates energy, changes in any
of these processes, if large enough, will require adjustments in the
earth's energy budget and could lead to long-term climatic change, not
only regionally, but perhaps globally.

At this point we could identify several other possible climatic consequences of increased coal burning (such as thermal pollution, increased evaporation to remove waste heat, changes in atmospheric chemistry), but instead we shall limit ourselves to a few problems that could cause fairly direct global effects. We shall begin by focussing on how the carbon dioxide released through fossil fuel combustion might lead to increased temperatures near the earth's surface. This problem is expected to be important globally over time scales of decades to millenia and is presently seen as the most troublesome climatic consequence of coal burning. After discussing CO_2, we shall present a general description of the global budgets of other pollutants (such as NO_x and SO_2) associated with coal combustion, and we shall indicate what impact they might have on other chemical species as well as on visibility and climate. A summary of remote sensing devices for monitoring the aerosol content or the purity of the air will conclude the chapter.

II. THE CO_2 PROBLEM

The problem of coal's influence on climate via carbon dioxide release can be divided into two parts: How will the atmospheric CO_2 concentration depend on fossil fuel combustion and how will changes in CO_2 concentration affect climate? It appears likely, as we shall discuss later, that continued growth in combustion of fossil fuels will increase the CO_2 concentration in the atmosphere, possibly to levels high enough to raise average global surface temperatures in the next century or so by at least a few degrees Celsius; more extreme changes could be expected in the polar latitudes. The primary climatic consequence of carbon dioxide is due to the CO_2 "greenhouse effect": CO_2 is relatively transparent to incoming sunlight (short-wave radiation) but intercepts some terrestrial (long-wave) radiation that would otherwise escape directly to space. This absorption of radiation energy makes it more difficult for the earth to cool itself and therefore makes the surface warmer than it would otherwise be. The effectiveness of the "atmospheric blanket" depends in part on the amount of CO_2 present. The "greenhouse effect" causes surface temperatures to increase when CO_2 concentrations increase.

CO_2 Concentrations

Measurements indicate that over the last century or so the atmospheric CO_2 concentration has risen by about 18% or an average increase of about .2% per year. In the past decade the rate of increase has accelerated to about .3% per year (Lowe et al., 1979) in response to increased combustion of fossil fuels. If all the carbon dioxide known to be released during fossil fuel combustion had remained in the atmosphere (and in the absence of other sources or sinks of carbon), we would expect these percentages to be double their actual values. About half the CO_2 released into the atmosphere must have somehow been removed. To predict future atmospheric CO_2 concentrations, then, we must determine what other processes contribute or remove CO_2 from the atmosphere; we need to know the overall characteristics of the carbon cycle (Bolin, 1979).

Although a significant contribution of atmospheric CO_2 can be attributed to fossil fuel combustion, the rate of addition has been relatively slow compared with the rate at which CO_2 is removed by green plants during photosynthesis and released by plants and animals as they respire or decay. Before man's intervention, the naturally high rates of removal and release of CO_2 in the atmosphere were thought to be nearly in balance. In fact, if the mass of all organically derived matter (including fossil fuels) were not changing, the two rates would be exactly equal. During different seasons of the year the rates of photosynthesis and decay are, of course, not equal (more growth in spring and summer; more death in fall and winter) and therefore the atmospheric CO_2 concentration varies seasonally within each hemisphere. Measurements indicate that in the Northern Hemisphere the magnitude of the effect is about a 2% variation within the year, with smaller variations in the Southern Hemisphere (Lowe et al., 1979), due ultimately to the influence of the larger oceans there.

There is recent evidence that, like the fossil fuel reservoir,
the size of the biosphere reservoir of carbon may have decreased
significantly over the last few centuries as a result of man's influence.
Bolin (1977), Woodwell (1978), and others estimate that the destruction
of many of the world's forests (to harvest the wood and to clear land
for agricultural purposes) has, during recent decades, led to a carbon
dioxide injection of comparable magnitude to that by fossil fuel
combustion. Recalling that the atmospheric carbon dioxide build-up
is observed to be increasing at only half the rate of fossil fuel
emission and thus perhaps a quarter the rate of anthropogenically
caused total emission, the question arises as to where the excess
carbon dioxide goes. In the long term we know that most of it becomes
dissolved in the oceans, but it is thought that, except in the shallow
surface layer, this is a very slow process (centuries to millenia).
It is currently thought that the oceans have indeed dissolved at least
half of the unaccounted for carbon dioxide, but it is not yet certain
how the remaining extra CO_2 is disposed of. It may be that the
estimates of forest clearing are too large (see Broecker et al. (1979)
for a discussion of uncertainties) so that there is actually no extra
CO_2 and the budget can be balanced. If, however, the current estimates
are accepted, it could be that the rate of mixing of carbon into the
deep ocean reservoir is faster than presently believed. Clearly, if
the mixing rate were large enough, the budget could be balanced. A
still speculative mechanism for increasing the mixing efficiency of
the ocean is that dead organic matter and carbonate shells sink to the
bottom of the ocean without decaying as much as previously believed
(cf., Wong, 1978). This would remove some of the CO_2 from the surface
layers of the ocean, allowing more atmospheric CO_2 to dissolve and thus
lowering atmospheric concentrations.

Another process that enhances the ability of the ocean to dissolve
atmospheric CO_2 is a buffering mechanism in sea water. With this
buffering, the oceans' capacity to dissolve CO_2 is less sensitive to
changes in temperature than it would otherwise be. It is likely that,
as more CO_2 becomes dissolved in the ocean, the buffer factor will

change, making it more difficult to further dissolve CO_2.

The uncertainties in the carbon cycle must be erased before accurate predictions can be made for changes in atmospheric CO_2 concentrations. Nevertheless, projections for future CO_2 concentrations have been made using our present limited knowledge. Kellogg (1978), for example, estimated that if fossil energy trends continue, the CO_2 concentration will increase by about 20% by the year 2000 and perhaps double within the following decade or two. Keeling and Bacastow (1977) have also made projections by assuming that most of the known reserves of gas, oil and coal will be burned in the next century or two. They find that maximum CO_2 concentrations of about eight times the present concentration could be reached by the 22nd century. Bolin (1979) believes that this prediction is probably not overestimated by more than a factor of two.

Climatic Effects

At present no measurable changes in global temperatures can be unambiguously attributed to changes in CO_2 concentration. The average global temperature changes that apparently occurred over the last century (less than .5°C) can be explained by internal fluctuations of the climate system and need no "external" cause such as increased atmospheric CO_2 concentrations.

With no hard observational evidence for CO_2-induced climatic change, scientists have turned to climate models in which conditions can be controlled and CO_2 concentrations can be increased to high enough levels to cause significant temperature changes. In these climate models most of the processes found in the natural climate system cannot be treated in their full complexity, and therefore the model predictions could prove to be in error. Schneider (1975) reviewed the state-of-the-art of modelling and concluded from the modelling results that his "best guess" was that a doubling of the current CO_2 concentration could change the average surface temperature by 1.5 to 3°C. Nothing published since his review has seriously cast doubt on his conclusions, although the uncertainty of "several-fold" admitted by Schneider should leave one a little skeptical.

The estimate given above is strengthened somewhat because it has
been supported by diverse models of varying degrees of complexity
(GARP, 1975): three-dimensional general circulation models (Manabe
and Wetherald, 1975), two-dimensional statistical dynamic models (e.g.,
Ohring and Adler, 1978), and radiative-convective models (e.g.,
Ramanathan, 1975). All the models are of course much simpler than the
natural climate system. The treatment of oceans and ice sheets and
their interactions with the atmosphere are particularly crude, even
though these interactions may be of primary importance in effecting
climatic change.

Although predicting the average global change in surface temperature induced by a doubling of CO_2 is a difficult enough task, we really
need to know in more detail how the global climate might change. We
would like to know, for example, how the growing season in the grain
belts would be affected by CO_2 changes. Although we are far from this
goal, conclusions from the modelling experiments provide preliminary
indications that the warming effect of CO_2 would be substantially
greater in the polar regions than in the tropics, the vertical stability
of the atmosphere would be decreased leading possibly to changes in the
synoptic weather patterns, and a doubling of CO_2 would accelerate the
hydrologic cycle (both precipitation and evaporation) by several percent.
Overall it would appear that if predictions of a doubling of the CO_2
concentration in the next century prove true, the climatic consequences
could be severe. Large changes in polar temperatures could partially
melt ice caps and lead to higher sea levels, thereby flooding low-
lying continental areas. For now such predictions are speculative, but
should not be ignored completely.

III. NO$_x$ AND PHOTOCHEMICAL PERTURBATIONS

Budget

Nitrogen oxides, trace constituents of the atmosphere, are
believed to play a major role in the tropospheric photochemistry and

control several parameters of environmental concern. In spite of this
central role and decades of study, many uncertainties remain in our
understanding of the NO_x system. Observations indicate that NO_x
varies from an abundance of several ppt (v/v) in remote marine
environments to several hundred ppt in remote continental regions
(Noxon, 1975; McFarland et al., 1978) to as much as 100 ppb in urban
environments. It is apparent that at least in some regions anthropogenic
activities have caused a significant increase in NO_x concentrations.
However, there is no evidence indicating that the global abundance of
NO_x has been directly affected by anthropogenic emissions thus far.

Table 1 summarizes the global sources and sinks of tropospheric
NO_x, as they are presently known. While it appears that lightning
discharges are the largest producers of NO_x on a global scale, anthropogenic emissions may be responsible for a significant fraction
(perhaps 30%) of the total source. Of the anthropogenic source in
the U.S., approximately 40% is from stationary sources, with 20%
being emitted from power plants. It is estimated that, for a given
amount of energy production, approximately 15% more NO_x is produced
from a coal-fired plant as compared to an oil-fired electric generating
plant and twice as much NO_x is produced by coal as compared to a gas-
fired power plant (Gartrell, 1977). Thus, in the absence of emission
control, the anthropogenic NO_x source will likely grow with a major
increase in coal burning. The major sink for NO_x is conversion to
HNO_3 vapor

$$NO_2 + OH + M \rightarrow HNO_3 + M$$

and removal of HNO_3 in rain. Because of uncertainties in the NO_x
budget as well as uncertainties in the future rate of anthropogenic
emission of NO_x, it is unclear if the global NO_x system will be
perturbed in the coming decades due to increased coal usage and other
anthropogenic activities. Nevertheless, the consequences of such a
perturbation could be severe, as outlined below.

Table 1. *Estimated Global Sources and Sinks of NO_x (10^{10} molecules $cm^{-2} s^{-1}$)* *

Sources	Rate	Reference
Anthropogenic (in 1975)	0.7	Robinson and Robbins (1970)**
Lightning	1.0	Chameides (1979)
NH_3 Oxidation	1	This work
Soils	uncertain	
TOTAL	2	
Sinks		
Heterogeneous Removal of HNO_3	2.4	This work
Deposition of NO_x	0.01	This work
TOTAL	2.4	

* 10^{10} molecules NO_x $cm^{-2} s^{-1}$ = 36 x 10^{12} g N yr^{-1}

** Extrapolated from Robinson and Robbins' estimate using data from 1965 and assuming a 5% annual rate of increase

Tropospheric Photochemistry

Figure 1 illustrates the key mechanisms of the tropospheric photochemical system. The photolysis of O_3

(R1) $O_3 + h\nu \rightarrow O(^1D) + O_2$

followed by

(R2) $O(^1D) + H_2O \rightarrow 2OH$

lead to the production of OH, which is the prime sink of many of the reduced gases emitted by the biosphere (i.e., CO, CH_4, CH_3Cl, NH_3, etc.).

(R3) $CO + OH \rightarrow CO_2 + H$

is a major sink for both CO and OH and is the major source of HO_2 via

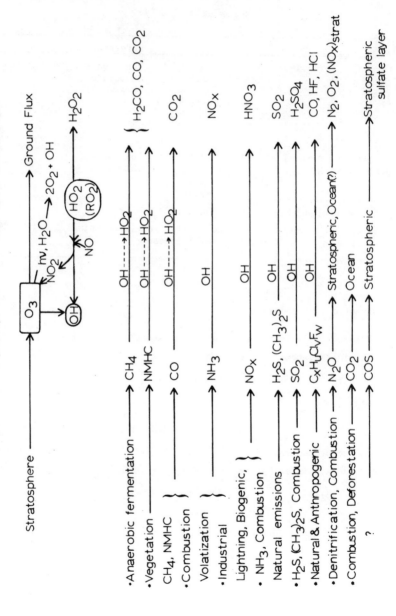

Fig. 1. Global Tropospheric Photochemistry

(R4) $H + O_2 + M \rightarrow HO_2 + M$.

The recycling of OH from HO_2 via

(R5) $NO + HO_2 \rightarrow NO_2 + OH$

is a major source of OH and is also believed to lead to the local
production of tropospheric O_3. The reaction

(R6) $HO_2 + O_3 \rightarrow 2O_2 + OH$

along with (R1) followed by (R2) acts as a photochemical sink for
tropospheric O_3.

Via reaction (R5), NO_x acts as an important source for OH and O_3
and thus an increase in its abundance could lead to a perturbation in
the $OH-O_3$ system. Figure 2 outlines the series of reactions that could
occur in the atmosphere as a result of such an increase in NO_x. (We
also include the effect of a CO increase, since CO is also a product
of anthropogenic activities and its increase tends to oppose that of
NO_x.) Detailed descriptions of the mechanisms summarized in Figure 2
have been presented by Sze (1977), Chameides et al. (1977), Liu (1977),
Logan et al. (1977), Butler et al., (1979), and Crutzen and Fishman
(1977). Our present understanding of tropospheric photochemistry is
that an increase in NO_x could ultimately lead to a temperature decrease
as well as an increase in the column O_3 abundance.

Perturbations to CH_4 and Temperature

An increase in tropospheric NO_x leads to an enhancement of OH
via (R5). Since

(R7) $CH_4 + OH \rightarrow CH_3 + H_2O$

is the major sink of CH_4, which is produced by microbial, anaerobic
fermentation, an increase in OH will lead to a decrease in the CH_4
abundance. As CH_4 is an absorber of terrestrial radiation in the
7.66 μ window region and produces stratospheric H_2O, which also
absorbs terrestrial radiation and contributes to the greenhouse effect
(Wang et al., 1976), an increase in NO_x may ultimately alter the

Anthropogenic NO_x [and CO] Effect

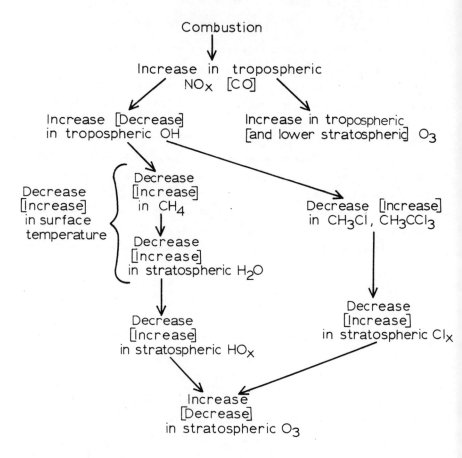

Fig. 2.   Anthropogenic NO_x [and CO] Effect

atmosphere's radiative-energy budget in such a way as to decrease the
atmospheric temperature. Increases in CO, on the other hand, tend
to oppose the above effects of NO_x by leading to a decrease in OH,
an increase in CH_4, and thus an increase in temperature.

Thus, the effects of anthropogenic emissions on the tropospheric
photochemistry and temperature depend upon the rate at which NO_x and
CO emissions increase. An upper limit temperature decrease due to NO_x
emissions alone (and neglecting other effects) of about 0.1°C by the
year 2010 is implied by simple one-dimensional model calculations.
However, calculations also indicate that inclusion of the effects of
CO_2 increases, which were discussed in the preceeding section, would
likely overwhelm the temperature perturbation implied by NO_x (and CO)
increases. Nevertheless, the results are indicative of the intricate
coupling of climate to atmospheric composition and thus to anthropogenic emissions.

Perturbations in Atmospheric Ozone

As indicated in Figure 2, increases in NO_x leading to changes
in OH abundance may eventually cause a perturbation in the column
ozone abundance. A steady state, one-dimensional model of the
tropospheric and stratospheric photochemistry (cf. Butler et al.,
1979) has been used to indicate the response of ozone to increases
in anthropogenic emissions. Using present-day rate data we find that
increases in the ground-level fluxes of NO_x and CO may result in an
increase in the total column abundance of O_3 of as much as 8% by the
year 2025 if other factors remain fixed. The calculations also
indicate that the perturbation on ozone due to chlorofluoromethane
emissions will likely swamp out the smaller effects caused by NO_x
and CO emissions; however, at present these two competing effects
appear to be comparable. The calculations indicate the need for
studying all chemical effects when attempting to infer the past and
possible future variations in atmospheric ozone, and once again
emphasize the highly complex nature of the atmosphere's climatic
and chemical system.

Conclusion

Sensitivity studies indicate that increases in the NO_x tropospheric abundance may lead to significant changes in tropospheric OH, CH_4, CH_3Cl, CH_3CCl_3, and therefore possibly perturb atmospheric temperature and ozone density. However, many uncertainties remain in our understanding of atmospheric photochemistry and its relation to anthropogenic sources of NO_x, as well as that of CO. Long-term theoretical and measurement programs are needed to better understand the implications of continued anthropogenic emissions of NO_x and CO upon atmospheric photochemistry.

IV. SULFUR OXIDES

SO_2 Budget

The elucidation of the atmospheric sulfur oxide budget and the chemistry which controls the ultimate removal of sulfur from the atmosphere remains a major challenge of atmospheric chemistry. Approximately 70 M tons (10^{12} g) S yr^{-1} as SO_2 are presently produced by anthropogenic processes (Friend, 1973). On the other hand, natural emissions of H_2S and CH_3SCH_3 probably lead to the production of SO_2, via photochemistry, at a rate of about (100-200) M tons S yr^{-1}. A small source of about 10 M tons S yr^{-1} of SO_2 may result from the oxidation of CS_2 (Sze and Ko, 1979); i.e.

$$CS_2 + OH \rightarrow CS + SOH \rightarrow COS + HS \rightarrow \ldots \rightarrow SO_2 \; .$$

While it is believed that the oxidation of stratospheric COS is a major source of the stratospheric sulfate aerosol (Crutzen, 1976), it is not known if anthropogenic emissions of COS are an important global source.

Recently reported global SO_2 data suggest natural SO_2 concentration levels of about 100 ppt (Maroulis and Bandy, 1979) with much higher levels near urban areas. At present too little data exist to ascertain what impact anthropogenic emissions might be having on SO_2 abundances either on a regional or a global scale. However, in view of the large rates of SO_2 production associated with coal burning (about 50% more

SO_2 is produced by a coal-fired electric power plant, as compared to an oil-fired plant (Gartrell, 1977)), the predicted increased usage of coal may soon cause the anthropogenic source to become competitive with the natural sources.

Sulfate Formation

The fate of atmospheric SO_2 is conversion to $SO_4^=$ and H_2SO_4 and eventual removal via wet and dry deposition. The processes by which SO_2 is converted to $SO_4^=$ is presently the subject of intense scientific investigation. Observations indicate that in plumes containing enhanced SO_2 abundances, on the average about 1-2% of the SO_2 is converted to $SO_4^=$ per hour. The rate controlling step in the conversion of SO_2 to $SO_4^=$ is believed to be the initial SO_2 oxidation, the subsequent steps rapidly following the first reaction. Four schemes for oxidation of SO_2 to $SO_4^=$ are believed to be of potential importance. These processes are:

a) The oxidation of SO_2 initiated by a gas-phase reaction of SO_2 with OH and possibly other oxidizing radicals such as HO_2 and CH_3O_2 (Sander and Seinfeld, 1976; Calvert et al., 1978; Davis et al., 1978): Since high radical concentrations are influenced by hydrocarbon and NO_x emissions during day-light conditions, it is likely that this process directly links $SO_4^=$ production with photochemical smog. Studies indicate that for sunny days this process can lead to conversion rates of as much as 2-4% hr^{-1}.

b) The catalyzed oxidation of SO_2 dissolved in droplets by transition metals: This process is thought to be of potential importance near stack plumes, and perhaps urban fogs, where transition metal concentrations may be large (Husar, 1978).

c) The oxidation of SO_2 in the liquid phase by strong oxidants also in the liquid phase such as H_2O_2 and possibly O_3 (Eggleton and Cox, 1978).

d) The surface catalyzed oxidation of SO_2 upon collision with particles (especially soot) (Novakov et al., 1974).

Both SO_2 and $SO_4^=$ are removed from the atmosphere by wet and dry precipitation. The residence time of SO_2, determined by the combined rates of conversion to $SO_4^=$ (1-2% hr^{-1}) and removal in precipitation (2-5% hr^{-1}), is believed to be about 15-30 hours. The $SO_4^=$ residence time is generally about 3-5 days. Assuming a typical horizontal wind velocity in the boundary layer of about 500 km/day, we find that SO_2 may be transported on average 500 km, while $SO_4^=$ may be transported 3000 km before removal. The effects of sulfur oxide emissions may therefore be experienced over 3000 km from the source, thus indicating the large scale of sulfur oxide pollution. In fact, satellite pictures indicate that, under the appropriate synoptic meteorological conditions, regional sulfate episodes occur in which large portions of the midwest and/or eastern seaboard of the United States are covered by a continuous sulfate haze which significantly limits visibility and perhaps has serious health impacts (R. Hussar, private communication, 1978). Such events effect non-urban areas as well as urban areas. A more detailed discussion of the sulfate aerosol and its impact upon visibility is presented in Section III.

Acid Rain

The release of both NO_x and SO_2 usually result in the deposition of HNO_3 or H_2SO_4 in rainwater. If the quantities of NO_3^- and $SO_4^=$ are sufficiently large, the incorporation of HNO_3 and H_2SO_4 into rainwater can overwhelm the CO_2-buffered pH of 5.6 normally observed in rainwater, leading to acid rain. Precipitation with low pH can have deleterious effects upon materials, agriculture, and other ecological systems.

Measurements reveal that over much of the eastern U.S. and western Europe, the emissions of NO_x and SO_2 have caused a significant lowering of the pH of rain (Likens et al., 1979). Generally most of the excess acidity is due to H_2SO_4, with about 30% from HNO_3, although the HNO_3 contribution has increased significantly over the past 20 years (cf., Cogbill and Likens, 1974). It is quite likely that extensive increases in coal utilization, without a major effort to control SO_2 emissions, will cause an intensification of acid precipitation over

major portions of the continental U.S., with the potential for serious
ecological and economic fallout.

Conclusion

While much remains to be learned of the sulfur system, it appears
that increases in SO_2 emissions can exacerbate regional problems
associated with the conversion of SO_2 to $SO_4^=$ and the production of
hazes and the scavenging of H_2SO_4 in rainwater leading to acid
precipitation. Even if 70-90% of the gaseous sulfur is removed in new
coal-based power plants, by use of scrubbers, it is estimated that SO_2
emissions in the U.S. will increase by as much as 25% over the next 20
years. Further experimental and modelling work is needed to determine
the potential scope of this problem and the impact intense coal
utilization will have.

V. THE GLOBAL DETERIORATION OF VISIBILITY

The increasing build-up of aerosols in the atmosphere, a major
environmental problem associated with the burning of fossil fuels in
general and coal in particular, and the consequent deterioration of
visibility along with the possible impact of the aerosols upon the
climate are matters of concern. These topics were dealt with at a
recent conference on Environmental and Climatic Impact of Coal
Utilization (Singh and Deepak, 1979). Here we will present an overview
of the visibility problem from the standpoint of the increment associated
with the replacement of gas and oil by coal. This incremental change
will depend largely upon the increase or possible decrease of sulfur
emission associated with the transition to coal.

Many studies (Castleman et al., 1975; Waggoner et al., 1976;
Charleson et al., 1974; Castleman, 1974; Castleman et al., 1976;
Harrison et al., 1976; Larson and Harrison, 1977) have implicated
sulfate aerosols as an important component of submicron-size aerosols
distributed over a large geographic area in the midwest and the south.
This large-scale optical dominance by sulfates in the aerosols suggests

that it is not due to local sources. Sulfate hazes have occurred over
much of the eastern United States and extend over large areas in western
and northern Europe, particularly in Sweden. These hazes play a
significant role in the optical properties of the atmosphere since they
are predominantly in the optically efficient size range (0.1-1 micron).
The oxidation rate of atmospheric SO_2 to sulfate aerosols has been
examined in many of the above studies and the characteristic reaction
times can be as short as a few minutes or as long as a few days.
Homogeneous processes between gaseous SO_2 and reactive free radical and
heterogeneous processes between dissolved sulfates and oxygen or ozone
are at different times likely to be rate controlling. Both catalysis
by metal ions and inhibition by organic molecules readily occur at
concentrations as low as 10^{-7} molar. Ambient ammonia has a strong
influence. The temperature dependence of heterogeneous rates and the
dependence upon partial pressures of O_2, O_3 and NO_2 are poorly known,
especially in the presence of metal catalysts. Turko et al. (1979)
have studied the contribution of carbonil sulfide (COS) to the formation
of stratospheric aerosol particles and the effects of these particles
on the earth's radiation balance and surface temperature. Recent
measurements of the rate of reaction of COS with OH place new limits
on the likely atmospheric lifetime of COS. Worldwide measurements show
an ubiquitous distribution. Detailed models of the transformation
involving COS and aerosols have been formulated. These models include
the physical processes of nucleation, growth and evaporation, coagulation,
sedimentation, and fusion.

The reactions involving ozone and the sulfur-bearing compounds
are of particular interest in aerosol problems since ozone is one of
the important pollutants, particularly in sunbelt areas. Although the
homogeneous reaction rate between SO_2 and O_3 is slow, the heterogeneous
reaction is potentially very important, leading under certain circumstances to characteristic rates of transformation to aerosols of 4-5%
per hour. With higher concentrations of ozone this would go up in
proportion. Detailed analyses have been made of the transformation of
sulfur dioxide to atmospheric H_2SO_4 aerosols or products of this acid

by neutralization with ammonia before the gas molecules have substantially drifted distances of 100 kilometers or so. These studies
indicate that very detailed analyses would be needed before determining
the conversion rates within such distances. The possibility of characteristic lifetimes of the order of minutes indicates that fast
conversion should be considered in connection with the utilization of
tall stacks for ameliorating SO_2 concentrations around electric power
plants.

The direct emission of fly ash and other atmospheric particulates
might also be an important cause for the deterioration of visibility.
However, electrostatic precipitators and filters of particulates have
been developed to a high degree of efficiency and this, plus the
advantages obtained by pre-coal cleaning, should handle the problem.
Indeed, mine-mouth coal cleaning would not only reduce the aerosol
build-up but would also assist the gas desulfurization in the coal
burner itself.

Geographic factors have an important influence in the impact of
power plant-generated aerosols upon visibility. Coastal plants, for
example, can take advantage of the greater pollution absorbing capacity
of the ocean during periods of offshore winds to reduce their use of
high-grade coal or the use of scrubbers. Offshore coal burning plants
analogous to offshore nuclear plants makes a certain degree of sense
in the context of air pollution problems. Most of the effluents would
end up in the ocean, whose capacity is far greater than lakes.

The sun belt is especially vulnerable to sulfate formation and
other hazards to visibility. First, the high ultraviolet level in the
sun belt makes its troposphere especially suited to atomic oxygen
production by the photodecomposition of O_3. High ozone load levels
and humidity levels also promote production of aerosols.

One of the big societal problems related to visibility deterioration
is the question of assigning a dollar value which might be used in
benefit/cost analyses such as, for example, that discussed in Chapter
15. In the Florida Sulfur Oxide Study, the Stanford Research Institute
(SRI) Report (FSOS, 1978) showed evidence that visibility deterioration

might be one of the major benefits to offset the cost of pollution
control. The dollar value which SRI placed on visibility was based
upon a survey in the Four Corners region of New Mexico which was
translated into a price per mile of visibility lost. However, it is
probable that the very long distances involved in the New Mexico
study (25-75 miles) may have artificially deflated the price per mile
assigned to visibility. For example, in Pasadena or San Bernardino,
California the value of a view depends upon whether one can see the
nearby mountains which are only five to ten miles away. Thus, the
dollar equivalent of, say, 60 miles loss of visibility in regions of
California where the hills and mountains are much closer is greater.
With this interpretation the per mileage visibility costs used could
be a factor of five to ten greater, a factor which would greatly
influence the assessment of the benefits of pollution reduction.

For areas of the country which have no mountains the visibility
question might well be academic. However, from a psychological point
of view blueness of the sky could have some type of dollar value equivalent to the value of good visibility. Smoggy and hazy skies are grey
or brownish and have a different psychological impact than a cheerful
blue sky. It is noteable that the highest suicide rates in the world
occur in grey-sky regions. There can be no doubt that people value
blue sky in a similar fashion to the way they value a scenic view.
However, how can we quantify the value of a blue sky? One might
connect visibility to atmospheric turbidity or optical depth or then
to the blueness of the sky through the following considerations.
According to the classic works of H. Green (1932), Knoll et al. (1946),
Middleton (1952) (see also McCarthy, 1976) the visual range in kilometers
is given approximately by

$$R = 3.9/k$$

where k is the scattering cross section in kilometers^{-1}. It can be
shown (Green-FSOS, 1978) that this is equivalent to

$$R = 1.7 \ h/\theta$$

where h is the mixing height of the haze or smog layer and θ is the
atmospheric turbidity as measured at 500 nm with a Volz-type sun
photometer (Volz, 1973; Flowers et al., 1969). Observation suggests
that for very blue skies $\theta \approx 0.02$, for blue skies $\theta \approx 0.05$, whitish
blue $\theta \approx 0.1$, and white $\theta \approx 0.2$. If one assigns the value of a blue sky
comparable to a 30 mile loss of visibility in the Four Corners region,
then the value of a blue sky, certainly to retirees and tourists in
the sunbelt region, takes on a large dollar value. Accordingly, one
of the burning issues related to the increased use of coal is whether
the value assigned to visibility loss, or loss of blueness of the sky,
would pay for the costs of coal cleaning and scrubbers necessary to
avoid such losses.

VI. REMOTE SENSING AND THE USE OF SATELLITES

There are technical problems in measuring visibility, blueness of
the sky, or aerosol content of the atmosphere. Point sampling
devices of aerosol densities and size distributions have come to a
high level of technological sophistication (Lundgren et al., 1979).
Most methods use in-situ techniques in which the particles are drawn
into a laboratory instrument which provides some technique of counting
particles and sorting them as to sizes and shapes. Various remote
sensing techniques for measuring aerosol content which are closely
related to the question of visibility have been reviewed by Green
(1979). The sun radiometer effectively measures the direct solar light
intensity at a central wavelength and compares this value with the light
intensity expected for a perfectly clean atmosphere. The ratio may be
used to calculate what is called the atmospheric turbidity. A national
network of sun photometers with two wavelength channels (Flowers, 1969)
have yielded substantial data on turbidity available throughout the
country. A multi-channel instrument (Volz, 1973) and the use of a
compact spectrometer (Doda and Green, 1976) have the potential for
giving the turbidity and the aerosol size distribution as well as the
water vapor content. Such instruments can be utilized to follow

visibility trends following a transition to greater coal use.

The aureole method by Green et al. (1971) is a very sensitive technique for measuring the aerosol loading of the atmosphere. Figure 3 illustrates this technique. The sun is photographed through a neutral density 4 filter so that the film registers the sun's image attenuated by 10,000 and the unattenuated image of the surrounding sky in the first 15 or so degrees surrounding the sun. Figure 3c illustrates a densitometric trace obtained in such an experimental arrangement. Here t(b) refers to the film densities in a trace going above (t) or below (b); the sun and $\ell(r)$ indicates the traces in a left (right) cut through the sun itself. When the sky is clean, the sky trace falls off very rapidly away from the sun. When the sky is loaded with aerosols, the trace falls off slowly to an extent depending upon the aerosol size distribution and content. The analysis of this technique has been further developed by Deepak et al. (1977) both theoretically and experimentally. A number of electrical methods are available for exploiting the information content of the sun's aureole (Green et al., 1971; Twitty, 1975; Deepak and Box, 1978; Green, 1979) and it appears that measuring the aureole might provide a helpful method for monitoring possible aerosol build-up associated with the increased use of coal.

Other radiometric techniques are available. For example, for monitoring an overall buildup of aerosols when the sky is aerosol-free, direct sunlight falling on a flat plate and the diffuse component will have a certain ratio which is calculable by radiative transfer techniques. Departures of the ratio from this value can be used as a measure of the aerosol content of the sky. To get the diffuse light (sky light) one simply uses a disc or ball to shadow the direct sun.

The use of satellites for remote sensing and particularly for monitoring pollution has been developed extensively over the past 20 years. Rather than attempt to recount the history of these developments, we will simply give an example of a modern day satellite which has concentrated on looking at the earth's atmosphere. We refer to Nimbus-7, a satellite launched in October 1978, and placed in a 955

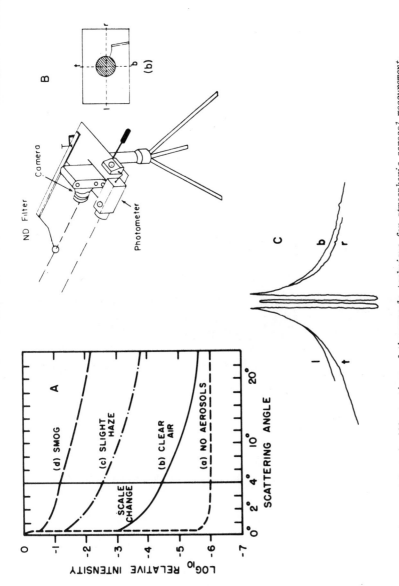

Fig. 3. Schematic illustrations of the aureole technique for atmospheric aerosol measurement. A shows relative intensity vs angle for (a) no aerosols, (b) clean air, (c) hazy air, (d) smog. B shows obscuring disc camera technique of Green et al. (1971). C shows densitometric trace of aureole photograph with attenuating disc for various scans; r(right), l(left), t(top), b(bottom).

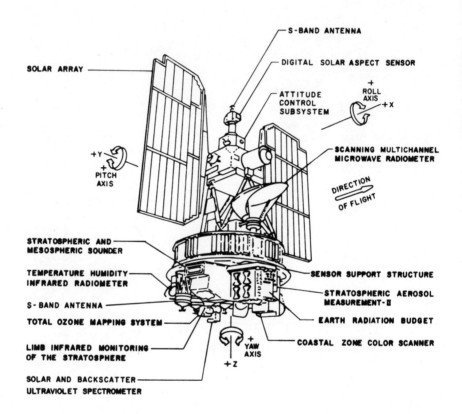

Fig. 4.   Nimbus 7 Atmospheric Observatory

kilometer sun synchronous polar orbit. Figure 4 shows the Nimbus-7
Satellite with its sun panels unfolded. The sensors are mounted below
in a torus structure; the altitude control and solar cell array are
supported above the torus by a truss. The Nimbus-7 mission involves
a variety of experiments in air pollution, and oceanographic and
meteorological disciplines. The mission refines the sounding and
atmospheric structure measurement capability developed in previous
Nimbus observatories and in other satellite programs. The scope of this
satellite system is illustrated in Table 2, which gives the various team
members, the various experiments, and their purposes. Of particular
interest to the coal burning question is ERB, the earth radiation budget
instrument. This instrument determines over a period of one year the
radiation budget of the earth on both synoptic and planetary scales by
simultaneously measuring the incoming solar radiation and the outgoing
earth-reflected short-wave radiation and emitted long-wave radiation.
Such data should be invaluable for the quantitative pursuit of the
question of climate change which has been addressed in Section II.

The stratospheric aerosol measurement (SAM II) has the objective
of mapping the concentration and optical properties of stratospheric
aerosols as a function of altitude, latitude and longitude. When no
clouds are present in the instantaneous field of view, tropospheric
aerosols can also be mapped. The objective of the tropospheric and
mesospheric sounder (SAMS) is to measure the vertical concentration of
H_2O, CH_4, CO, NO and CO_2 species that are very much involved in coal
burning issues and climate change questions.

The objective of the Solar Backscatter Ultraviolet/Total Ozone
Mapping (SBUV/TOMS) device is to determine the vertical distribution
of ozone, map the total ozone, and monitor the incident solar ultraviolet irradiance and the ultraviolet radiation backscattered from the
earth. Anthropogenic impacts upon the ozone layer have received
perhaps the most extensive attention of all of the potential atmospheric
modification problems. While no direct impact of coal burning on the
ozone layer has been proposed, the possible indirect effects via reaction
rates that tie up NO_x or Cl compounds must be given further attention.

Table 2: *Nimbus-7 Atmospheric Observatory*

Experimental Team	Affiliation	Type of Device	Parameter Determinations
SBUV/TOMS - SOLAR AND BACKSCATTERED ULTRAVIOLET AND TOTAL OZONE MAPPING SYSTEM			
Heath, D.	Goddard SFC		
Mateer, C. L.	Atm. Envir. Service (Canada)	Sun and Earth Viewing	Solar UV Irradiance,
Belmont, A. D.	Control Data Corp.	Monochromators, Nadir	Ozone Profiles, Global
Miller, A. J.	NOAA, Nat'l Meteorological	Viewing and Nadir	Maps of Total Ozone
Green, A. E. S.	U. of Florida	Scanning	
Cunnold, D. M.	MIT		
Imhof, W. L.	Lockheed Space Sc. Lab.		
ERB - EARTH RADIATION BUDGET			
Jacobowitz, H.	NOAA		
Vonder Haar, T. H.	Colorado State University		
House, F. B.	Drexel University	Sun and Earth Viewing	Solar Irradiance, Flux
Coulson, K. L.	U. of California	Spectroradiometer,	and Radiance of Earth
Hickey, J. R.	Eppley Lab., Inc.	Nadir Viewing and	Reflected Short-wave and
Stowe, L.	NOAA/NESS	Nadir Scanning	Emitted Long-wave Radiation
Ingersoll, A. P.	Calif. Inst. of Technology		
Smith, G. L.	LaRC		
CZCS - COASTAL ZONE COLOR SCANNER			
Hovis, W.	GSFC		
Yentsch, C. S.	N. E. Research Foundation		Chlorophyll, Sediment
Clark, D.	NOAA/NESS		Distribution, Gelbstoffe
Apel, J. R.	NOAA/ERL	Earth Viewing Radio-	Concentration (Salinity),
El-Sayed, S. Z.	Texas A & M	meter Nadir Scanning	Coastal Water Temperature,
Gordon, H. R.	U. of Miami		Currents
Wrigley, R. C.	Ames		
Harris, T. F. W.	CSIR (South Africa)		
Newth, F. H.	EURASEP - CEC		
SMMR - SCANNING MULTICHANNEL MICROWAVE RADIOMETER			
Gloersen, P.	GSFC		Sea Ice, Sea Surface
Ramseier, R. O.	Dept. of Environment, Canada		Temperatures and Winds,
Staelin, D. H.	MIT	Earth Viewing	Cloud Liquid Water Content,
Campbell, W. J.	Dept. of Interior/Geol. Sur.	Microwave Radiometer,	Precipitation, Water Vapor,
Gudmandsen, P.	Tech. Inst. of Denmark	Nadir Scanning	Soil Moisture, Snow Cover
Ross, D. B.	NOAA/ERL		
Windsor, E. P. L.	British Aircraft Corp.		
Kunzi, K. F.	Univ. of Berne, Switzerland		
SAM II - STRATOSPHERIC AEROSOL MEASUREMENT			
McCormick, M. P.	LaRC		
Pepin, T. J.	U. of Wyoming	Solar Extinction	Aerosols - Vertical
Grams, G. W.	NCAR	Photometer, Limb	Profiles
Herman, B. M.	U. of Arizona	Viewing	
Russell, P. B.	Stanford Research Institute		
LIMS - LIMB INFRARED MONITOR OF THE STRATOSPHERE			
Russell, J.	LaRC		
Gille, J. C.	NCAR	Limb Scanning Infrared	H_2O, HNO_3, NO_2, O_3,
House, F. B.	Drexel University	Radiometer	Temperature - Vertical
Remsberg, E. E.	LaRC		Profiles
Loevy, C. B.	U. of Washington		
Drayson, S. R.	U. of Michigan		
Fischer, H.	U. of Munchen, W. Germany	Limb Scanning Pressure	CH_4, CO, H_2O, NO, N_2O
Planet, W. G.	NOAA/NESS	Modulated Infrared	
Girard, A.	Onera, France	Radiometer	Temperature-Vertical
Harries, J. E.	National Physical Labs., U.K.		Profiles
THIR - TEMPERATURE AND HUMIDITY BY IR			
		Earth Viewing Infrared	Correlative Imagery for
		Radiometer, Nadir	Other Experiments, Sea
		Scanning	Surface Temperature

Nimbus-7 is the most sophisticated pollution monitoring satellite launched to date. Others have been launched or will be launched with more specialized purposes. However, a great state-of-the-art advance will occur when the space shuttle goes up. Then we will have a platform to monitor pollution much more closely and more accurately, since the scientists aboard will have the opportunity to correct instrument malfunctions and to recalibrate. Accordingly, we might reasonably anticipate that many of the issues related to possible atmospheric modification induced by coal burning will receive clarifying information when the space shuttle is operational.

References

Bolin, B., On the role of the atmosphere in biogeochemical cycles, Quart. J. Roy. Met. Soc., 105:25-42, 1979.

Bolin, B., Changes of land biota and their role for the carbon cycle, Science, 196:613-615, 1977.

Broecker, W. S. et al., Fate of fossil fuel carbon dioxide and the global carbon budget, Science, 206:409-418, 1979.

Butler, D. M., R. S. Stolarski, and W. L. Chameides, Effect of ground level emission of carbon monoxide and nitrogen oxides on tropospheric and stratospheric ozone. J. Geophys. Res., submitted, 1978.

Castleman, A. W., Jr., Nucleation processes and aerosol chemistry, Space Science Reviews, 15:547-589, 1974.

Castleman, A. W., Jr., R. E. Davis, H. R. Munkelwitz, I. N. Tang, and W. P. Wood, Kinetics of association reactions pertaining to H_2SO_4 aerosol formation, Inter. Jour. Chem. Kinetics, Symp. No. 1, John Wiley and Sons, Inc., pp. 629-640, 1975.

Castleman, A. W., Jr., and I. N. Tang, Kinetics of the association reaction of SO_2 with the hydroxyl radical, J. Photochemistry, 6:349-354, 1976-77.

Chameides, W. L., Effect of variable energy input on nitrogen fixation in instantaneous linear discharges, Nature, 277:123-125, 1979.

Chameides, W. L., S. C. Liu, and R. J. Cicerone, Possible variations in atmospheric methane, J. Geophys. Res., 82:1795-1798, 1977.

Charlson, R. J. et al., Sulfuric acid-ammonium sulfate aerosol: Optical detection in the St. Louis region, Science, 84:156-158, 1974.

Cogbill, C. V., and G. E. Likens, Acid precipitation in the northeastern United States, Water Reservoir Res., 10:1133-1137, 1974.

Crutzen, P. J., The possible importance of CSO for the sulfate layer of the stratosphere, Geophys. Res. Lett., 3:73-76, 1976.

Crutzen, P. J., and J. Fishman, Average concentrations of OH in the northern hemisphere troposphere and the budgets of CH_4, CO, and H_2, Geophys. Res. Lett., 4:321-324, 1977.

Davis, D. D., W. Heaps, D. Philen, and T. McGee, Boundary layer measurements of the OH radical in the vicinity of an isolated power plant plume: SO_2 and NO_2 chemical conversion times, Atmos. Env., 1978.

Deepak, A., and M. A. Box, Photogrammetry of the solar aureole, Applied Optics, 17:1120-1124, 1978.

Doda, D. D., and A. E. S. Green, Spectral sunphotometry using a compact spectrometer, Remote Sensing of the Environment, 7:97-104, 1978.

Eggleton, A. E. J., and A. Cox, Homogeneous oxidation of sulfur compounds in the atmosphere, Atmos. Environ., 12:227, 1978.

FSOS-ICAAS, Green, A. E. S. et al., An interdisciplinary study of the health, social and environmental economics of sulfur oxide pollution in Florida, ICAAS Report, Un. of Fla, Feb., 1978.

Flowers, E.C., R. A. McCormick and K. R. Kurfis, J. Appl. Meteor., 8:995, 1969.

Friend, J. P., The global sulfur cycle, in The Chemistry of the Lower Atmosphere, (ed. S.I. Rasool), Plenum Press, N.Y., 1973.

GARP-16, The physical basis of climate and climate modeling, WMO-ICSU Joint Organizing Comm., GARP Publ. Series 16, Geneva, 1975.

Gartrell, R. E., Power generation, in Air Pollution IV. Engineering Control of Air Pollution, (ed. A. C. Stern), Academic Press, 466-531, 1977.

Green, A. E. S., The aureole method, in preparation, 1979.

Green, A. E. S., A. Deepak, and B. J. Lipofsky, Interpretation of the sun's aureole based on atmospheric aerosol models, Appl. Optics, 10:1263-1279, 1971.
Green, H. N., The atmospheric transmission of coloured light, RAE Rept. E and I: 720, Royal Aircraft Establishment, Barnborough, England, 1932.
Husar, R. B., in Proc. Inter. Symp. on Sulfur in the Atmosphere, Dubrovnik, Yugoslavia, Sept. 1978.
Harrison, H., T. V. Larson, P. V. Hobbs, Oxidation of Sulfur Dioxide in the Atmosphere: A Review, The Institute of Electrical and Electronics Engineers, Inc., Annals No. 75CH1004-1 2301, 1976.
Keeling, C. D., and Bacastow, Impact of industrial gases on climate, in Energy and Climate, Studies in Geophysics, U.S. National Research Council, pp. 72-95, 1977.
Kellogg, W. W., Global influence of mankind on the climate, in Climate Change, J. Gibbin, ed., Cambridge University Press, 1978.
Knoll, H. A., R. Tousey, and E. O. Hulbert, Visual thresholds of steady point sources of light in fields of brightness from dark to light, J. Opt. Soc. Am, 36:480-482, 1946.
Larson, T. V., and H. Harrison, Acidic sulfate aerosols: Formation from heterogeneous oxidation by O_3 in clouds, Dept. of Atmos. Sci., Univ. of Washington, Seattle (preprint).
Liken, G. E., R. F. Wright, J. N. Galloway, and T. J. Butler, Acid rain, Scientific American, 241:43-51, 1979.
Liu, S. C., Possible effects on the tropospheric O_3 and OH due to NO emissions, Geophys. Res. Lett., 4:325-328, 1977.
Logan, J.A., M. J. Prather, S. C. Wofsy, and M. B. McElroy, Atmospheric chemistry: Response to human influence, Phil. Trans. Roy. Soc., 290:187-234, 1978.
Lowe, D. C., P. R. Guenther, and C. D. Keeling, The concentration of atmospheric carbon dioxide at Baring Head, New Zealand, Tellus, 31:58-67, 1979.
Lundgren, D. A., S. H. Franklin, Jr., W. H. Marlow, M. Lippmann, W. E. Clark, M. D. Durham (eds.), Aerosol Measurement, University of Florida Press, Gainesville, FL., 1979.
Manabe, S., and R. T. Wetherald, The effects of doubling CO_2 concentration on the climate of a general circulation model, J. Atmos. Sci., 32:3-15, 1975.
Maroulis, P. J., and A. R. Bandy, Measurements of tropospheric background levels of SO_2 on project GAMETAG, EOS, Trans. Amer. Geophys. Union, 59:1081, 1978.
McCartney, E. J., Optics of the Atmosphere, John Wiley and Sons, N.Y., 408 pp., 1976.
McFarland, M., D. Kley W. C. Kuster, A. L. Schmeltekopf, and J. W. Drummond, NO, O_3, NO_2, and CO measurements made in the equatorial pacific region, EOS, Trans. Amer. Geophys. Union, 59:1077, 1978.
Middleton, W. E. K., Vision Through the Atmosphere, University of Toronto Press, Canada, 1952.
Noxon, J. F., NO_2 in the stratosphere and troposphere by ground based absorption spectroscopy, Science, 189:547-549, 1975.
Ohring, G., and S. Adler, Some experiments with a zonally averaged climate model J. Atmos. Sci., 35:186-205, 1978.
Ramanathan, V., A study of the sensitivity of radiative-convective models, 2nd Conf. of Atmospheric Radiation, Amer. Meteor. Soc., pp. 124-125, 1975.
Robinson, E., and R. C. Robbins, Gaseous nitrogen compound pollutants from urban and natural sources, J. Air Poll. Contr. Assoc., 20:303-306, 1970.
Sander, S. P., and J. H. Seinfeld, Chemical kinetics of homogeneous atmospheric oxidation of sulfur dioxide, Env. Sci. Tech., 10:1113, 1976.
Schneider, S. H., On the carbon dioxide-climate confusion, J. Atmos. Sci., 82:2060-2066, 1975.
Singh, J. J. and A. Deepak, Environmental and Climatic Impact of Coal Utilization, Williamsburg, Va., April, 1979 (Academic Press, to be published).
Sze, N. D., Anthropogenic CO emissions: Implications for the atmospheric $CO-OH-CH_4$ cycle, Science, 195:673-675, 1977.
Sze, N. D., and M. K. W. Ko, Is CS_2 a precursor for atmospheric COS?, Nature, 278:731-732, 1979.
Turco, R. P., et al., Carbonyl sulfide, stratospheric aerosols and terrestrial climate, Abst. Digest Symp. on Environmental and Climatic Impact of Coal Utilization, Williamsburg, Virginia, April, 1979.
Twitty, J. T., Inversion of aureole measurements to derive aerosol size distributions, J. Atmos. Sci., 32:584-591, 1975.
Volz, F., Photometer mit Selen-Photoelement zur spektrolung und zur Bestimmung der Wellenbangen abhangigkilt der Dunsttribung, Arch. Met. Geoph., 10:1 (with translation), 1954.
Waggoner, A. P., A. J. Vanderpol, and R. J. Charlson, Sulphate-light scattering ratio as an index of the role of sulfur in tropospheric optics, Nature, 261:120-122, 1976.
Wood, W. P., A. W. Castleman, Jr., I. N. Tang, Mechanisms of aerosol formation from SO_2, J. Aerosol. Sci., 6:367-374, 1975.
Woodwell, G. M., The carbon dioxide question, Scientific American, 238:34-43, 1978.
Wong. C. S., Atmospheric input of carbon dioxide from burning wood, Science, 200:197-199, 1978.

CHAPTER 12

SOLID WASTE AND TRACE ELEMENT IMPACTS

William E. Bolch, Jr.

I. INTRODUCTION

In addition to moisture, volatile matter, fixed carbon, ash, Al,
Ca, Fe, K, Mg, S, and Si, coal may contain a wide spectrum of trace
elements including As, Cd, Ce, Cr, Cu, Hg, Mn, Pb, Se, Sr, V, Zn, and
naturally occurring radioactivity, especially the uranium and thorium
series.

Ash and trace element content of coal is a direct result of its
geological history. Geologic age peat swamps were buried by deposition
of materials as the swamps subsided. Some coal may have thin layers of
deposited materials. Also the mining process itself may add inadver-
tant overburden or foreign matter from the bottom of the seam. Trace
elements may be derived from the original organic matter or from aqueous
input either during or after coalification.

The recent monograph, Trace Contaminants from Coal (Torrey, 1978a),
contains a most comprehensive review of the potential impacts from trace
elements, including radioactivity from coal burning. This book summarizes information, primarily from federal agency reports, on (1) the type
and amount of trace contaminants in coal ash, (2) analytical methods for
their measurement, (3) transport, transformation, and biological uptake
of trace contaminants, (4) health effects, (5) ecological effects, (6)
responses to emission control devices, and (7) factors to be considered
in ash disposal. Topics (3) and (7) suffer somewhat from lack of detailed technical treatment. A companion monograph, Coal Ash Utilization:

231

Fly Ash, Bottom Ash, and Slag (Torrey 1978b) focuses on the recovery of
these waste products as beneficial resources. These monographs were
published before the impact of the Resource Conservation and Recovery
Act was reflected in the proposed hazardous waste regulations issued in
late 1978 (EPA 1978). The Department of Energy Environmental Impact
Statement covers some of the topics not emphasized by Torrey. However,
health effects from combustion of coal are not quantified. Other broad
studies include some aspects of trace element contaminants or solid
waste disposal (USOTA 1979; Brandi et al., 1978).

A. Environmental Concerns

Air pollution became a major national concern during the past two
decades. Degradation of air quality by central power plants burning
fossil fuels, especially coal, resulted in massive monitoring efforts,
broad-ranging effect studies, conversion to cleaner fuels, and incorporation of sophisticated air pollution control devices. Special
attention was given to the oxides of carbon, nitrogen, and sulfur as
well as hydrocarbons and particulate emissions. Fuel switching, i.e.,
to petroleum and natural gas, was an acceptable solution providing
excellent air pollution abatement. Whenever the conversion solution
was unavailable, air pollution control devices often changed an air
quality problem to a water quality or solid waste disposal concern.

Starting with the National Environmental Policy Act and similar
legislation by states that required various forms of environmental impact analyses and continuing through the associated development of the
Environmental Protection Agency and the environmental regulations of the
last half of this decade, there has been an increasing emphasis on the
less obvious pollutants from fossil fuel power plants. The return to
coal, a defined national policy, results in the consideration of the
ultimate fate of trace elements in coal as a significant coal burning
issue.

B. Environmental Pathways

As existing power plants are converted to coal and new coal burning facilities are built, both owners and consumers will become keenly aware of the local environmental pathways through the site selection processes, the permitting requirements, and the pollution abatement facilities and their economic impact on power costs. The specific plant site pathways for trace elements in coal need to be addressed.

There are, however, many additional environmental contamination potentials associated with coal utilization, from the mine to the delivery of coal at the power plant. Off-site pathways include those environmental problems at the mine, the post mining drainage, reclamation of disturbed lands, coal cleaning operations, storage pile leachate, and transportation impacts.

C. Trace Elements in Coal

Direct measurements of the trace element composition of coal are limited. A number of sources estimate the concentrations in coal by back calculation from ash measurements and percentage of ash in the source. Figure 1 summarizes the data on trace elements in coal from two review sources. The data in Figure 1 have been arranged in descending order of the average concentration from the various sources. It is not always clear whether data are for raw or cleaned coal and no data referred to cores from unmined sources. It should be re-emphasized that any particular source of coal may have a value different by a factor of two.

Note that the ordinate spans a concentration range from about 100 ppm to 1 ppm. Barium is the most common trace element followed by fluorine, a volatile element. The elements Hg and Te were not graphed as they have concentrations of from 0.06 to 2 ppm and 0.03 to 0.5 ppm, respectively.

D. Radioactivity in Coal

Coal contains both carbon-14 and potassium-40--the former a function of its geologic age and the latter in direct correlation with the K content. Neither radionuclide has any significant impact on the environment. Both U and Th are found in coal deposits. Figure 2 presents these important radioactive series by location and type of coal.

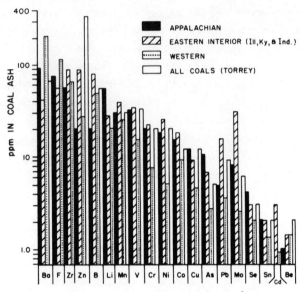

Fig. 1. Trace Elements in Coal

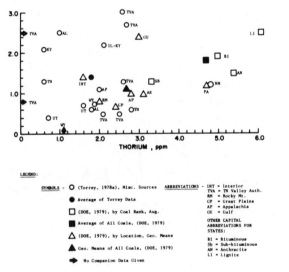

Fig. 2 Radioactivity in Coal by Location and Class

The data suggest that the U series is more easily quantified and
has a value of about 1.4 ppm regardless of the source and method of
combining data. The Th data may be more affected by analytical difficulties, whether data are combined as means or geometric means, and
the inclusion of lignite data. It is interesting to note that all
sedimentary rocks average about 1.2 ppm U while Th ranges between 5.9
and 11 ppm. Thus, coal appears to be depleted in Th when compared to
other sedimentary rocks.

A concentration of 1.4 ppm U is equal to 0.47 pCi/g if uranium-238
and each of its 13 daughters in equilibrium with this parent radionuclide. Thus, important species of the series such as thorium-230 and
radium-226 will also be at 0.47 pCi/g. Similarly a Th concentration of
2.7 ppm converts to 0.40 pCi/g of thorium-232 and each of its nine
daughters. Measureable radiation is also produced by the 10-daughter
series from uranium-235 although the parent represents only 0.73%
of natural uranium.

Some small deposits in the Western United States contain U in
rather high concentrations--50 to 100 ppm with at least two reported in
the 6300 to 7300 ppm range. Important U ore resources are in the range
of about 1000 to 3000 ppm. The concentration of U and its many
daughters in the unburned residues of coal is, of course, at higher
concentrations and thus their dispersal or disposal become a coal
burning issue.

II. ASH FROM COAL COMBUSTION

Ash is the noncombustible residue that remains when coal is burned.
Ash is formed by chemical changes in extraneous mineral matter or in
inorganic and organic plant material during the combustion of coal.

Figure 3 diagrams the fractionation of coal into various components
in a typical power plant. Both inorganics and unburned organics may be
contributed to the volatile vapors and ash. Dense ash may settle as
bottom ash while the finer particles must be controlled in a suitable
filter or scrubbing device. Either of these solid residues is removed
by dry mechanical means or be wet slurry.

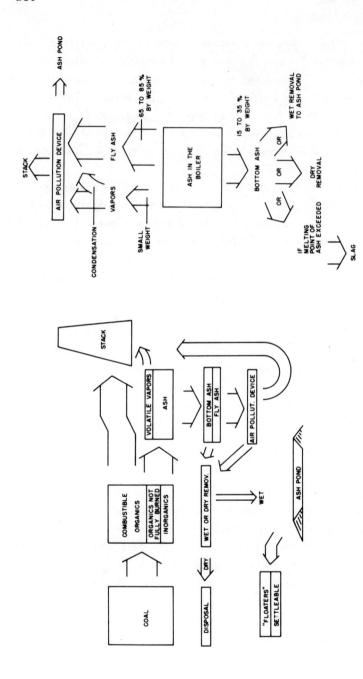

Fig. 4. Transformations of Ash in Coal Combustion

Fig. 3. Fraction of Coal in a Typical Power Plant

Figure 4 depicts other transformations and weight distributions
that ash may undergo in the boiler systems. Note the production of a
glassy slag in the event that the melting point of the bottom ash is
reached. Many elements may be vaporized within the boiler, but some
fraction apparently condenses upon fly ash particles within the gaseous
stream or the air pollution device. There are distinct differences in
ash fraction from a similar coal, depending upon the boiler feed design. Cyclone designs produce a much higher percentage of bottom ash.

A. Trace Elements in Ash

Figure 5 shows the concentration of 19 trace elements in coal ash,
arranged in order of the average concentration for each rank that exceeds 10 ppm (anthracite, high volatile bituminous, low volatile
bituminous, and medium volatile bituminous). Note that the concentrations range from about 1000 ppm to 10 ppm. The actual enrichment
factor for a particular element depends upon numerous factors. The
value for each element-category represents from 4 to 24 measurements.
The variability found in any single sampling of a coal ash.

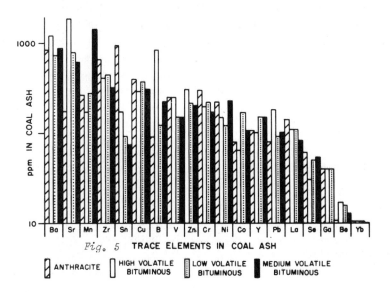

Fig. 5   TRACE ELEMENTS IN COAL ASH

The chemical constituents in Figure 5 may be grouped into four general classes:

Class I.   Elements volatilized not in the combustion zone but instead from a rather uniform melt that becomes both fly ash and slag: Ba, Sr, Mn, Sn, Co, La, Sc, and Ti.

Class II.   Elements that are volatilized on combustion and condense or absorb on the fly ash as the flue gas cools, leading to depletion from the slag and concentration in the fly ash, include Cu, Zn, Pb, Ga, As, Se, and Cd.

Class III.   Elements that remain almost completely in the gas phase: F, Hg, Cl, and Br.

Class IV.   Elements for which little data is available: Zr, B, Li, V, Cr, Ni, Mo, and Ba.

Lignite ash, when compared to the trends in Figure 5 has a somewhat different distribution, with Ba, Mn, Cu, and B contents higher (1.3x to 5.4x) than the average within the category, while all other element concentrations shown in Figure 5 are about one-half for lignite ash.

Some of the trace elements are known to be toxic to man and to animal and plant life. In addition to the trace inorganic elements, a broad spectrum of other toxic or potentially toxic substances are released into the environment during coal combustion. Table 1 is an abbreviated version of a more complete listing given by Torrey (1978a).

Table 1: *Selected Toxic and Potentially Toxic Substances Released During Coal Burning*

Group	Example
Anhydrides	Meleic Anhydride
Aminos	Aliphatic Aminos
Carbonyl Compounds	Formaldehyde
Heterocyclics	Pyridines
Hydrocarbons	Benzene
Phenols	Cresols
Polycyclic Aromatic Hydrocarbons	Benzo/a/pyrene & Dibenzofluorene
Sulfur Compounds	Thiophenes
Organometallics	Tetraethyl Lead
Other Particulates	Tar
Cyanides	Hydrogen Cyanide

B. Radioactivity in Ash

As stated earlier, the major radionuclides in coal are ^{238}U, ^{232}Th and their daughters, and ^{40}K. Since coal is approximately 10% ash, the concentrations of these noncombustible nuclides in ash will be about 10 times greater. More specifically, uranium, lead, and polonium tend to concentrate in the fly ash while thorium, radium, and potassium concentrate in the bottom ash. Other radionuclides in these series have not been adequately tested.

A hypothetical mass balance and radioactivity fractionation can be seen in Figure 6. These data are for coal having concentrations of 1.4 ppm U and 2.7 ppm Th. The noble gases ^{222}Rn and ^{220}Rn will continue to decay through their daughters in the environment. Any fly ash that escapes the air pollution control devices will be at the fly ash concentration shown. The solid wastes will be a combination of bottom and fly ash. More data is needed on the radionuclide fractionation versus boiler type.

C. Physical Characteristics of Ash

Table 2 outlines the physical characteristics of coal ash. These parameters are important to both waste disposal considerations and to the potential for any utilization of the ash as a resource.

D. Coal Ash Utilization Potentials

By 1985, the electric utility industry in the United States may burn nearly 1200 million tons of coal for the production of electrical energy. The corresponding ash production will be 180 million tons, of which 120 million tons will be fly ash and the remaining 60 million tons will be dry bottom ash and boiler slag. Utilization is both a resource recovery and a solution to a waste disposal problem. Table 3 summarizes some of the present successful uses of ash and a number of minor and potential applications. About 68% of slag is currently utilized and 8% is removed at no expense to the company. Bottom and fly ash utilization, including removal at no cost, approaches 15%.

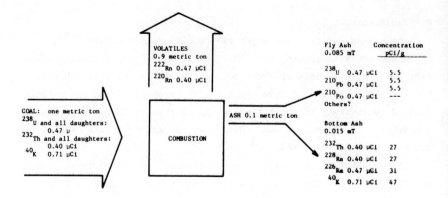

Fig. 6. *Postuated Mass Balance of Radioactivity in Coal Combustion*

Table 2: *Physical Characteristics of Ash*

Bottom Ash

    General:
        Coarse, dense particles; 1 to 40,000 microns (size range
        analogous to medium sand to fine gravel);
        angular shape with porous surface

    Slag:
        Angular shape with glass-like appearance

Fly Ash

    General:
        Spherical; 0.5 to 100 microns (size range analogous to
        medium clay to find sand); pH of pulverized fly ash
        plus water: mostly basic (pH 8 to 12)

    Cenospheres:
        Very lightweight particles (20% by volume)
        Silicate glass filled with nitrogen and carbon dioxide
        will float in water (0.8 g/cc), 20 to 200 microns

    Cyclone Fired Units:
        About 90% of the fly ash, 10 microns

Air Pollution Device and Fly Ash

    Electrostatic Precipitators:
        50% of the particles are less than or equal to 2 microns

    Mechanical Collectors:
        50% of the particles are less than or equal to 20 microns

---

Source: Torrey 1978b

A comprehensive review of the current and potential recovery
of this waste product resource is contained in Coal Ash Utilization:
Fly Ash, Bottom Ash and Slag (Torrey 1978b). He discusses in considerable detail many of the topics in Table 3.

E. Ash Disposal

Two types of ash must be handled: bottom ash and fly ash. Specific treatments are site dependent; however, the following discussion is

Table 3: *Present and Potential Usages of Coal Ash*

Present Usage

    Mixed with raw material before forming cement
    Pozzolan cement
    Partial replacement of cement in
        concrete products
        structural concrete
        dams and other massive concrete
    Lightweight aggregate
    Fill material for roads, construction sites, etc.
    Stabilizer for road bases, parking lots, etc.
    Filler in asphalt mix

Potential and Limited Usage

    Agricultural fertilizer
    Mine spoil and refuse bank reclamation
    Minor traffic surfacing (for bicycle paths, hiking trails,
        camping and picnic areas)
    Brick filler, color and texture amendments
    Sanitary landfill to improve compaction or contribute a
        neutralizing effect
    Abrasive materials for skid resistant surfaces, sand blasting,
        cleaning powders
    Thermal insulation and fireproofing of steel
    Water waste treatment to aid coagulation, growth of flocs,
        sludge-conditioning and absorb organics
    Fire fighting, fire lands, ice control
    Kitty litter, deodorization of animal waste
    Neutralization, normally basic with limited buffering capacity,
        flue gas desulferization

typical of the industry (U.S. EPA 1973, Alachua 1978). Bottom ash settles
from the furnace to the ash hopper where it is cooled by recirculating
slurry water. The cooled bottom ash is then sluiced to ash ponds where
it settles out, leaving bottom sediments and slurry water which can be
used again. Periodically, the bottom ash sediments are removed and
placed in on-site, above-ground landfill cells.

Large fly ash particles are removed by mechanical dust collectors
while smaller particles are removed by electrostatic precipitators. The
collected ash can then be handled either wet or dry. The wet process is
similar to that for bottom ash, while dry fly ash is transported directly
to the landfill sites. Ash remains there until the sites become full,

at which time they are covered with soil and revegetated to within ten
feet of the working face. New hazardous waste regulations will have a
considerable impact on the design and monitoring of landfills for the
purpose of ash disposal (U.S. EPA 1978).

III. HEALTH EFFECTS

    Fatalities and serious injuries from mining operations and the
transportation of coal have a historical baseline and thus are easy to
extrapolate for increased coal use. It has been estimated (DOE 1979)
that by the year 1990 some 16 deaths per year will occur from mining and
cleaning operations and 62 deaths per year will be attributed to the
transportation phase as rail traffic and other modes attempt to move
enormous quantities of coal. It is interesting to note that the impact
statement (DOE 1979) on the Fuel Use Act refused to quantify the health
impacts of combustion. The National Environmental Policy Act, however,
demands that the agencies "identify and develop methods and procedures
which will assure that presently unqualified environmental values will
be given appropriate consideration in decision making." Certainly,
environmental transport and dose-to-risk models exist that can attempt
to place the health effects of coal burning in proper perspective with
other energy sources. The major effects, such as SO_2, O_3, particulates,
and oxides of nitrogen are adequately addressed in other chapters. The
impact of trace elements added to the environment by coal combustion
is nontrivial and remains a coal burning issue.

A. Trace Element Effects

    Table 4 summarizes some of the potential effects of trace element
releases to the environment. These contaminants may be directly transferred to air by stack releases, to water via slurry removal techniques,
and to the soil by landfill of solid wastes. Indirect pathways involve
mechanisms such as deposition, resuspension, runoff, leaching, flooding,
aquatic and terrestrial uptake, and human exposure by breathing, eating
and drinking of contaminated media. It is fully recognized that a
quantitative assessment of public ill health would have a high degree of

Table 4: *Summary of Potential Health Effects of Trace Elements*

Element	Human Target Organs	Location of Suspected or Known Cancers	Toxic Effects on Humans other than Cancer	Other Effects
Ag (silver)	Skin	--	Mucous membrane and skin discoloration	Toxic to marine organisms. High CF.
As (arsenic)	RS, GI, CNS, liver, skin, blood, endocrine system	Mouth, larynx, esophagus, bladder	Poisoning of oxidative phosphorylation system	CP, chronic effects on both aquatic and mammalian species. Coal burning may add to other sources, i.e., smelting, pesticides, herbicides.
B (boron)	GI	--	--	EE for plants, but toxic at water concentrations 1 mg/l.
Ba (barium)	GI, RS, CNS, CVS, skin	--	Highly toxic when ingested	Toxic for soluble compounds, high CF. INH pathway.
Be (beryllium)	RS, skin, bone	Lungs, bones	Skin damage	INH pathway, strongly absorbed by soils & clays.
Cd (cadmium)	GI, RS, CNS, CVS, liver, skin, kidney	Lungs	Hypertension, Pulmonary emphysema, kidney damage	CP for fish and mammals, may effect growth and photosynthesis in plants.
Co (cobalt)	GI, RS, CNS, CVS, endocrine system	--	--	EE for plants and animals, but range between nutritional and toxic dose is small.
Cr (chromium)	RS	RS (Cr^{+6})	Emphysema, pneumonia	Effects depend upon valance state, some reduction in crop growth documented.
Cu (copper)	GI, blood	--	--	EE for most organisms, but readily concentrated by marine organisms to toxic levels.
F (fluorine)	Bone	--	--	Damage to wide range of plants, livestock fluorisis.
Fe (iron)	GI, RS, CNS, CVS, blood endocrine system	--	--	EE
Ga (gallium)	Kidney, liver muscles	Linked to tumor formation	Neuromuscular poison, kidney	Little known
Ge (germanium)	GI, RS, CVS	--	--	Low toxicity.
Hg (mercury)	RS, CNS, CVS, liver, kidney	--	CNS poison	Toxicity depends upon chemical form. Natural methylation can occur in bottom sediments.
Mn (manganese)	RS, CNS, kidney	--	Affects CNS	Toxic to very low concentrations to crops, EE to mammals.
Mo (molybdenum)	Liver, blood, kidney, bone	--	--	EE, toxicity only in areas of high soil Mo.
Ni (nickel)	RS, CNS, skin	Various (nickel carbonyl)	Respiratory disorders, dermatitis	Toxic to citrus fruits, nickel salts results in fish toxicity. Nickel carbonyl extremely toxic to man.
Pb (lead)	GI, CNS, blood, kidney	--	Hemolytic and renal poison	Little is known of biological effects on plants and marine organisms.
Sb (antimony)	GI, RS, CNS, liver, skin	--	Implicated in heart disease	Little is known of ecological effects.
Se (selenium)	GI, CNS, liver, skin	Liver	Degeneration of liver and kidney. Pneumonia.	May be beneficial in some areas but deterious in others. Narrow range between nutritional and toxic levels. Synergistic effect with Hg, As, and Tl.
Sn (tin)	GI, CNS	--	--	Organic SN is toxic but little is known about inorganic effects.
Sr (strontium)	CNS, teeth, bone	--	--	EE for mammals.
Te (tellerium)	CNS, liver, kidney	--	Kidney and liver degeneration	Potentially dangerous, but little known.
Tl (thallium)	GI, RS, CNS, liver, kidney, blood, bone	--	Cumulative hemolytic poison	Thallium oxide (released by coal combustion) toxic to rats when administered orally.
U (uranium)	Kidney	--	Kidney disease	Soluble uranyl ion may result in acute renal damage irrespective of its radioactivity.
V (vanadium)	RS, CNS, skin, kidney	--	--	May be an EE.
W (tungston)	Bone	--	--	No reports of toxicity.
Zn (zinc)	GI, skin, bone	--	--	Long-term sublethal effects in marine water suspected.

CF = Concentration Factors   EE = Essential Element
CNS = Central Nervous System   GI = Gastrointestinal Tract
CP = Cumulative Poison   INH = Inhalation
CVS = Cardiovascular System   RS = Respiratory System

Adapted from Table 5.5 and pp. 73-208 of Torrey 1978(a).

uncertainty, but a preliminary calculation would serve as a stimulant
of interest in its refinement and as a basis of comparison to other
energy sources.

B. Radiological Effects

The health effects from the radionuclides contained in coal ash may
occur from three major pathways. First, the utilization of ash and slag
in building materials and other places enhances the potential for continued exposure. An example is the current moratorium on the use of TVA
coal slag in concrete blocks for home construction due to the elevation
of radon daughters within the closed spaces and the resulting increase
in the lung cancer risk of occupants.

A second pathway is less well-documented and presents a long-term
hazard. This pathway is that of water contamination from leachate from
either ash disposal or utilization practices. The Resource Conservation
and Recovery Act attempts to address this kind of problem.

Perhaps the most significant radiological impact of coal burning is
that of the airborne emissions of radioactivity. Even if the off-site
health effects of radioactivity are small in comparison to those produced
by the major releases (i.e., SO_2, particulates, etc.), the relative
radiological impact of equivalent-sized, identically-sited nuclear and
coal-fired plants is still at issue. A most comprehensive comparison
was presented by McBride et al. (1977). Table 5 summarizes some of the
results for three (1000 MWe) power plant types at a hypothetical midwestern site. Most of the doses shown are small in comparison, for
example, to either natural background or to man-made contributions by
diagnostic x-rays. The real issues raised by the data include (1) the
fact that the public and industry accepts a high degree of waste treatment for the nuclear plants and yet often resists high-cost pollution
control devices and siting restrictions for the coal plants; and (2) the
fact that the trace elements, including radioactivity, each have a potential for an impact that should be quantified in terms of some health
effect index and integrated with the major releases for a total environmental impact.

Table 5   *Dose Commitments from Airborne Radioactivity*
           Releases at Power Plants

Type, Units	Coal Fired Plant	Boiling Water Reactor	Pressurized Water Reactor
Maximum Individual (mrem/yr)			
Whole body	1.9	4.6	1.8
Bone	18.2	5.9	2.7
Thyroid	1.9	36.9*	3.8
Surrounding Population (man-rem/yr)			
Whole body	19	13	13
Bone	192	21	20
Thyroid	19	37	12

*Assumes site boundary dairy cow.

IV.   EFFECTS OF PRE-COMBUSTION OPERATIONS

A. Mining

Depending upon geological conditions and characteristics of the
coal seam, either surface mining or underground mining is employed when
excavating coal from the earth. Underground mining is of primary concern when dealing with trace elements. The reason for this is the high
amount of acid mine drainage and related trace elements associated with
underground mines. Acid drainage results from the leaching process of
high sulfur-bearing materials. Iron disulfides undergo oxidation to a
series of hydrous iron sulfates which, in the presence of water,
generate large amounts of iron and sulfate leading to acid mine drainage.
The extent of this process is a function of water-sulfide contact time,
hydrological, geological, and topographic mine and terrain features,
type of mine operation, and activity status of mining. In general,
inactive underground mines are the primary source of acid mine drainage.
Heavy metals, such as Fe, Al, Mn, and Z, are found in acid mine
drainage at concentrations of 1 to 10 ppm. Uranium also forms
soluble sulfate complexes with sulfur acid.

Another potential health hazard related to coal mining is the inhalation of radon gases (^{220}Rn and ^{222}Rn). Early work shows no significant danger from these radionuclides (DOE 1979, EPA 1977).

B. Coal Cleaning

To minimize transportation costs, to minimize effluent and bottom sediments after combustion, and to improve the Btu content, coal cleaning is performed to remove unwanted sulfur and other noncombustible materials (ash) from coal by crushing it and washing it with water. The wastes generated by coal cleaning are pyritic materials, shales, clays, and various silicates termed GOB which are stacked in GOB piles. Trace elements exist at toxic levels in these piles, presenting an environmental and health hazard as water leaches through them. In general, though, GOB piles from coal cleaning do not post as great a danger as acid mine drainage from coal mining (DOE 1979, EPA 1977(a)).

C. Storage Piles

Two types of coal storage piles exist at power plants--live and reserve. Live piles are kept small and used quickly to minimize oxidation which reduces the Btu content of freshly cut coal. Reserve piles maintain about a 100-day supply with oxidative processes decreasing with time. Again, acid runoff from the piles presents environmental and health hazards. Al, Cu, Mn, Zn, Fe, Mg, and sulfates dissolve at low pH, which degrades streams and aquifers. The concentrations of trace elements is dependent upon drainage quality and quantity, rain and humidity, area of pile, and type of coal (DOE 1979, EPA 1977(a)).

D. Coal Conversion

With decreasing gaseous and liquid fuel reserves, attention has been directed toward gasification and liquefaction of coal--the former in practice and the latter under research. Gasification is a complex process which concentrates trace elements such as As, Be, Hg, Se, Cd, and Pb in byproduct chars. The main health problem is cancer, developed through contact or inhalation from coal dust or other pollutants generated at the gasification plant.

In 1976 the Department of the Interior and the Energy Research and
Development Administration published a draft Environmental Statement for
a Synthetic Fuel Commercialization Program. In this statement, even
with the known carcinogenic effects from exposure to coal and trace
elements, little attention was directed toward health effects of coal
conversion. The National Institute for Occupational Safety and Health
(NIOSH) strongly recommends this document be revised to address adequately the environmental and occupational health effects problem
(DOE 1979; EPA 1977(a); and Young, 1979).

V. NEW STANDARDS AND REGULATIONS

Philbin (1979) has presented an excellent review of the possible
implications of new legislation to the electric power industry. The
Resource Conservation and Recovery Act (RCRA) is the key legislation
with respect to solid waste and trace element impacts of coal burning.
A brief review and analysis follows.

The National Environmental Policy Act has been the tool by which
compliance with applicable regulations are shown to have been met via
the environmental impact statement (EIS) process. The Federal Water
Pollution Act (WPCA) of 1948, with the Federal Water Pollution Control
Act Amendment of 1972, the Clean Water Act Amendments of 1977, and
the Clean Air Act (CAA) of 1970 and its 1970 amendments are keys to
the Environmental Protection Agency (EPA) requirements for electric
power stations. The EPA's permit programs, especially the National
Pollution Discharge Elimination System (NPDES) for water and the Prevention of Significant Deterioration (PSD) for air quality, have a
major impact on coal-fired plants. These requirements have the potential for making the EIS process for coal burning stations nearly as
complicated as for nuclear stations. Note that the CAA lists radioactivity as a pollutant.

The new laws, namely, The Toxic Substance Control Act (TOSCA) and
its companion RCRA, represent a significant impact on industry. RCRA
is the solid waste analog of FWPA and CAA. Utility waste (not only ash
but also sludges generated by air pollution control devices and other
solid wastes generated on site) may have to undergo prescribed leachate
and toxicity tests. Under RCRA, ash with a concentration of ^{226}Ra in

excess of 5 pCi/g would be considered a hazardous waste. If these and
other definitions stand, waste from coal burning plants will require
numerous tests, carefully planned disposal sites, periodic monitoring,
and long term administration. Specific regulations and requirements
for RCRA are under development and remain a coal burning issue.

References

Brandi, R., P. Chan, et al., Interim regional report for national coal utilization assessment,
    Impacts of Future Coal Use in California, 1978.
McBride, J. P. et al., Radiological impact of airborne effluents of coal and nuclear plants,
    Science, Vol. 202, No. 4372, Dec. 1978.
Philbin, T. W., Coping with future engironmental regulations, presented at the Atomic Industrial
    Forum Conference on Environmental Regulation: Looking Ahead, Monterey, California, June 13,
    1978.
Torrey, S., Trace contaminants from coal, Pollution Technology Review, No. 50, Noyes Data
    Corporation, N.J., 1978.
Torrey, S., Coal ash utilization; fly ash, bottom ash, and slag, Pollution Technology Review,
    No. 48, Noyes Data Corporation, N. J., 1978.
U.S. Congress, The direct use of coal, Office of Technology Assessment, Washington, D.C., 1979.
U.S. DOE, Fuel use act, Final Environmental Impact Statement, U.S. Dept. of Energy, Washington,
    D.C., April 1979.
U.S. EPA, Development document for proposed effluent limitations guidelines and new source
    performance standards for the steam electric power generating point source category,
    EPA 440/1-73029, March 1974.
U.S. EPA, Characterization of ash from coal-fired power plants, EPA-600/7-77-010, January 1977.
U.S. EPA, Potential radioactive pollutants resulting from expanded energy program, Office of
    Research and Development, EPA-600/7-77-082, August 1977.
U.S. EPA, Hazardous waste, Federal Register, Environmental Protection Agency, Vol. 43,
    No. 243, December 18, 1978.
Young, R. J., Occupational health and coal conversion, Occupational Health and Safety, 22-26,
    September 1979.

Bibliography

Beck, H. L., C. V. Gagolak, K. M. Miller and W. M. Lowder, Perturbations on the natural
    radiation environment due to the utilization of coal as an energy source, Presented at the
    DOE/UT Symposium on the Natural Radiation Environment III, Houston, Texas, April 23-28, 1978.
Coles, D. G. et al., Behavior of natural radionuclides in western coal-fired power plants,
    Environmental Science and Technology, Vol. 12, No. 4, April 1978.
Denslow, V. A., Resource Conservation and Recovery Act of 1976: Material Recovery and Con-
    servation Status and Future Impacts, Presented at American Institute of Chemical Engineers
    87th National Meeting, Boston, Mass., August 20, 1979.
Eisenbud, M., and H. G. Petrow, Radioactivity in the atmospheric effluents of power plants
    that use fossil fuels, Science, Vol. 144, April 1964.
Fisher, G. L., Potential health significance of coal fly ash, Symposium on Trace Metals and
    Toxic Substances from Coal Combustion, American Institute of Chemical Engineers 87th
    National Meeting, Boston, Mass., August 1979.
Kaakinen, J. W., R. M. Jorden, M. H. Lawasani and R. E. West, Trace element behavior in coal-
    fired power plant, Environmental Science and Technology, Vol. 9, No. 9, pp. 862, September 1975.
Klein, D. H. and P. Russell, Heavy metals: fallout around a power plant, Environmental Science
    and Technology, Vol. 7, No. 4, pp. 357, April 1973.
Klein, D. H. et al., Pathways of thirty-seven trace elements through coal-fired power plants,
    Environmental Science and Technology, Vol. 9, No. 10, pp. 973, October 1975.
Moon, D. K., RCRA and the hazardous waste generator, Paper #19C, Session No. 3090, AIChE
    National Meeting, Boston, Mass., August 20, 1979.
Singth, J. J., Measurement of trace metals in coal plant effluents - a review, Symposium on
    Environmental and Climatic Impact of Coal Utilization, Williamsburg, VA. April 17-19, 1979.
Sloth, E. and M. Locke, A fossil plant environmental impact study, Power Engineering, April, 1974.
Styron, E. et al., Assessment of the radiological impact of coal utilization, MLM-2514 Monsanto
    Research Corp., Mound Facility EPA/DOE Interagency Agreement IAG-D5-#681, February 1979.

CHAPTER 13

AGRICULTURE

Shreve S. Woltz and Jimmy J. Street

I. INTRODUCTION

Increased utilization of coal as an alternative to oil or natural gas will result in increased environmental impact on agriculture and wildlife ecology. Most of the effects will unfortunately be adverse without proper assessment and avoidance techniques. The objectives of this chapter are: (1) to evaluate the prospective disruptions and minimize damage to agriculture from coal use; and (2) to consider utilization of coal ash in agriculture and forest soils.

The impact of coal residues on the soil and plant environment depends on the partitioning of this material between captured and non-captured residues. The emitted flue gases and particulates that escape to the atmosphere have direct impact on the surrounding environment. The magnitude of this impact depends on the nature of the gases and particulates and their interaction with the environment. The coal ash residues left at the site of coal combustion, which constitutes about 90% of the total residues, exert an impact on the environment only when discharged or disposed directly onto the land. The following sections deal with the effect of escaping gases and emissions with primary emphasis on sulfur dioxide and the potential use of captured residues as a soil amendment. The geographic area emphasized in these discussions is the southeastern United States. Agriculture in the southeast is highly energy intensive and, to function effectively, needs petroleum products. Agriculture is therefore interested in alternate energy uses to save

249

agri-business as we know it.

II. SULFUR DIOXIDE EFFECTS ON AGRICULTURE

For agriculture and ecology, the principal deterrent to steadily increasing coal-burning activities is the sulfur present in coal. Biologically, sulfur is indispensable, being a major constituent of amino acids and thus protein and performs a variety of roles in metabolic intermediaries. Sulfur is required by plants at about one-tenth the level of nitrogen, principally as a component of the amino acids methionine, cysteine, and cystine. The sulfur requirement is acquired principally as sulfate from the soil, but also as other organic and inorganic sulfur compounds. Significant amounts of sulfur are obtained from assimilation of airborne sulfur compounds. Animals obtain most of their sulfur as protein constituents.

Sulfur dioxide is somewhat similar to carbon dioxide in physicochemical qualities. It is a heavy gas, readily absorbed by porous materials, and easily liquefied. It dissolves readily in water, producing a relatively weak acid, although stronger than carbonic and hydrofluoric acids. The acid produced, sulfurous acid H_2SO_3, is poorly characterized in terms of physical structure. SO_2 acts with water, both as an oxidizing and as a reducing agent. Biochemically, SO_2 may be either oxidized or reduced to a variety of sulfur-containing compounds. SO_2 is extremely soluble in fat solvents such as acetone, which indicates a lipophilic quality which aids in penetration of plant cuticle. An affinity for water as well as lipids makes the molecule readily mobile in plant tissue, having relatively easy access to most cell types. Plant material effectively scrubs SO_2 from the air (Amer. Soc. for Hort. Sci., 1970) up to certain concentrations. The geometry of the plant and successive plant contacts of a moving mass of air determines the scrubbing efficiency. Thus, successive rows of trees or other plants receive decreasing SO_2 dosages from a traveling air mass. Plant uptake of airborne SO_2 is primarily into the leaves via the stomates or "breathing pores" (Elkley and Ormrod, 1979; Kondo and Sugahara, 1978). Uptake via the inter stomatal epidermal cells via the cuticle is also significant, but not as damaging. SO_2 absorbed by the epidermal cells contributes significantly to increased foliar sulfur content with less damage than uptake via the stomates.

The degree of stomatal aperture is of such importance in SO_2 damage to leaves during episodes of high level fumigation ($>2600\mu g/m^3$) that in short exposures of 1 to 4 hrs. on susceptible plants, the pattern of acute injury coincides approximately with the pattern of leaf areas possessing functioning, open stomates. The youngest and oldest leaves frequently escape damage because stomates are immature and not yet functional in young leaves and closed because of senescence in old leaves. Individual plants that are suffering from water deficiency and therefore have closed stomates generally escape acute damage from high level, short duration exposure to SO_2. Environmental factors that favor increased stomatal aperture (Majernik and Mansfield, 1972) also enhance susceptibility to fumigation damage from SO_2 through the stomatal opening mechanism. Higher relative humidities (above 70-80%) increase susceptibility to acute SO_2 damage, whereas low relative humidity (below 30-40%) clearly aids in the protection of plant life exposed to atmospheric SO_2. Stomatal opening on many plant species corresponds to higher diurnal periods of enhanced photosynthetic activity. This happens for most plants during the morning hours on sunny days when temperatures are optimal for the plant species involved. Heavy SO_2 exposure at this time is generally much more damaging than at other times of the day. Plant nutritional status also clearly alters both SO_2 sensitivity and stomatal opening patterns (Leone and Brennan, 1972). Increased levels of nitrogen lower SO_2 susceptibility apparently by increasing sulfur nutritional needs for protein synthesis. High levels of fertilization with sulfates increase the damage from SO_2 air pollution, especially from long-term, low-level fumigation.

There is an extremely wide range in susceptibility-tolerance categorization for plant species in response to airborne SO_2 effects (Zimmerman and Hitchcock, 1956; National Academy of Sciences, 1978). Information on the relative tolerance of native and cultivated species can be used to advantage in land use planning for locating coal burning facilities and in selecting species for planting within the range of influence of SO_2 emissions. Tolerant species that effectively scrub SO_2 from the air may be used to minimize air pollution problems, especially if large acreages surrounding a facility could be planted to something like Euca- lyptus trees that grow very rapidly in the southeastern United States and can be harvested for pulp operations. The value of preventive planting would be determined by an evaluation of distribution patterns of SO_2 emissions into the atmosphere. If a source emitted SO_2 at 300 ft. elevation, the planting effects would not be effective locally, but would be if the SO_2 were moving at ground level.

Indicator plants are used in diagnosing plant damage caused by SO_2 air pollution (Oshima, 1974; Seidman et al., 1956; and van Raay, 1969). If symptoms on susceptible species are representative of SO_2 damage and the range of effects on other species correlates well with their toler- ances to SO_2, this correlation should lead to a further investigation into the cause of apparent SO_2 damage. Included should be air sample and meterology monitoring relative to potential emission sources of SO_2. Also, within the limitations of plant analysis and with due attention to other sources such as fertilizers and sprays, sulfur analysis of plant material may give supplemental indications of the cause (Linzon et al., 1979; Ratsch, 1974; and Sidhu and Singh, 1977). Plant analysis alone is not reliable in SO_2 damage evaluation.

The physiological, metabolic, and anatomical bases for SO_2 toler- ance-susceptibility have not been well explored (Cohen et al., 1973; Lamoreaux and Chaney, 1978; Priebe et al., 1978; and Silvius et al., 1976). As such information becomes available it will be of value in ex- plaining plant responses, in avoiding damaging situations, and in de- signing plant breeding programs to include considerations of resistance or tolerance to ubiquitous air pollutants. This information may be used in urban landscaping where higher levels of air pollution may be a con- tinuing problem. Therapeutic measures such as foliar applications of protective chemicals have been evaluated and found useful in protection against SO_2 and other atmospheric pollutants. Species with high gas exchange capacity are usually more susceptible to damage from SO_2. Spe-

cies with slow growth rates, low rates of photosynthesis, and low gas
exchange characteristics are much less susceptible to SO_2 damage.
Shade plants (those which grow best at low sunlight intensities and
very poorly at high light intensities) are adapted to low gas exchange,
low rates of photosynthesis, and are generally less subject to SO_2 injury.

A. SO_2 Fumigation

Controlled fumigation of plants with SO_2 provides information on
the qualitative and quantitative effects on various species (McLaughlin
et al., 1976; Nriagu, 1978; Zimmerman and Hitchcock, 1956; and National
Academy of Sciences, 1978). Cultivated and native plants may be categorized for their response to airborne SO_2. Such information is useful
in (1) diagnostic technique for plant damage of unknown etiology; (2)
land use planning, to suit the plant species to the ambient SO_2 problems
as well as the reverse, to suit the SO_2 air pollution to the susceptibility of the surrounding plant populations; (3) for tracing the path of
SO_2 plumes or assistance in establishing isopleths of SO_2 effects around
an isolated emission source; and (4) establishment of bioindicator techniques in SO_2 biological air quality evaluation.

Since we are dependent on information obtained from fumigation procedures and need to be critical of conclusions derived therefrom, it is
necessary to list fumigation techniques followed by comments on the weaknesses and strengths of each procedure. The questions to be applied relate to the reliability of fumigation environment control and the applicability of such findings to agriculture and ecosystems. Methods of
fumigating plants are:

1. Controlled environment plant growth chamber
2. Greenhouse fumigation compartments
3. Fumigation gun, directed airstream
4. Filtered, directed airstream in polluted area
5. Greenhouses erected in field in polluted area,
    comparing ambient with filtered air

6. Open-top greenhouse compartments providing normal
    light and precipitation using airflow of desired
    pollutant level to surround plants with a controlled
    atmosphere

7. A plastic film bubble to enclose a branch or part
    of a plant *in situ* providing the desired atmosphere
    by air exchanges

8. Wind tunnel with standard polluted air passing at a
    controlled velocity over plant subjects

9. Container-grown, automatically irrigated plant
    placement in selected locations to sample the
    atmosphere as bioindicators which are fumigated
    in situ

10. *In vitro* fumigation of leaves in reaction vessels
    of manometric apparatus, such as Warburg, to
    measure metabolic response.

The diversity of apparatuses and procedures demonstrated by the
above list indicates the complexities associated with reproducing fumigation effects and relating them in a valid manner to that occurring in
the field (Mudd and Kozlowski, 1975). Plant growth chambers offer the
best control of atmospheric and general environmental conditions and reproducibility of results but are deficient in direct application to
field conditions. Greenhouses are not directly related to field conditions but come closer in regard to light quality (sunlight instead of
artificial light) and in the sizes and stages of plant growth that may
be accommodated; plant environment control for reproducibility of results suffers. Methods 3, 4, 6, and 9 have the advantage of normalized
field conditions of light, temperature, humidity, and precipitation.
Greenhouses erected in a polluted area generally answer the question as
to the occurrence of atmospheric pollution but do not escape the greenhouse effect. Enclosing a branch for fumigation or providing filtered
air is convenient but has the drawback of the effects of this type of
enclosure. A wind tunnel is useful but has the drawback of limited num-

bers and sizes of plants that can be accommodated at one time as well as
the drawback of the scrubbing effect of plant material on the windward
side which reduces pollutant levels on the leeward side. Bioindicator
plants placed at increasing distances from a pollution source may give
a practical, rapid assay of the severity of the problem. In vitro fumigation of leaves in appropriate apparatuses provides information as to
the time sequence of metabolic responses to a pollutant but the ultimate
applicability to in vivo situations awaits further exploration.

B. SO_2 Effects on Plants

The effects of SO_2 on plants have been under investigation for over
a century, SO_2 being the oldest recognized air pollutant, long associated with coal-burning and industry (Nriagu, 1978; National Academy of
Sciences, 1978; Mudd and Kozlowski, 1975; and Meyer, 1977). Possible
responses of plants to atmospheric SO_2 include (1) beneficial effects
due to a supply of an indispensable nutrient, sulfur; (2) hidden or invisible injury reflecting a reduced growth rate and impaired plant metabolism; (3) chronic injury visible as foliar chlorosis, stunting, and
possibly impaired produce quality; and (4) acute injury causing the
death of plant tissues, especially foliage, resulting in a characteristic interveinal "scorch" of leaf laminae (necrotic tissue between leaf
veins). Injury from the standpoint of agricultural interests and society at large may differ from the physiologic classification of plant
damage (Environmental Research and Technology, Inc., 1978). Depending
on the intended use of the plant, damage may be (1) economic, (2) aesthetic, or (3) ecologic. All plant damage is not perceived as having
adverse effects since damage may be minor and go unnoticed or may involve
unwanted plants such as weeds. Economic damage relates to decreased
yield, impaired quality, and increased cost of production. Aesthetic
damage could involve, for example, forests not intended for harvest, and
landscape plantings. Ecologic effects are not completely separable from
the others but include an impaired ability of vegetation to furnish habi-
tats for wildlife, to control soil erosion, to preserve rainfall infiltration and retention and to perform not-readily apparent biological bal-

ance roles. Inroads by SO_2 air pollution in damaging native vegetation
are autocatalytic in increasing damage since they impair a natural
scrubbing system that filters out airborne SO_2. The effects of SO_2 in
the atmosphere are direct and indirect in managed and natural plant communities and ecosystems. They include altered populations, biological
balance changes, and changes in plant tolerance to biotic and abiotic
factors. Biological susceptibility is quite variable. It is the combined effect of planned use and biological susceptibility of plant populations at risk that determines potentials for adverse effects.

     The interpretation of SO_2 effects on plants is usually linked to
the degree of visible leaf damage, i.e., percent of leaf area destroyed
(Brisley et al., 1959; Spugel et al., 1977; Tingey et al., 1971; Benedict, 1973; and Heggestad, 1968). Although this is a tangible and usually accepted criterion for economic damage estimation it does not take
into account certain types of hidden, chronic injury. While most emphasis should be on visible injury, some allowances need to be made for
invisible injury. Many changes have been found in metabolic and physiologic processes due to SO_2 but have not been directly applicable to
plant damage estimates. Effects on plants and microorganisms are of
ecological significance due to changes in vigor, structure and composition of the flora. SO_2 and SO_3 cause changes in soil pH nutrient providing capacity and microorganism population which, in turn, affect
higher plant population stability and resistance to stress (Irving and
Miller, 1977). These effects will be mentioned in greater detail under
the acid rain discussion.

C. SO_2 Dose-Response Relationship for Plants

     The effects of a given concentration of SO_2 for a specific time
period is the dose-response relation. Atmospheres may be controlled
with standardized fumigation or, when warranted, ambient atmospheric
SO_2 levels may be closely monitored and the effects under each individual condition documented. A compilation of the effects of varied concentration x duration may then be studied for generalizations as to relations between the two features of exposure (Temple, 1972; Benedict,

1973; Jones et al., 1973; and Heggestad, 1968). Controlled fumigation
has the greatest merit for accurate description and reproducibility of
exposure conditions while monitored ambient fumigation has merit for
direct applicability to field SO_2 emissions. Also, results of controlled and monitored ambient fumigation episodes may be compared. Controlled fumigation is usually a steady state process, whereas monitored
ambient fumigation fluctuates by gradient changes from level to level
and also is characterized by peaks and valleys of exposure of poorly
defined duration. It is clearly evident (Table 1) that short duration
but high concentration exposure is much more damaging than low concentrations of SO_2 over extended exposure periods. A basic principle encountered in the peculiar effects of SO_2 is that detoxification of SO_2
may prevent acute damage if the atmospheric level is below a critical
level wherein SO_2 is metabolized into less toxic compounds before acute
damage occurs. If atmospheric SO_2 is above the critical level for the
environmental conditions, acute damage will result. Differences in susceptibility of various plant species are illustrated by the data in
Table 1.

Table 2 lists concentrations at indicated durations of exposure
for which no visible (acute) foliar damage was reported. Some of these
values were set in the reported research as thresholds for injury levels, above which visible injury did occur. These data may be compared
with those in Table 1 to support contentions that dose-response levels
for plants are not first-class criteria for prediction of damage. Alfalfa, for example, is reported as receiving injury (Table 1) at concentrations lower than some cases (Table 2) where injury did not occur.
This is attributable to differences in environmental conditions. The
data from Tables 1 and 2 indicate that plant response from SO_2 is not
accurately predictable, that plant dose-response predictions are more
relative than absolute. Relative comparisons have special meaning as,
for example, in comparing an assortment of plant species for SO_2 susceptibility in the same experiment under the same environment. Dose-
response data for one plant species for reliable predictions would have
to be relative, namely comparing the combinations of concentration and

Table 1: *Concentrations of SO_2 at which Foliar Injury Occurred
in Experimental Exposures of Plants*

Duration of Exposure	SO_2 Concentration, mg/m^3	Plant
1.5 min.	2,620	Tomato, tobacco
8 min.	165	Tomato
10 min.	157	Scotch pine
12 min.	22.3	Alfalfa
1 hr.	0.131	Eastern white pine
	0.655	Petunia
	2.10	Wheat, meadow fescue
	2.57	Swiss chard, Chinese cabbage
	2.83	Eggplant, endive
	10.5	Tomato, salvia, coleus
2 hr.	0.655	Eastern white pine, red pine, Scotch pine
	2.57	Radish
	2.75	Turnip, cos lettuce
	10.5	Ginkgo
3 hr.	1.73	Oats, rye
	2.16	Sweet pea
	3.28	Alfalfa
	4.98	Blueberry
	10.48	Coleus, salvia
4 hr.	1.31	Tomato, pinto bean
	2.02	Cucumber
	2.62	Tobacco
	3.46	Alfalfa
5 hr.	1.73	Alfalfa, tomato, oat, barley, rye, sweet clover
	2.57	Salvia
	2.62	Castor bean, salvia, pepper
	5.24	Tobacco, tomato, sugar beet
6 hr.	0.066	Eastern white pine
	1.31	Buckwheat, anemone, ixia
	1.83	Rose
	1.97	Turfgrass species, Crocus species, sparaxis
8 hr.	1.57	Barley
	2.15	Alfalfa
24 hr.	0.786	Begonia, dahlia, aster, zinnia, cucumber, violet, spiderwort
	1.31	Barley
	13.1	Lichens

Source: National Academy of Sciences, 1978.

Table 2: *Concentrations of SO_2 at which an Absence of Foliar Injury or "Threshold for Injury" was Reported in Plants*

Duration of Exposure	SO_2 Concentration, mg/m^3	Plant
0.25 hr.	21.000	Rye
0.50 hr.	1.310	Soybean
	13.000	Rye
1 hr.	2.148	Sweet clover
	7.000	Rye
	26.206	Winter Rape
2 hr.	1.310	Tobacco
	2.100	Crimson clover
	2.148	Sweet clover
	20.96	Pin oak
3 hr.	2.800	Oat
	3.300	Rye
	3.668	Sycamore maple
	12.838	Blueberry, high bush
4 hr.	1.050	Tomato, crimson clover
	1.153	Alfalfa
	1.310	Tobacco
	1.467	Cucumber
5 hr.	3.013	Alfalfa
	11.266	Lily, orchard species
6 hr.	2.000	Rye
	7.86	Pin oak
8 hr.	1.310	Jerusalem cherry, tulip, ixora, sorghum
	2.279	Alfalfa
	2.620	Tobacco
	5.24	Ginkgo, pin oak
12 hr.	0.524	Tobacco
	2.620	Broadbean
1 day	0.200	Bean
	0.786	Azalea, periwinkle, Swiss chard, pepper
	2.044	Barley
4 days	0.200	Bushbean
9 days	1.500	Tobacco
10 days	1.965	Pin oak
13 days	1.500	Corn
15 days	1.00	Sunflower
20 days	1.00	Radish
21 days	1.310	Blueberry, petunia, pepper, coleus
77 days	0.208	Perennia ryegrass

Source: National Academy of Sciences, 1978.

duration of exposure under the same environmental conditions. For each
there may be a different response to a given dose. This does not represent an insurmountable obstacle but reinforces the concept of relativity
and the need for comparative observations between species and between environments. Guidance is to be obtained from data in Tables 1 and 2 on
the significance of concentration x duration or dosage-expected response,
but only in broad terms. The variability in response relative to environment may be partially attributed to changes in the physiological basis of the response as described in Table 3

Table 3: *Physiologic Effects of SO_2 on Plants*

Concentration mg/m^3	Duration of Exposure	Effect
0.070	10 min.	Decreased stomatal resistance, broadbean
0.131	5 min.	Decreased stomatal resistance, broadbean, corn
0.524	10 days	Decreased CO_2 fixation, alfalfa
1.31	30 min.	Decreased[14] CO_2-derived photosynthate, barley
	8 hr.	Decreased regeneration of leaves, moss
1.78	3 hr.	Threshold for effect on potassium-efflux, lichen
2.62	1 hr.	Decreased CO_2 fixation, European larch
	4 hr.	Decreased regeneration of leaves, moss
	5 hr.	Decreased CO_2 fixation, Norway spruce
	24 hr.	Breakdown of chlorophyll, bryophytes
3.80	30 min.	Decreased CO_2 fixation, crimson clover
7.074	6 min.	Decreased CO_2 fixation, Scotch pine
8.00	15 min.	Inhibition of net CO_2 assimilation, mosses and lichens
13.1	5 hr.	Increased O_2 uptake, bryophytes

Source: National Academy of Sciences, 1978

Comparative fumigation of plant species reveals differences and
similarities among species. This is impor tant to the diagnosis of specific plant problems, which potentially include airborne SO_2 as the
principal etiological agent. Plant species exposed simultaneously to

the same SO_2 concentration-duration exposures with comparable environments may be compared for their response to SO_2 fumigation. Compilation of data from various sources permits comparisons, but they are of
less definitive nature since fumigation conditions affect the results
to a highly significant degree. Greenhouse fumigations have been frequently used to categorize plant species into 3 or more susceptibility-
resistance classes for SO_2 fumigation effects. Examples of such classifications representing plants of interest in the southeastern United
States are shown in Tables 4 and 5.

D. Genetic Adaptations to Airborne SO_2

Atmospheric SO_2 has an effect on the genetic makeup of native populations of perennial ryegrass (Horsman et al., 1978) and on Geranium
carolinianum (Taylor, 1978-a). The selection pressure of SO_2 air pollution results in the evident changes in native populations in affected
geographic areas such that the populations from polluted areas are statistically superior in SO_2 resistance over populations from nearby unpolluted geographic areas. There is a precedent for this in the genetic
response of native populations in that increased tolerance occurs to
pollution with heavy metal residues. The demonstrated evolution of increased native resistance to an environment-degrading factor has wide
implications for ecosystems in that it holds hope that nature will adapt
to man's changes in the environmental influences impinging on native ecosystems and biota.

SO_2 has a roughly predictable rate of deposition over various ter-
rains and various vegetation covers (Table 6). Acid soils do not accumulate as much SO_2 as alkaline soils. SO_2 trapping obviously is favored
by the formation of alkaline sulfites (Lockyer et al., 1978). The
scrubbing action of vegetation is described in a very approximate manner in Table 6. Research is needed to characterize plant retention of
SO_2. Much of the information available is expressed in terms of plant
response and sulfur content (Linzon et al., 1979).

Table 4: *Comparative Susceptibility of Plant Species to SO_2 Fumigation*

Susceptible	Intermediate	Resistant
Chicory	Azalea	Jerusalem cherry
Spanish needles	Honeysuckle	Gladiolus
Nightshade	Plantain	Tulip
Celery	Grape, wold	Sorghum
Tomato	Rose	Ixora
Cucumber	Galinsoga	Stevia
Pumpkin	Lamb's-quarters	Corn, field, and sweet
Pigweed	Taxus	Deutzia
Alfalfa	Bean	Lily
Clover, sweet		Iris
Coleus		Apple
Geranium		Cotton
Eggplant		
Blueberry		
Hibiscus		

Source: Zimmerman and Hitchcock, 1956

Table 5: *Comparative Susceptibility of Plant Species to SO_2 Fumigation*

Susceptible	Intermediate	Resistant
Eucalyptus cineria	Bald cypress	Sweetcorn
Eucalyptus amplifolia	Eucalyptus viminalis	Turkey oak
Eucalyptus camaldulensis	Soybean	Eucalyptus
Bushbean	Gloxinia	robusta
Aster	Petunia	Box elder
Geranium	Rosewood	Live oak
Zinnia	Squash	Tobacco
	Tomato	Red mangrove
	Pepper	Bahiagrass
	Cat's claw	Sorghum
	Buttonwood	Kalenchoe
	Slash pine	
	Blackeyed pea	
	Begonia	

Source: Woltz, 1979

E. Acid Rain

Acid rain and acid mist (H_2SO_4 + HNO_3) have several types of effects on vegetation, soil, bodies of water, and ecological balance (Nriagu, 1978; National Academy of Sciences, 1978). These effects have not yet become as pronounced in the southeast as they have in the northeastern United States, Canada, and Northern Europe (Likens, 1976). The problems of acid rain are associated with the long-range transport of sulfur dioxide and acid sulfates and attributable to the practice of exporting SO_2 from tall stacks instead of trapping the gas locally. Rainwater in equalibrium with atmospheric CO_2 has a pH of 5.6 due to CO_2 acidification. SO_x + NO_x may lower the pH to 4.0 or occasionally less which is potentially quite harmful, biologically. In 1972–73, the principal rainwater pH range for the southeast was 5.0 to 4.4, whereas in 1955–56, the range was 5.6 to 4.7 (Likens, 1976). The effects of acid rain on vegetation include spotting of upper leaf surfaces by deposited fine droplets of acid mist, especially on beet, Swiss chard, and alfalfa; decreased productivity of forests affected by acidified rain; reduction in the incidence of certain plant diseases; inhibition of nitrogen fixation; erosion of leaf cuticle, and increased disease susceptibility after acid rain damage to leaves; and leaching of nutrient cations from leaves. Soil acidification by acid rain results in impairment in soil microbiology, nitrogen fixation, and leaching losses of soil nutrient cations. Toxicity and enhanced uptake of undesirable elements also may occur, as for example, enrichment of plant produce with cadmium. It was suggested that acid rain altered cellwall and membrane structures, making plants more susceptible to other pollutants and to plant parasites. If standards for sulfates are not prescribed and enforced, forests in affected areas will be degraded. The combined effects of acid rain in cuticular (epidermis outer layer) erosion, foliar lesions, leaching of foliar and soil-contained nutrients, and disturbances in host-parasite and pollutant relationships will threaten the profitability of agriculture and degrade the native ecosystems.

The acidification of lakes and streams by acid rain and SO_2 effects may cause severe stress on the aquatic ecosystems if the waters are not

Table 6:   *Sulfur Dioxide Deposition Velocities Over Soil
            And Vegetation in Humid Areas, Such as South-
            eastern United States*

| | Deposition velocity[a] (V_d), cm/s | |
	Range	Typical
Soil, pH > 7, dry	0.3 - 1.0	0.8
Soil, pH > 7, wet	0.3 - 1.0	0.8
Soil, pH ∿ 4, dry	0.1 - 0.5	0.4
Soil, pH ∿ 4, wet	0.1 - 0.8	0.6
Grass, 0.1 m height	0.1 - 0.8	0.5
Crops, 1.0 m height	0.2 - 1.5	0.7
Forest, 10.0 m height	0.2 - 2.0	Uncertain

Source: National Academy of Sciences, 1978

buffered by alkaline geological source materials such as limestone. In
the future, efforts will be made to compensate for acidification by
treatment with lime and other alkaline substances. This will improve
the situation but may not solve the problem of excessive levels of dissolved sulfur compounds. The specific effects of acidification of natural waters are many: acidified reservoirs will have an increase in metal
concentrations that can exceed public health limits for drinking water;
acid lakes will not support fish populations and will become visually un-
attractive due to dense growth of filamentous algae and sphagnum mosses;
and unique communities of aquatic populations as well as individual species may be in danger of extinction.

F. Trace Elements

    Trace elements from coal-burning find their way into the biosphere
in various ways (Phung et al., 1979; Klein et al., 1975; and Gladney
et al., 1978). The amounts that impact agriculture and ecology are not
of a large magnitude nor are the elements usually distributed over large
geographical areas. The pathways of 37 trace elements have been traced
through a coal-fired power plant (Klein et al., 1975). The present discussion will emphasize three trace elements as some of the agriculturally

and ecologically important by-products of coal-burning, to illustrate the
present concern for the distributional fate and action of coal-derived
trace elements.

Cadmium is a common contaminant of coal but the amounts are generally low. Cadmium that finds its way into the soil in which plants are
growing can be detrimental to plant growth (Woltz and Chambliss, 1979),
damaging soybean, bushbean, tomato, and pepper at levels of 3 to 50 mg/kg
of soil. Soybean leaves accumulated 0.5 to 9 ppm Cd while bahiagrass accumulated much more, 3 to 100 ppm Cd, both in terms of dry weight of
leaves. Soil-derived cadmium represents a food-chain hazard (although
poorly defined quantitatively) because of its common association with hy-
pertension in animals (Goyer and Mehlman, 1977).

Boron is evolved in coal-burning (Gladney et al., 1978) and may affect crop plants adversely or beneficially, depending on the rate of deposition in balance with fertilizer applications of boron, crop removal,
and loss of boron by leaching. Tolerance and requirements by crop plants
is quite variable according to plant species. The lower level of tolerance of total boron supply is about 1.2 lbs. and the upper level is about
4.8 lbs. per acre per year distributed in increments over the year, assuming leaching losses prevent excessive carry-over from year to year.

Fluoride is evolved from coal in varying amounts according to burning process and coal source but was not cited as a significant coal contaminant quantitatively (Tourangeau et al., 1977). Airborne fluoride has
characteristic effects on plant material (Woltz, 1964; Woltz and Leonard,
1964; Woltz and Waters, 1978-a and 1978-b) but the amounts from coal are
not likely to be an air pollution problem and may not be a soil or water
problem (Woltz et al., 1971). The balance of the element in soils and
the biosphere warrants long-term observation.

III. COAL ASH UTILIZATION ON AGRICULTURE AND ON FOREST SOILS

The burning of coal in power plants results in a waste residue composed of the inorganic mineral constituents of the coal and any uncom-
busted organic matter. The inorganic mineral portion, which is called
ash, makes up from 3 to 30% of the coal. Upon burning, this ash is dis-

tributed into two fractions, <u>bottom ash</u> (collected from the bottom of the
boiler unit) and <u>fly ash</u> (collected by air pollution control equipment
through which the stack gases pass). A third residue, vapors, is that
part of the coal volatilized in the furnace. Most of the vapors escape
to the atmosphere in the stack gas.

Fly ash is the predominate waste material and may compose between 10
and 85% of the coal ash residue. Fly ash occurs as spherical particles
usually ranging in diameter from 0.5 to 100 microns (Davison <u>et al</u>., 1974).
The chemical and physical characteristics of fly ash depend on the composition of the parent coal, conditions during burning, kind and efficiency
of emission control devices, storage and handling of the ash and climate.
Ashes from anthracite, bituminous and lignite coals may have considerable
variations in their chemical make-up.

The mineralogical composition of fly ash indicates that between 70
and 90% of the particles are glassy spheres with the remainder consisting
of the following minerals: quartz (SiO_2), mullite ($3Al_2O_3 \cdot 2SiO_2$), hema-
tite (Fe_2O_3) and magnetite (Fe_3O_4) and small amounts of unburned carbon
(Hodgson and Holliday, 1966). In addition to these minerals, calcite
($CaCO_3$), tourmaline, borax, rhodium boride, boron arsenite and boron
phosphate have been identified (Gal <u>et al</u>., 1978). Fly ash from eastern
U.S. coals consistently showed a larger percentage of amorphous material
than that of the western U.S. coals (Fisher <u>et al</u>., 1976).

Chemically, almost all naturally existing elements can be found in
fly ash (Klein <u>et al</u>., 1975; Kaakinen <u>et al</u>., 1975). The major matrix
elements in fly ash are Si, Al, and Fe together with a small percentage
of Ca, K, Na, and Ti (Rees and Sidrak, 1956; Natusch <u>et al</u>., 1975). The
elemental compositions of coal, fly ash, and soil are compared in Table 7.
Although elemental compositions of fly ashes may vary widely, they usually
contain higher concentrations of essential plant nutrients, except N,
than do common cropland soils.

Bituminous and subbituminous coals are located mostly in the eastern
and mideastern regions of the U.S., whereas the lignites are predominant
in the western U.S. The former are characteristically high in S content
and consequently produce ashes generally low in pH. On the other hand,

Table 7: *Range in Amounts of Trace Elements Present in Coal Ash and Soil (ppm)*

Element	Anthracites[1]		High volatile bituminous[1]		Soils[2]	
	Max.	Min.	Max.	Min.	Max.	Min.
Ag	1	0	3	1	---	----
B	130	63	2800	90	100	2
Ba	1340	540	4660	210	100	3000
Be	11	6	60	4	---	----
Co	165	10	305	12	1	40
Cr	395	210	315	74	5	3000
Cu	540	96	770	30	2	100
Ga	71	30	98	17	15	70
Ge	20	20	205	20	---	----
La	220	115	270	29	30	----
Mn	365	58	700	31	100	4000
Ni	320	125	610	45	10	1000
Pb	120	41	1500	32	2	100
Sc	82	50	78	7	10	25
Sn	4200	19	825	10	---	----
Sr	340	80	9600	170	500	4000
V	310	210	840	60	50	1000
Y	120	70	285	29	---	----
Yb	12	5	15	3	---	----
Zn	350	155	1200	50	10	300
Zr	1200	320	1450	115	---	----

[1]From Torrey (1978b)

[2]From Allaway (1968) and Lisk (1972)

lignites have lower S but higher Ca and Mg contents and produce ashes characteristically high in pH (Furr et al., 1977; Bern, 1976; Natusch et al., 1975). The micro-elements content of ashes from western states have a higher B content but are lower in other trace elements (As, Cd, Co, Cr, Pb, Sb and Zn) when compared with ashes of eastern and midwestern coal deposits (Abernathy, 1969; Natusch et al., 1975).

The key to management of industrial waste, such as coal ash, is to recycle it in a beneficial manner with a minimum impact on the environ-

ment. It is estimated that less than 10% of the annual production of fly
ash is utilized by recycling (Torrey, 1978a). The major use of this
waste material has been in road construction, building materials, etc.
The disposal of fly ash is becoming a potential problem that may be
solved to some degree with applications onto selected agricultural and
forest soils.

The chemical nature of a waste material such as coal ash must be ascertained before the soil can be effectively utilized as a disposal medium. The chemical and physical alterations resulting from the addition of
coal ash to soils and the subsequent effect on vegetative growth have
been investigated recently on a small scale with greenhouse and field
studies. However, caution should be used when limited data on coal waste
ash disposal onto specific soils are extrapolated to make broad recom-
mendations as to the environmental consequences of using the soil as a
disposal site. The following discussions are a review of existing data
on the effects on plants and soils of fly ash from coal combustion.

A. Plant Effects

There are sixteen (16) elements that have been shown to be essential
for plant growth. Thirteen (13) of these elements must be absorbed from
the soil solution in contact with plant roots. In most agricultureal
soils all these elements are present, but usually at very low levels;
hence there is need for addition of plant nutrients in the form of chemical fertilizers. The ash from coal combustion also contains all or most
of the essential plant nutrients and if this material could be utilized
as a source of plant nutrients, then there exists a novel solution to the
waste problem of coal-fired power plants.

Some apparent problems accompany application of coal ash to soils.
These include the possibility of introduction of undesirable or toxic
elements into the soil matrix that subsequently may be taken up by
plants. Addition of coal ash can change the physical and chemical nature of soils and spoil the soil as a plant growth medium. There also
exists the possibility that water percolating through ash-amended soils
may lower the ground water quality to an unacceptable level. To deter-

mine if the addition of coal ash will have a beneficial or detrimental
effect on a particular soil requires a careful evaluation of the chemical and physical properties of both the coal ash and the soil, as well
as the subsequent mixture.

Preliminary studies have shown that many chemical constituents in
fly ashes may be beneficial to plant growth and can improve agronomic
properties of the soil (Chang et al., 1977). Effects on plants are induced primarily by changes in the chemical nature of the soil solution
upon addition of fly ash. For this reason, the low-pH fly ash of the
eastern states would have a different effect from that of the fly ash
with an alkaline pH range. Weathering fly ash in storage lagoons will
affect the chemical characteristics of the ash and may render the material less hazardous to plants when applied to soils. Interactions between plants and fly ash-amended soils are further complicated by varying edaphic factors, plant species factors, and many others.

Concentration of the essential plant nutrients K, S, Mo, B and Zn
has been shown to increase when fly ash containing these elements was
added to the soil (Martens et al., 1970; Mulford and Martens, 1971;
Schnappinger et al.,1975; and Elseewi et al., 1978b). However, the concentration of the non-essential trace elements Al, As, Ba, Cs, Se, Sr,
Rb, W and V have also been shown to increase with the addition of fly
ash to the soil (Connor et al., 1976; Bradford et al., 1978; Adriano
et al., 1978b).

There have been reports of increased plant dry matter yields where
fly ash has been added to soils deficient in macro- or micro-nutrients.
Greenhouse studies using unweathered western U.S. fly ash at rates rang-
up to 8% (by weight) produced higher yields of several crops (Page et
al., 1979a). The yield increases were attributed to increased plant
availability of S from fly ash, while reductions in lettuce (Lactuca
sativa) yields were due to high salinity and excessive B (Elseewi et
al., 1978a, b). Similarly, increases in alfalfa (Medicago sativa)
yields were attributed to an alleviation of B deficiency by field application of fly ash (Plank and Martens, 1974). Fly ash applied on acidic
stripmine soils in several states increased yields of several crops ap-

parently caused by increased plant nutrient availability (Fail and Wochok, 1976; Kovacic and Hardy, 1972; Capp and Engle, 1967).

The micro-nutrients Mn, Zn, Cu and Fe from fly ash may be available
to plants depending on the pH of the fly ash amended soil. The increases
in soil pH caused by alkaline fly ash application are likely to induce
deficiencies of these nutrients. However, Zn and other micronutrient
deficiencies can be corrected by incorporation into the soil of acidic
fly ash containing these elements. Zinc in acidic fly ash has been
shown to be highly available to plants (Schnappinger et al., 1975). The
application of high rates of alkaline fly ash may induce a Zn, Mn, or Cu
deficiency for some plants and the addition of acidic fly ash to alkaline
soils may alleviate a deficiency for some plants.

The availability of S, Mo, B and Se to plants has been shown to increase with the addition of fly ash. Elseewi et al. (1978b) concluded
that yield increases of alfalfa and bermuda grass were due to a correction of sulfur deficiency in soils and also demonstrated that fly ash-
derived S was equally as effective as fertilizer-derived S.

Doran and Martens (1972) showed equal plant availability of fly ash-
Mo and $Na_2MoO_4 \cdot 1H_2O$ and suggested that high fly ash imputs to soil could
increase Mo in forage crops to levels potentially toxic to animals feeding on the forage.

Boron in fly ash is also readily available to plants and numerous
investigators (Holliday et al., 1958; Cope, 1962; Hodgson and Townsend,
1973; Townsend and Gillham, 1975) considered B as a major limiting factor for successful cropland utilization of ashes, especially when ashes
are not fully weathered. Since B compounds in soils are quite water-
soluble the weathering of fly ash, by allowing adequate drainage, should
reduce detrimental effects of B to plants.

Selenium concentrations in plant tissues have been shown to increase
with fly ash addition to soil. Its increase was shown to be related to
the amount of fly ash applied and the Se content of the ash (Furr et al.,
1976; Furr et al., 1977; Straughan et al., 1978). Although many trace
elements in fly ash are considered potentially detrimental to plants,
only B has been associated with any significant reductions in crop pro-

duction. Cadminum, Co, Cr, Cu, F, Ni, Tl and V in fly ash have not been shown to be deleterious to plants. Molybdenum and Se are essentially non-toxic to plants at input levels expected from fly ash but their un-controlled increase in plant tissues would be a potential hazard. Al-though Be, Cd, F, Hg, Ni and Tl are considered potentially hazardous to animals, their significant uptake from fly ash by plants has not been demonstrated.

B. Soil Effects

The response from plants indicates the physical and/or chemical properties of the soil must have been altered by the addition of fly ash. Essentially all the physical properties of soil included in Table 8 are affected by fly ash application. For most soils, fly ash input would reduce the bulk density of the soil mixture. Fly ash by itself did not appear to be effective in retaining water. However, when added into soils at as little as 8% by weight, it significantly increased the water-holding capacity of the soil mixture (Chang et al., 1977). Fly ash also reduced the modulus of rupture (cohesiveness of soil particles) in all soils tested.

Perhaps the most limiting factors in fly ash utilization on land are unfavorable changes in the chemistry of the soil such as increases in pH, salinity, and addition of certain toxic elements. Lignite ashes, which are characteristically high in oxides of Ca and Mg, may increase the pH when added to the soil.

Laboratory studies showed that an alkaline fly ash was chemically equivalent to approximately 20% of reagent-grade $CaCO_3$ in reducing soil acidity and supplying plant Ca needs (Phung et al., 1978). Other studies have shown an increase in the pH of soils of the eastern U.S. when alka-line fly ash was applied (Plank et al., 1975).

Soil salinity increased five- to six-fold with application of 8% unweathered fly ash over a short period of time (Page et al., 1979). Similar observations were also reported by other investigators (Mulford and Martens, 1971; Elseewi et al., 1978a; Phung et al., 1978; Adriano et al., 1978b). However, weathering of fly ash before application should

Table 8: *Potential and Observed Effects of Fly Ash on Physical, Chemical, and Microbiological Properties of Soils Compared with that of a Non-Treated Typical Agricultural Soil*[1]

Soil properties	Typical agricultural soil	Soil treated with	
		weathered ash	unweathered ash
Bulk density	1.3 (average)	Lower	Lower
Aeration	High	Higher	Higher
Water holding capacity	High	Higher	Higher
Modulus of rupture	High	Lower	Lower
Wind erosion	Resistant	More susceptible	More susceptible
Water erosion	Resistant	More susceptible	More susceptible
Nutrient content	All nutrients present	Very low N; othrs present	Very low N; others present
pH	6.0 - 7.5	< 6.0 to ≈ 8.0	< 6.0 to ≈ 12.0
Cation exchange capacity	Medium to high	Lower	Lower
Toxic salts	None	None	B and soluble salts of Ca, Mg, Na, and K
Salinity	Low	Moderate	High but diminished after 2-3 years
Temperature	Adequate	Higher	Higher
Microbial activity	High	No effect	Initially low

[1]From Adriano et al., 1979a

significantly reduce the impact of soil salinity. Lagooning, stockpiling, and leaching considerably reduced the soluble salt and B contents of fly ash (Townsend and Hodgson, 1973; Townsend and Gillham, 1975). Under normal storage conditions, the complete stabilization would take several years (Plank and Martens, 1974). Of the various soil impacts of fly ash—high B, salinity and certain soluble salts are often linked with yield reductions.

C. Reclamation

It has been estimated that a million hectares of U.S. land remain as virtual wastelands due to damages resulting from stripping operations in coal mining (Fail and Wochok, 1977). The spoil areas are acidic and infertile and do not support good vegetation which subjects them to severe erosion. Fly ash has been demonstrated effective in reclaiming these areas (Capp and Gillmore, 1973; Fail and Wochok, 1977).

In reclaiming spoil areas, the quantities of fly ash applied usually exceed the amounts applied to croplands. A stabilization period of about 1 year was required for uniform plant growth on revegetated sites where unweathered fly ashes were used. Fly ash added to soils not only chemically stabilized but also increased the water-holding capacity of the soils affected.

D. Potential Utilization

The chemical and physical properties of coal ash have been characterized and show promise as a material that can be utilized in several different areas (Torrey, 1978a). Perhaps the area of agriculture and forestry are potentially the most resourceful and ideal from the standpoint of recycling. The soils of the southeastern United States have both chemical and physical properties that, in their natural state, reduce productivity; however, the use of coal ash in some cases, could be used to correct or alleviate the low productivity.

The use of coal ash alone or in combinations with other waste such as sewage sludge, mine tailings, dredge spoil material, peat or most any waste material of biological origin is not yet tested and deserves early evaluation. From a chemical and physical analysis of the waste materials, a compatibility could be worked out and tested. Soils could be selected, based on their properties, that make ideal disposal sites and at the same time are made more productive for forestry, agronomic, or ornamental crops. In utilizing the soil as a disposal site for coal ash, a consideration must be given to the long-term effects on the environment and on food and fiber products.

References

Abernathy, R. F., Spectrochemical analysis of coal ash for trace elements, Investigations 7281, U.S. Dept. of Interior, Washington, D.C., 1969.

Adriano, D.C., A. L. Page, A. A. Elseewi, A. C. Chang, and I. Straughan, Utilization and disposal of fly ash and other coal residues in terrestrial ecosystem: a review, J. Environ. Quality (in press), 1979a.

Adriano, D. C., T. A. Woodford, and T. G. Ciravolo, Growth and elemental composition of corn and bean seedlings as influenced by soil application of coal ash, J. Environ. Quality, 7, 416-421, 1978b.

Allaway, W. H., Agronomic controls over the environmental cycling of trace elements, Adv. Agron., 20, 235-274, 1968.

American Society for Hort. Science, Pollutant impact on horticulture and man, Hort, Science, 5, 4, 1970.

Benedict, H. M., Atmospheric sulfur dioxide - its effects on vegetation, Northwest Workshop Air Pollution and How it Affects Plants, 8-21, 1973.

Bern, J., Residues from power generation: processing, recycling, and dispersal. In Land Application of Waste Materials, Soil Conserv. Soc. Amer. Ankeny, Iowa, 226-248, 1976.

Bradford, G. R., A. L. Page, I. R. Straughan and H. T. Phung, A study of the deposition of fly ash on desert soils and vegetation adjacent to a coal-fired generating station. In D. C. Adriano and I. L. Brisbin (ed.) Environmental Chemistry and Cycling Processes, CONF-760429, 383-393, U.S. Dept. of Commerce, Springfield, Va., 1978.

Brisley, H. R., C.· R. Davis, and J. A. Booth, Sulfur dioxide fumigation of cotton with special reference to its effect on yield, Agronomy Journal, 1959.

Capp, J. P. and D. W. Gillmore, Soil-making potential of power plant fly ash in mined-land reclamation. In Proceedings Research and Applied Technology Symposium on Mined-Land Reclamation, Pittsburgh, 1973.

Capp, J. P., and C. F. Engle, Fly ash in agriculture, Bur. Mines Info. Circular 8348, 210-220, U.S. Dept. Interior, 1967.

Chang, A. C., L. J. Lund, A. L. Page, and J. E. Warneke, Physical properties of fly ash-amended soils, J. Environ. Quality, 6, 257-270, 1977.

Cohen, H. J., R. T. Drew, J. L. Johnson, and K. V. Rajagopalan, Molecular basis of the biological function of Molybdenum. The relationship between sulfite oxidase and the acute toxicity of bisulfite and SO$_2$, Proc. Nat. Acad. Sci., 70, 3655-3659, 1973.

Conner, J. J., J. R. Keith and B. M. Anderson, Trace-metal variation in soils and sagebrush in the Powder River Basin, Wyoming and Montana, J. Res. U. S. Geol. Survey, 4, 49-59, 1976.

Cope, F., The development of soil from an industrial waste ash, Trans. Comm. IV, V. Intl. Soc. Soil Sci. (Palmerston, New Zealand), 859-863, 1962.

Davison, R. L., D. F. S. Natusch, J. R. Wallace, and C. A. Evans, Jr., Trace elements in fly ash: dependence of concentration on particle size, Environ. Sci. Technol. 8, 1107-1113, 1974.

Doran, J. W., and D. C. Martens, Molybdenum availability as influenced by application of fly ash to soil. J. Environm. Quality, 1, 186-189, 1972.

Elkley, T., and D. P. Ormrod, Leaf diffusion resistance responses of three petunia cultivars to ozone and/or sulfur dioxide, J. of Air Pollution Control Assoc., 29, 622-625, 1979.

Elseewi, A. A., F. T. Bingham, and A. L. Page, Availability of sulfur in fly ash to plants, J. Environ. Quality, 7, 69-73, 1978b.

Environmental Research & Technology, Inc., Environmental effects, Florida sulfur oxides study, Environmental Research & Technology, Inc., Atlanta, Ga., 1978.

Fail, J. L., Jr., and Z. S. Wochok, Soybean growth on fly ash-amended strip mine spoils, Plant Soil, 48, 472-484, 1977.

Fisher, G. L., D. P. Y. Chang, and M. Brummer, Gly ash collected from electrostatic precipitators: microcrystalline structures and the mystery of the spheres, Science, 192, 553-555, 1976.

Furr, A. K., W. C. Kelly, C. A. Bache, W. H. Gutenmann, and D. J. Lisk, Multielement uptake by vegetables and millet grown in pots on fly ash amended soil, J. Agric. Food Chem., 24, 885-888, 1976.

Furr, A. K., T. F. Parkinson, R. A. Hinrichs, D. R. Van Campen, C. A. Bache, W. H. Gutenmann, L. E. St. John, I. S. Pakkala, and D. J. Lisk, National survey of elements and radioactivity in fly ashes: absorption of elements by cabbage grown in fly ash-soil mixtures, Environ. Sci. Technol., 11, 1194-1201, 1977.

Gal, M., A. L. Page, and I. Straughan, Mineralogical composition of clay fractionated from coal and fly ash, Agron. Abstract, 25, 1978.

Gladney, E. S., L. E. Wangen, D. B. Curtis, and E. T. Jurney, Observations on boron release from coal-fired power plants, Environmental Science & Technology, 12, 1084-1085, 1978.

Goyer, R. A., and M. A. Mehlman, Toxicology of Trace Elements, 303 pp., Hemisphere Publ. Corp., Washington, 1977.

Heggestad, H. E., K. L. Tuthill, and R. N. Stewart, Differences among poinsettias in tolerance to sulfur dioxide, Hort Science, 8, 337-338, 1979.

Hodgson, D. R., and R. Holliday, The agronomic properties of pulverized fuel ash, Chem. Inc. (May issue), 785-790, 1966.

Hodgson, D. R., and W. N. Townsend, the amelioration and revegetation of pulverized fuel ash, 247-271. In R. J. Hutnik and G. Davis (eds.) Ecology and Reclamation of Devastated Land, Vol. 2, Gordon and Breach, London, 1973.

Holliday, R., D. R. Hodgson, W. N. Townsend, and J. W. Wood, Plant growth on fly ash, Nature, 181, 1079-1080, 1958.

Horsman, D. C., T. M. Roberts, and A. D. Bradshaw, Evolution of sulphur dioxide tolerance in perennial ryegrass, Nature, 276, 493-494, 1978.

Irving, P. M., and J. E. Miller, Response of soybeans to acid precipitation alone and in combination with sulfur dioxide, Annu. Rep. Argonne Natl. Lab. Radiol. Environ. Res. Div. ANL-77-65(Pt. 3), 24-27, 1977.

Jones, H. C., J. R. Cunningham, S. B. McLaughlin, N. T. Lee, and S. S. Ray, Investigation of alleged air pollution effects on yield of soybeans in the vicinity of the Shawnee Steam Plant, TVA Division of Environmental Planning, Aug., 1973.

Kaakinen, J. E., R. M. Jorden, M. H. Lawasani, and R. E. West, Trace element behavior in coal-fired power plant, Environ. Sci. Technol., 9, 862-869, 1975.

Klein, D. H., A. W. Andren, J. A. Carter, J. F. Emery, W. Fulkerson, W. S. Lyon, J. C. Ogle, Y. Talmi, R. I. Van Hook, and N. Bolton, Pathway of thirty-seven trace elements through coal-fired power plant, Environmental Science & Technology, 9, 973-978, 1975.

Kondo, N., and K. Sugahara, Changes in transpiration rate of SO_2-resistant and sensitive plants with SO_2 fumigation and the participation of abscisic acid, Plant & Cell Physiol., 365-373, 1978.

Kovacic, W., and R. G. Hardy, Progress report: Utilization of fly ash in the reclamation of coal mine spoil banks in southeastern Kansas, Kansas Geol. Survey Bull., 204, 29-31, 1972.

Lamoreaux, R. J., and W. R. Chaney, Photosynthesis and transpiration of excised silver maple leaves exposed to cadmium and sulfur dioxide, Environ. Pollut., 17, 259-268, 1978.

Leone, Ida A., and E. Brennan, Sulfur nutrition as it contributes to the susceptibility of tobacco and tomato to SO_2 injury, Atmospheric Environ., 6, 259-266, 1972.

Likens, G. E., Acid precipitation, Chem. Eng. News, 54, 29-44, 1976.

Linzon, S. N., P. J. Temple, and R. G. Pearson, Sulfur concentrations in plant foliage and related effects, J.A.P.C.A., 29, 520-525, 1979.

Lisk, D. J., Trace metals in soils, plants, and animals, Adv. Agron., 24, 267-325, 1972.

Lockyer, D. R., D. W. Cowling, and J. S. Fenlon, Laboratory measurements of dry deposition of sulfur dioxide on to several soils from England and Wales, J. Sci. Fd. Agric., 29, 739, 1978.

Majernik, O., and T. A. Mansfield, Stomatal responses to raised atmospheric CO_2 concentrations during exposure of plants to SO_2 pollution, Environ. Pollut., 3, 1-7, 1972.

Martens, D. C., M. G. Schnappinger, Jr., and L. W. Zelazny, The plant availability of potassium in fly ash, Soil Sci. Soc. Amer. Proc., 34, 453-456, 1970.

McLaughlin, S. B., V. G. Schorn, and H. C. Jones, A programmable exposure system for kinetic dose-response studies with air pollutants, Jour. of the Air Pollution Control Assoc., 26, 132-135, 1976.

Meyer, B., Sulfur, Energy, and Environment, 448 pp., Elsevier Scientific Publ. Co., N.Y., 1977.

Mudd, J. B., and T. T. Kozlowski (eds.), Responses of Plants to Air Pollution, 383 pp., Academic Press, New York, 1975.

Mulford, F. R., and D. C. Martens, Response of alfalfa to boron in fly ash, Soil Sci. Soc. Amer. Proc., 35, 296-300, 1971.

National Academy of Sciences, Sulfur Oxides, 209 pp., Natl. Acad. of Sci., Wash. D.C., 1978.

Natusch, D. F. S., C. F. Baver, H. Matusiewicz, C. A. Evans, J. Baker, A. Loy, R. W. Linton, and P. K. Hopke, Characterization of trace elements in fly ash, 553-575. In Proceedings of Intl. Conf. on Heavy Metals in the Environment, Toronto, Canada, Vol.2, Part 2, 1975.

Nriagu, J. O., Sulfur in the Environment, 482 pp., John Wiley & Sons, New York, 1978.

Oshima, R. J., A viable system of biological indicators for monitoring air pollutants,

Journal of Air Pollution Control Assoc., 24, 576-578, 1974.

Page, A. L., A. A. Elseewi, and I. Straughan, Physical and chemical properties of fly ash from coal-fired power plants with reference to environmental impacts, Residue Review, 7, 83, 1979.

Phung, H. T., L. J. Jund, and A. L. Page, Potential use of fly ash as a liming material, 504-515. In D. C. Adriano and I. L. Brisbin (ed.). Environmental Chemistry and Cycling Processes, CONF-760429, U.S. Dept. Commerce, Springfield, VA, 1978.

Phung, H. T., L. J. Lund, A. L. Page, and G. R. Bradford, Trace elements in fly ash and their release in water and treated soils, J. Environ. Anal., 8, 171-175, 1979.

Plant, C. O., and D. C. Martens, Boron availability as influenced by application of fly ash to soil, Soil Sci. Soc. Amer. Proc., 38, 974-977, 1974.

Plank, C. O., D. C. Martens, and D. L. Hallock, Effect of soil application of fly ash on chemical composition and yield of corn (Zea mays L.) and on chemical composition of displaced soil solutions. Plant Soil, 42 465-476, 1975.

Priebe, A., H. Klein, and H. J. Jager, Role of polyamines in SO_2-polluted pea plants, Journal of Experimental Botany, 29, 1045-1050, 1978.

Ratsch, H. C., Sulfur content of Douglas-Fir foliage near a paper mill, National Environmental Research Center, EPA-660/3-74-018, 1974.

Rees, W. J., and G. H. Sidrak, Plant nutrition on fly ash, Plant Soil, 8, 141-159, 1956.

Schnappinger, M. G., Jr., D. C. Martens, and C. O. Plank, Zinc availability as influenced by application of fly ash to soil, Environ. Sci. Technol., 9, 258-261, 1975.

Seidman, G., I. J. Hindowi, and W. W. Heck, Environmental conditions affecting the use of plants as indicators of air pollution, J.A.P.C.A., 15, 168-170, 1965.

Sidhu, S. S., and P. Singh, Foliar sulfur content and related damage to forest vegetation near a linerboard mill in Newfoundland, Plant Dis. Reptr., 61, 7-11, 1977.

Silvius, J. E., C. H. Baer, S. Dodrill, and H. Patrick, Photoreduction of sulfur dioxide by spinach leaves and isolated spinach chloroplasts, Plant Physiol., 57, 799-801, 1976.

Spugel, D. G., J. E. Miller, R. N. Muller, H. J. Smith, and P. B. xerikos, Effect of sulfur dioxide fumigation on development and yield of field-grown soybeans, Annu. Rep. Argonne Natl. Lab. Radiol. Environ. Res. Div. 1977 ANL-77-65 (Pt. 3), 7-11, 1977.

Taylor, G. E., Jr., Genetic analysis of ecotypic differentiation within an annual plant species, Geranium Carolinianum L., in response to sulfur dioxide, Bot. Gaz., 139, 362-368, 1978.

Temple, P. J., Dose-response of urban trees to sulfur dioxide, Journal of the Air Pollution Control Assoc., 22, 271-273, 1972.

Tingey, D. T., W. W. Heck, and R. A. Reinert, Effect of low concentrations of ozone and sulfur dioxide on foliage, growth and yield of radish, Jour. of Amer. Soc. Hort. Sci., 97, 369, 1971.

Torrey, S., Coal ash utilization - fly ash, bottom ash, and slag, Pollution Tech. Review No. 48, Norgen Data Corp., Park Ridge, New Jersey, 1978a.

Torrey, S., Trace contaminants from coal, Pollution Tech. Review No. 50, Norgen Data Corp., Park Riege, New Jersey, 1978b.

Townsend, W. N., and D. R. Hodgson, Edaphological problems associated with deposits of pulverized fuel ash, 45-56. In R. J. Hutnik and G. Davis (ed.). Ecology and Reclamation of Devastated Land, Vol. 1, Gordon and Breach, London, 1973.

Townsend, W. N., and E. W. F. Gillham, Pulverized fuel ash as a medium for plant growth, 287-304. In M. J. Chadwick and G. T. Goodman (ed.), The Ecology of Resource Degradation and Renewal, Blackwell, Oxford, 1975.

Tourangeau, P. C., C. C. Gordon, and C. E. Carlson, Fluoride emissions of coal-fired power plants and their impact upon plant and animal species, Fluoride, 10, 47-53, 1977.

Woltz, S. S., Translocation and metabolic effects of fluorides in gladiolus leaves, Proc. Fla. State Hort. Soc., 77, 511-515, 1964.

Woltz, S. S., and C. G. Chambliss, Phytotoxicity of cadmium to bushbean, tomato, and other plants, Proc. Fla. State Hort. Soc., 92, in press, 1979.

Woltz, S. S., and C. D. Leonard, Effect of atmospheric fluorides upon certain metabolic processes in Valencia orange leaves, Proc. Fla. State Hort. Soc., 77, 9-15, 1964.

Woltz, S. S., and W. E. Waters, Airborne fluoride effects on some flowering and landscaping plants, Hort. Science, 13, 430-432, 1978.

Woltz, S. S., and W. E. Waters, Airborne fuoride effects on some foliage plants, Hort. Science, 13, 585-586, 1978.

Woltz, S. S., W. E. Waters, and C. D. Leonard, Effects of fluorides on metabolism and visible injury in cut-flower crops and citrus, Fluoride, 4, 30-36, 1971.

Zimmerman, P. W., and A. E. Hitchcock, Susceptibility of plants to hydrofluoric acids and sulfur dioxide gases, Contrib. Boyce Thompson Inst., 18, 263-279, 1956.

CHAPTER 14

HEALTH EFFECTS OF AIR POLLUTION
RESULTING FROM COAL COMBUSTION

Evelyn H. Schlenker and Marc J. Jaeger

Coal combustion, either in a power plant or in a home stove releases a mixture of gases and particulates into the air. The products of combustion include large amounts of fly ash, carbon dioxide, and of sulfur oxides and, lesser amounts of nitrogen oxides, carbon monoxide, hydrocarbons, radionuclides and traces of numerous elements. The interaction of meteorological factors such as humidity, sunlight, and temperature with these substances results in the production of secondary pollutants such as photoxidants (ozone) and aerosols (Calvert, 1973, also see Chapter 9). The quantity and quality of primary and secondary pollutants depend on the type of coal burned and on the geographical and seasonal variations in meteorological factors of a particular area. Extensive coal burning may therefore result in different kinds of pollution in different parts of our country. The population is exposed to a complicated mixture of substances, which may act additively, competitively, or synergistically both in environment and within the body.

We intend first to describe the various pollutants and their physical, chemical, and biological properties. This information is mostly derived from laboratory studies with animals and cell cultures. The other two sources of our knowledge of an association of pollution with health effects are controlled human exposures and epidemiological studies. Each of these topics is treated separately. The National Primary Standards for the main pollutants are listed in Table 1 for comparison with the health effects described in this chapter.

277

Table 1:   *Primary and Secondary Ambient Air Quality Standards*

Pollutant	National Primary	National Secondary ***
Sulfur Oxides	80 $\mu g/m^3$ (0.03 ppm)* ann. arith. mean 365 $\mu g/m^3$ (0.14 ppm) max. 24-hr. avg. **	60 $\mu g/m^3$ (0.02 ppm) ann. arith. mean (Repealed, 1977) 260 $\mu g/m^3$ (0.1 ppm) max. 24-hr. avg. (Repealed, 1977) 1300 $\mu g/m^3$ (0.5 ppm) max. 24-hr. avg.
Particulate Matter	75 $\mu g/m^3$ annual geometric mean 260 $\mu g/m^3$ maximum 24-hr. avg.	60 $\mu g/m^3$ annual geometric mean 150 $\mu g/m^3$ maximum 24-hr avg.
Carbon Monoxide	10 mg/m^3 (9 ppm) max. 8-hr avg. 40 mg/m^3 (35 ppm) max. 1-hr avg.	Same as primary Same as primary
Photochemical Oxidants	235 $\mu g/m^3$ (0.120 ppm) max. (Revised 1979) 1-hr avg.	Same as primary
Hydrocarbons	160 $\mu g/m^3$ (0.24 ppm) max. 3-hr avg. (6-9 a.m.)	Same as primary
Nitrogen Oxides	100 $\mu g/m^3$ (0.05 ppm) ann. arith. mean	Same as primary

*   ppm given on volume basis

**   all maximum averages are not to be exceeded more than once

*** adapted as Florida's Ambient Air Standards

Source: Code of Federal Regulations Title 40 –
         "Protection of the Environment."
         Parts 50-59. U.S. Government. Printing
         Office, Washington, D.C. 1978.

I. BIOCHEMICAL AND BIOPHYSICAL PROPERTIES

A. Particulate Matter

Suspended particulate matter may be solid or liquid. It may be
composed of inert or reactive substances which may act as catalysts of
atmospheric reactions. The scrubbers installed in power plants remove
primarily large particulates. Relatively large amounts of small particulate matter of < 2μm are released into the atmosphere (Office of
Technology Assessment, "The direct use of coal", 1979, p. 191).

Most particulate matter enters the body through the nose and
mouth. The amount of ventilation and the site of deposition of particulate matter in the lungs depends upon its size, shape, hydroscopicity,
and on the ventilation rate (Stuart, 1973). Large particulate matter
is impacted in the nasopharynx and is cleared by nose blowing, sneezing,
coughing, and swallowing (where they may act on the gastrointestinal
tract). Smaller particulate matter is deposited in the tracheobronchial
tree by sedimentation. From there particles may be cleared by coughing,
mucociliary action, and, to a small degree, by the bronchial blood flow.
The smallest particles (< 0.1μm) are deposited in the alveolar zone.
Clearance of these particles is slow. It involves phagocytosis by
macrophages (amoeba - like cells) and transport into the lymph and
blood. Particles may stimulate irritant receptors in the upper airways
and produce, by reflex action, a bronchoconstriction and increased mucous secretion. Particulate matter may also act directly on mast cells
located in large numbers in the lower airways. Mast cells release mediators which cause bronchoconstriction and increase the capillary permea-
bility, thereby causing edema.

It may be useful at this point to describe briefly some anatomical
and physiological features of the respiratory system. The airways com-
prise a system of tubes that branch, giving rise to bronchi, which get
smaller and smaller, the smallest being less than 1 mm in diameter.
The walls of bronchi contain muscle fibers, which may shorten and decrease the bronchial caliber (bronchoconstriction measured as increased
airway resistance). The bronchi are lined by cells with a border of
cilia, which beat regularly and transport a covering layer of mucous

and particles toward the mouth. The transport of oxygen and of toxic
gases (carbon monoxide) into the blood occurs in terminal sacs of 0.5
mm diameter, which are surrounded by a network of capillaries. Macro-
phges, which pick up particles, patrol the whole lining of the lungs.
Mast cells, which release physiologically active substances, are found
in the wall of the bronchi (Said, 1978). The lung tissue has elastic
properties, which can be measured (so-called lung compliance).

Other actions of small particles depend on their composition. Fly
ash is largely composed of oxides of carbon, silica, aluminum, nickel,
and iron (Chap. 12). The distribution of trace elements found in coal
combustion residues is given in Chap. 12. The concentration of these
elements is particularly high in small particles (Office of Technology
Assessment, 1979). Some of these elements have been shown to cause
inflammatory reactions that may lead to pulmonary fibrosis and cancer
if the elements are inhaled in high concentration over long periods
of time. A consequence of such an exposure is silicosis found in coal
miners. National Standards for fine particles (< 2µ) do not yet exist.

B. Gases

Gases produced by coal combustion include large amounts of carbon
dioxide (CO_2), sulfur dioxide (SO_2), carbon monoxide (CO), and nitrogen
oxides (NO_x). The site and fractional uptake of gases by the lungs is
determined by gas solubility, gas reactivity, and by the depth and rate
of the individual's ventilation (Frank, 1970 and Miller, 1977).

SO_2 is highly soluble and is primarily absorbed in the nose and
upper airways where it is converted into sulfurous acid (Frank, 1970).
These acids may irritate nerve endings located in the throat and in the
trachea, causing coughing and bronchoconstriction. SO_2 may be converted
into other gases such as organic sulfones, H_2S, and H_2SO_4 mists in the
presence of other pollutants (McJilton et al. 1973). These substances
may adhere to fine particles and may be transported deep into the lungs.
They may also be neutralized by ammonia that is present in the oral region (Larson et al. 1977).

Nitrogen oxides (NO_2, NO, etc.) and ozone (O_3) are less soluble

gases. Unlike SO_2, they penetrate deeply into the lung where they may,
even in small concentrations, have toxic effects. Ozone and NO_2 oxi-
dize cell membranes, interfer with metabolic processes, and compromise
defense mechanisms (Mustafa and Tierney, 1978, and Roekm et al., 1971).
Unlike SO_2, they can reach and stimulate mast cells directly and release mediators. Ozone is a product of air pollution; it is the result
of a chemical reaction of primary pollutants with air under the catalytic action of sunlight. Ozone levels are high even in some non-polluted
areas because of high solar irradiation.

Carbon dioxide and CO are non-irritating gases. CO is taken up in
the alveolar region. It binds avidly to hemoglobin and forms carboxy-
hemoglobin. This results in physiological effects similar to those
of anemia. In individuals who are anemic or whose cardiopulmonary systems can no longer compensate for an O_2 deficit, even relatively small
amounts of CO (100 ppm) may be health threatening (Arrow et al. 1973).
CO is particularly toxic in pregnancy (Longo, 1977). In addition, carbon monoxide may result in subtle reductions of altertness, vigilance,
concentration, and similar pyschological functions (Horvath, 1973, and
Bartlett, 1973). It may promote the production of arteriosclerotic
heart disease (Astrup et al. 1967) and exacerbate underlying coronary
heart disease (Goldsmith and Aronow, 1975). Only 2% of the CO emissions
in the U.S. are due to combustion of coal and oil, the rest are mostly
the result of automobile exhaust (Waldblott, 1978). Presently the levels
of CO in our environment are low and pose little threat to our health,
however, high density traffic lanes and improperly vented stoves in homes
may result in dangerous accumulations of CO.

Carbon dioxide at levels encountered in the environment today
(440 ppm) has no known health effect.

II. ANIMAL EXPERIMENTATION

Animal and cellular exposures to air pollutants are conducted to
evaluate dose-response relationships; pathophysiolocical as well as
anatomical and biochemical changes have been observed. The susceptibility of different animal species to toxic and subtoxic levels of pollutants

varies. Results can not therefore be directly extrapolated to humans.
In research, animals are exposed to high levels of pollutants for short
periods of time. This may not be equivalent to a life long exposure
to low levels of the same substances. Animal experiments, however, are
useful in understanding the underlying anatomical, biochemical, and
biophysical effects pollutants have on biological systems.

A. SO_2

Animals exposed for weeks to SO_2 levels of 0.14 to 1 ppm did not
develop any morphological biochemical alterations (Alarie et al. 1972).
Higher levels (5 to 30 ppm) did result in markedly increased airway
resistance and a decrease in compliance of the lungs even when exposures
lasted only minutes (Balchum et al. 1960).

B. Sulfuric Acid

Sulfuric acid aerosols, which are secondary products of air pollution, may be transported deeply into the respiratory system. Monkeys
exposed to high levels of H_2SO_4 (2.5 - 4.8 $\mu g/m^3$) for 78 weeks showed
both histological and pulmonary function changes (Alarie et al. 1973).
Sackner and co-workers (1978) did not find any effect of 14 $\mu g/m^3$ of
H_2SO_4 aerosols on the cardio-pulmonary system of sheep. Sackner's exposure lasted only 20 minutes.

C. Particulate Matter

The particles primarily associated with coal combustion are fly
ash and sulfates. These particles penetrate deep into the lungs. They
affect the pulmonary system by depressing the immunological defense
system, by increasing airway resistance, and by decreasing lung compliance (Fenters et al. 1979, Wehner et al. 1970, and Amdur et al. 1978).
The magnitude of the response is related to the concentration, the composition, and the size of particles.

Elements found in trace quantities in fly ash and polycylic hydrocarbons can produce fibrosis and cancer of the lung. In general, these
effects occur only if large amounts of these substances are inhaled
(Waldbot, 1978).

D. Nitrogen Dioxide

Animals exposed for long periods of time to 2 ppm (3760 $\mu g/m^3$)
show distinct morphologic alterations of the alveoli and bronchioles
(Freeman et al. 1968 and 1968a). The structure of alveolar walls is
altered and ciliated cells in bronchi diminish in number. Continuous
exposures to NO_2 result in the production of emphysematous lesions in
mice at concentrations of approximately 2 ppm. NO_2 disrupts the body's
defense mechanisms, and an increased mortality in mice was noted with
either a continuous (1.5 ppm) or intermittent (3.5 ppm) exposures
(Ehrlich, 1966). For a concise review of the effects of NO_2 on animals
see Morrow's 1975 paper.

E. Ozone

O_3 is toxic in small concentrations. It diffuses with oxygen (O_2)
into the blood stream and affects not only linings of the lungs but also the blood (Buckley et al. 1975). Schwartz and co-workers (1976)
exposed mature rats to various levels of O_3 (0.2, 0.5 and 0.8 ppm) for
either 8 or 24 hours a day for 7 consecutive days. They found: 1)
damage in small airways with exposures to 0.2 ppm (about 2x the Present
One Hour Primary National Standard) and 2) a dose-dependent response to
the 3 levels of ozone that could be quantified by alterations of biochemical marker enzyme activity. Barlett et al. 1974 using similar
levels of O_3 for 30 days in growing rats noted a decrease in lung elasticity and in growth rate. Animals exposed to ozone at low level can become adapted to exposures to ozone at higher levels (Frank et al. 1979).
However, associated with this tolerance is a decrease of lung elasticity.

III. HUMAN EXPERIMENTATION

Studies using human subjects are restricted by law in length of
exposure and concentration of pollutant for ethical reasons. Exposures
are limited to responses that are expected to be reversible and easily
measureable. All functional changes described in the next section were
of a transient nature. The most important information obtained from
such studies is the acute response of human beings to well-defined levels

of single pollutants or combinations of pollutants. Human exposures
also provide information concerning effects of pollutants on individuals
who suffer from a chronic illness or who may be particularly sensitive
or reactive.

A. SO_2

The response of humans to gaseous SO_2 is dose-dependent. In a
study by Frank and co-workers (1962), healthy volunteers were exposed
to 1, 5, and 13 ppm of SO_2 for 10 to 30 minutes. No significant response to SO_2 occurred at 1 ppm, but airway resistance increased at the
two higher levels. Some subjects showed a decrease of lung compliance
at higher concentrations. Using sensitive lung function tests, Jaeger
and co-workers (1979) found a small but significant decrease in small
airways functions in asthmatics exposed to 0.5 ppm SO_2 for 3 hours.
Some effects were delayed. The patients wore nose clips and the scrubbing properties of the nose were therefore by-passed. Normals exposed
to the same levels of SO_2 showed no statistically significant response.
As early as 1957, Sim and Pattle noted that subjects who were especially
sensitive to the "London Fog" also had greater reactions to SO_2 exposures. Lawther and co-workers (1975) found significant increases of
specific airway resistance in normals exposed to 1 ppm if the subjects
hyperventilated. Hyperventilation brought about either voluntarily or by
exercise seems to enhance the effect SO_2 has on lung function.

B. H_2SO_4

Few studies exist in which humans have been exposed to sulfuric
acid mists. Published effects range from very small to significant decreases in small airways function (Amdur et al. 1952; Newhouse et al.
1978; Sackner et al. 1978). Sackner and co-workers found no effect of
exposing asthmatic subjects to 1 mg/m^3 for 10 minutes.

C. Particulate Matter

The deposition of particles in the human respiratory system depends
primarily on size. Seventy percent of small particles; which are less

the 3μm, are deposited in the small bronchi and alveoli. Clearance
mechanisms remove most of the deposited particles. These mechanisms
include phagocytosis, ciliary movement, and coughing. The deposition
fraction may be different in subjects suffering from chronic bronchitis
or in smokers (Stuart, 1973, and Goldberg and Lourenco,   1973).   In
these subjects particles may be unevenly distributed   because of the
mosaic-like pattern of disease process. Moreover, clearance is often
impaired (Cohen et al. 1979). Our knowledge of the health effects of
some particles is derived from information of occupational exposures of
workers in various plants or mines. In general, occupational exposures
involve only a small number of agents such as coal, silica, asbestos,
etc., to which an individual is exposed at a very high concentration.

Functionally, the exposure of normal individuals to inert dusts
may result in a transient increase of airway resistance. In asthmatics
and bronchitics the effect is more pronounced (Norris et al. 1966).

D. Nitrogen Dioxide

NO_2 exposure (2.5 ppm) results in a delayed effect in man (30 minutes). Neding and Kreheler (1971) concluded that NO_2 acts on mast cells
and not on vagal irritant receptors, since the increase in airway resistance can be prevented with an antihistamine drug. Orehek and co-workers
(1976) showed that in some asthmatics NO_2 acts in synergism with bronchoconstriction drugs.

E. Ozone

Both healthy and asthmatic human subjects have been experimently
exposed to levels of ozone present in certain American cities such as
Los Angeles and Philadelphia. Kagawa and Toyama (1975) found that 0.9
ppm O_3 slightly increased airway resistance after a 5 minute exposure.
However, when the same subjects exericsed moderately, the same exposure
caused an appreciable response. Healthy subjects exercising intermit--
tantly for 2 hours and exposed to 0.75 ppm O_3, showed a decrease of lung
compliance, an increase of airway resistance, and an uneveness of distribution of ventilation (Bates et al. 1972). Folincebee et al. (1977)

found great decrements in lung function in 14 non-smoking males exposed
to 0.5 ppm O_3, exericse, and heat. Lower levels of O_3 (0.25 ppm to
0.37 ppm) affect lung function significantly only in some individuals
such as smokers and visitors to polluted cities. When low levels of
O_3 (.37 ppm) are combined with low levels of SO_2 (.37 ppm), all normal
subjects react (Hazucha and Bates, 1975). Recently Golden, Nadel and
Boushey (1978) found that the reactivity of normal subjects to histamine
may be enhanced by O_3. This phenomenon is similar to observations mentioned earlier in connection with NO_2.

The fact that pollutants may combine to produce enhanced or even
potentiated effects is presently not recognized by the National Standards
which are set for each pollutant separately (Brain, Proctor, and Reid,
1977).

IV. EPIDEMIOLOGICAL STUDIES

Epidemiological studies are the yardstick of air pollution research.
Only through the evaluation of past and present effects of air pollution
on large population groups can the long-term action of low levels of
possibly toxic substances on humans be determined. Health effects can
occur because of the accumulation of inhaled substances in the body if
no excretory mechanisms exist. Pollutants may also cause chronic irri-
tation of tissues. Both mechanisms are well-documented in medical
toxicology (e.g. lead poisoning and skin cancer). Unfortunately, even
though epidemiological studies are of prime importance in the evaluation
of air pollution, they are also the most difficult to conduct.

Epidemiological studies can be subdivided according to the nature
of the health considered: some authors attempt to related ambient air
pollution levels to mortality rates, others to morbidity rates, and
still others to cancer rates.

A. Mortality Studies

The clearest evidence for an association of SO_2 and/or particulates
to death occurred in a number of episodes in which air pollution reached
catastrophic levels such as in Belgium (Meuse Valley, 1930), Donora

(Pennsylvania, 1948), London (1952 and 1962), and New York City (1953).
Each of these episodes has been studied extensively. Together they resulted in approximately 5000 deaths; the London episode of 1952 alone
resulted in 4000 deaths. The mortality rates were estimated most carefully by trying to compare the deaths occurring at the time of the pollution episode with the seasonally adjusted death rate. The main conclusions drawn from those observations are the following:

1) The excess mortality was found particularly in people 45 years
    old and older who suffered from chronic cardiorespiratory dis-
    ease. It contained for several weeks after the episode.

2) SO_2 and particulate matter each reached levels in excess of
    1000 $\mu g/m^3$ (.38 ppm) (American Thoracic Society, 1978). This
    figure may be questionned, since it is based on extrapolation
    rather than on direct measurements.

3) Photochemical oxidants (ozone and NO_2) have not been linked
    with any similar catastrophies.

Other authors have tried to relate the daily variation of SO_2/
particulate level with the daily variation of mortality (Martin et al.
1960; Martin, 1964; Boyd, 1960; Buechley et al. 1973; Lebowitz, 1973).
Most of these studies found a definite relationship between pollution
and mortality. Buechley in a study covering 5 years found a 2% increase
in the mortality of New York City residents when SO_2 levels were above
500 $\mu g/m^3$ (.2 ppm). The levels of particulate matter varied in a similar
fashion to those of SO_2. It was, therefore, not possible to attribute
the increased mortality to either SO_2 or to the particulate matter.
Buechley's results are typical for studies done in this country and
abroad (Brasser, 1967; Lindeberg, 1968; and Watanbe 1971).

Schimmel et al. (1977) however, reported data that contradict those
mentioned above and failed to find any decrease of mortality in New York
City from 1963 through 1972, i.e., during a period characterized by a
significant decrease of SO_2 levels.

Several studies have failed to find any correlation between mortality rates and the level of photochemical oxidants in Los Angeles. Even
the mortality of elderly people and of residents of nursing homes seemed
unaffected by waves of photochemical pollution. These results are

astounding because controlled exposures, of both animals and of humans
have documented beyond doubt the toxcity of O3 and NO_2. It has been
pointed out that the mortality rates are actually at their seasonal
low when oxidant concentrations peak in the summer months in Los Angeles.

The ambient levels of CO may also have an effect on mortality as
suggested by studies in Los Angeles by Cohen et al. (1969) and Hexter
et al. (1971). These studies point to the special significance of CO
pollution even at moderate level for heart patients. This aspect of
pollution is not elaborated upon here, since it is not an expected result of coal utilization in power plants. However, it might be an important problem in interior pollution caused by improperly maintained
heating equipment.

B. Morbidity Studies

The possible contribution of air pollution to the development of
chronic lung disease, such as chronic bronchitis and emphysema, can be
studied by two means: questionnaires and lung function studies. Both
methods have been appreciably improved and standardized by early British
work (Medical Research Council, 1960). Recently, in this country,
Ferris (1978) summarized the standardization procedures. Most authors
now use the very same questionnaire and identical techniques and procedures which make various studies comparable. Such studies seem so important that we reproduce a table that lists 22 papers with a succinct
description of the results from a study sponsored by the American
Thoracic Society (Tables 2a and 2b). The majority of the authors find
an increased incidence of symptoms indicative of a chronic lung infection such as cough, and/or production of phlegm, and/or a decline of the
pulmonary function in populations exposed to SO_2 and particulates.

C. Confounding Factors

The evaluation of most epidemiological studies of mortality and
morbidity mentioned above is difficult because of one or more of the
following factors:

1) In several studies the levels of SO_2 and of particulate
    matter were measured at one rather than several points in the
    communities under consideration. Since some of the cities
    were very large (e.g. New York City) a more detailed descrip-
    tion of the pollution in various sections would have made the
    analysis more meaningful.

2) Smoking habits of the population often were not known. Since
    smoking is a prominent cause of respiratory symptoms, the lack
    of this information might invalidate some of the epidemiologi-
    cal studies. (Smoking is referred to in Section 5 of this
    chapter).

3) Most studies neglect to include the effect of seasonal changes
    of temperature on mortality and morbidity. This omission con-
    founds the analysis, since the seasonal temperature variation
    has itself an important influence on health, and since high
    pollution levels often accompany extremes of temperature.

4) Poor neighborhoods show a high prevalence of contagious res-
    piratory disorders because of crowding and poor hygiene. Peo-
    ple in low socioeconomic communities also smoke more heavily
    and may live close to factories and power plants. These fac-
    tors interact with air pollution as reported earlier.

5) It is often difficult to separate community health effects due
    to air pollution from those related to occupational exposures.

6) The migration of people into and out of communities selected
    as representative for either polluted or clean air areas fur-
    ther confounds statistics.

An overall judgment of how these confounding factors influence each
of the epidemiological studies is difficult and beyond the scope of
this paper. In comparing the work of various authors one finds that
"the more adjustments are made to account for confounding factors the
weaker the observed (residual) association of mortality or mobidity with
air pollution . It becomes then difficult to determine whether the effect
of air pollution is really quite small or whether the effects were over
adjusted for some of the confounding factors" (American Thoracic Society,

1978, p. 17).

The overall tenor of these studies and critical reviews is that the effects of air pollution on health are significant enough to be observed, i.e., to prove that a cause-and-effect relationship between pollution and health exists. However, the level of pollution is in general so low and the magnitude of the immediate health effects so small that a quantitative dose-response relationship seems difficult to establish beyond doubt. In other words, the air pollution-to-health relationship seems sometimes to disappear in the background noise, the other known and unknown causes of injury to our health.

D. Susceptible Subjects

It was mentioned earlier that individuals with chronic cardiorespiratory ailments are particularly susceptible to episodes of unusually high air pollution. Persons with asthma represent another group of highly susceptible subjects. Several studies have indicated that there is an increase of respiratory symptoms in asthmatics during periods of high pollution (Schrenk et al. 1949, Glasser et al. 1967, Chiaramonte et al. 1970, Cohen et al. 1972a and b, Finklea et al. 1974, Goldstein et al. 1974). Other authors have failed to find the same relationship (Greenburg et al. 1964, Weill et al. 1964, Rao et al. 1973). The confounding factors that afflict all epidemiological evaluations, obviously play a role here too. Asthmatics are usually susceptible to several factors such as rapid temperature changes, humidity, dust, and pollens, all of which may interplay with air pollution. The information gained from asthma studies is valuable because it points to a section of the population that is particularly susceptible to injury by air pollution. Not only are asthmatics a population at increased risk; other kinds of reactive or sensitive subjects also fall into this category. Results of acute exposures to controlled levels of pollution (see sections II & III) support this view. One might speculate whether responses to air pollution found in the general population are due to the susceptibility of a selected few or whether they are the result of a random occurrence.

Table 2a

Studies of Air Pollution and Prevalence of Chronic Respiratory Symptoms *

Study	Characteristics	Findings
Fairbairn and Reid (1958)	Comparison of respiratory illness among British postmen living in areas of heavy and light pollution.	Sick leave, premature retirement, and death due to bronchitis or pneumonia were closely related to pollution index based on visibility.
Ferris and Anderson (1962) Anderson and Ferris (1965)	Questionnaire and ventilatory function survey of random sample of adult population in Berlin, N.H., and Chilliwack, British Columbia, 1961-1963.	No apparent excess in symptom prevalence in association with air pollution. Diminished pulmonary function in residents of more polluted community, although differences in occupation, climate and ethnic factors may account for these findings.
Mork (1962)	Questionnaire and ventilatory function tests of male transport workers 40-59 yrs of age in Bergen, Norway and London, England.	Greater frequency of symptoms and lower average peak flow rates in London. Differences were not explained by smoking habits or socioeconomic factors.
Petrilli et al., (1966)	1961/1962 study of respiratory symptoms in nonsmoking women 65 yrs of age and over who had not worked in industry and were long-time residents of their neighborhood in Genoa, Italy.	Strong association between chronic respiratory symptom prevalence and area gradient for particulates and SO_2. Comparability between residents of surburban and industrialized areas was not assured by exclusion of persons on social welfare assistance.
Cederlof (1966) Hrubec et al., (1973)	Chronic respiratory symptom prevalence in large panels of twins in Sweden and in U.S. Index of air pollution based on estimated residential and occupational exposures to SO_2, particulates, and CO.	Increased prevalence of respiratory symptoms in twins related to smoking, alcohol consumption, socioeconomic characteristics, and urban residence, but not to indices or air pollution.
Bates et al., (1962, 1966) Bates (1967)	Comparison of symptom prevalence, work absences, and ventilatory function in Canadian veterans residing in 4 Canadian cities	Lower prevalence of symptoms and work absences and better ventilatory function in veterans living in the least polluted city.
Bates (1973)	10 yr follow-up study of Canadian veterans initially evaluated in 1960, and followed at yearly intervals with pulmonary function tests and clinical evaluations.	Least decline in pulmonary function with age in veterans from least polluted city.

*Reprinted with permission from the American Thoracic Society, 1978.

Table 2b

Lambert and Reid (1970)	Postal survey of nearly 10,000 British residents, 35-40 yrs of age using standardized questionnaire.	Among smokers, prevalence of chronic respiratory symptoms increased with increasing air pollution. Lesser symptom/pollution association in smokers.
Reichel (1970) Ulmer et al., (1970)	Respiratory morbidity prevalence surveys of random samples of population in 3 areas of West Germany with different degrees of air pollution.	No different in respiratory morbidity, standardized for age, sex, smoking habits, and social conditions, between populations living in the different areas.
Ferris et al., (1971, 1973 1976)	Repeat survey in 1967 of residents of Berlin, N.H., 6 yrs after the original 1961 study and again in 1973, 12 yrs after the original study. TSP concentrations decreased from 180 μg per m^3 in 1961 to 90 μg per m^3 in 1973.	Slightly lower prevalence of chronic respiratory symptoms and improved ventilatory function tests in the 1967 survey after standardization for age, sex and smoking habits. Results were attributed to decrease in air pollution values, although some of the effects may be associated with differences in temperature and season between the 1961 and 1967 surveys. No differences in chronic respiratory symptom prevalence found in 1973, even though air quality continued to improve.
Tsenetoshi et al., (1971)	Prevalence survey (Medical Research Council questionnaire) of 9 areas of Osaka and Hyogo prefectures, Japan, residents 40 yrs. of age and over.	Multiple regression analysis revealed increasing prevalence of chronic bronchitis, adjusted for age, sex, and smoking, corresponding to the area gradient of air pollution.
Cohen et al., (1972a)	Symptom and ventilatory function survey of nonsmoking Seventh Day Adventists 45-64 yrs, of age in Los Angeles and San Diego.	No association of symptom prevalence or ventilatory function with differences in peak oxidant readings of the 2 areas.
Chapman et al., (1973)	1970/71 prevalence survey of chronic respiratory symptoms in adult parents of school children in New York, Salt Lake City area. 5 Idaho-Montana communities, and Chattanooga; and of military inductees in the Chicago metropolitan area.	Increased prevalence of chronic respiratory symptoms in nonsmokers and smokers residing in more polluted communities. Effect of cigarette smoking exceeded that of air pollution. No association of symptoms with nitrogen oxide pollution in Chattanooga.
Comstock et al., (1973)	Repeat survey in 1968/69 of east coast telephone workers and of telephone workers in Tokyo.	After adjustment for age and smoking, no significant association of respiratory symptom prevalence with place of residence.
Becklake et al., (1975)	Prevalence of respiratory symptoms and measurement of ventilatory function in adults and children living in 3 areas of Montreal having different degrees of particulate air pollution (84 to 130 per m in 1974).	No difference in symptom prevalence or ventilatory function related to the degrees of air pollution in the 3 communities.

E. Acute Respiratory Infections

Air pollution has been linked to acute respiratory infections.
It has been shown by numerous authors that adults living in polluted
areas are more susceptible to respiratory infections of the lower respiratory tract, such as bronchitis and pneumonia (Dohan et al. 1960;
Ipsen et al. 1969, Angel et al. 1965, Verna et al. 1969, Lebowitz,
et al., 1972, Love et al. 1974, Finklea et al. 1974a). Socioeconomic
and seasonal temperature variables were taken into account. The
findings obtained from adult populations were confirmed in studies of
children (Reid, 1969; Lunn et al. 1967; Douglas et al. 1966; Colley
et al. 1970; Irwig et al. 1975; Lunn et al. 1970). Unfortunately,
it has not yet been possible to identfy which pollutant is responsible
for this increased risk.

The importance of these findings for the general health of our
population is considerable. Indeed, recent studies have shown a relationship between acute illnesses of the lower respiratory tract in
childhood and young adulthood with progressive chronic obstructive
lung disease of later life (Burrows et al. 1977). Since air pollution
increases the susceptibility to such infections, it may be linked to
health effects occurring 10 to 20 years after the exposure.

F. Cancer

The relationship between smoking habits and lung cancer is well
recognized. It is also known that some atmospheric pollutants are
carcinogenic in animal experiments and bear strong resemblance to
polycylic hydrocarbons believed to be responsible for the carcinogenic
effect of cigarette smoke. On this ground alone air pollution might be suspected to cause cancer in the respiratory tract. In addition, it is well
recognized that some other compounds contained in coal such as chromates,
nickel, asbestos, arsenic, etc. are carcinogenic. Their relationship to
cancer is established from occupational exposures (Cole et al.1975). Thus,
much information would lead to the suspicion that air pollution might
contribute to an increase of the cancer rate in the population exposed.

Here, as in other epidemiological studies, however, the expected relationship of pollution with health has not been established beyond doubt. A major confounding factor is the overriding effect that smoking has on cancer (Hoffman et al. 1954; Buell et al. 1967; Hammond and Horn, 1958; Haenszel et al. 1964; Hammond, 1972).

V. SMOKING AND AIR POLLUTION COMPARED

Smoking has been mentioned in previous sections as a confounding factor in air pollution investigations. It may be useful to compare briefly these two forms of pollution.

Cigarette smoke contains many of the same substances found in coal emissions, e.g. gases such as CO, CO_2, NO_2, SO_2, carcinogenic hydrocarbons, and a host of organic compounds (see Table 3).

Table 3 : *Components of Tobacco Smoke.* *

Metals:	Organic Compounds:	Gases:
iron	aldehydes	SO_x
manganese	organic acids	NH_3
titanium	phenols	Hydrogen cyanide
zinc	acrolein	CO
lead	nicotine	CO_2
nickel	pyrenes	NO_x
arsenic	flourenes	particulates (< 1.0 μm)

* from Shephard (1978).

Quantitatively, the amount of these agents is quite different in cigarette smoke and in polluted air. Thus the particulate matter inhaled by smoking one unfiltered cigarette ($\sim$ 5mg) would approximate the amount of particulate matter inhaled over a period of 5 days if the subject is exposed to polluted air of 75 μg/m^3. Another comparison may be made with CO: the level of CO-hemoglobin in a heavy smoker may reach 15%; in a heavily polluted community the levels of CO-hemoglobin could reach 1.85%

on an average basis, with possible peaks in city traffic of 5%.

Smoking one cigarette affects lung function by causing an increase
in large and small airways resistance of 20 and 6%, respectively
(DaSilva and Hamosh, 1973; Nadel and Comroe, 1961). The effect is immediate and lasts up to 35 minutes. By comparison, Kreisman and co-
workers (1976) needed 3 ppm SO_2 to raise the resistance of small airways in normal subjects by 5.7%. Other air pollutants such as O_3 and
NO_2 at high levels may result in similar responses.

Chronic effects of cigarette smoking are well documented (Woolcock
and Bernard, 1977; Barter and Campell, 1976; Lambert and Reid, 1970).
Cigarette smoking is linked to low birth weights, lung and cardiovascular disease, and various kinds of cancer. Smokers have an altered defense system with a decreased mucociliary clearance (Cohen et al. 1979)
a preponderance of lymphocytes, an increase in the immunoglobulin IGA,
and an increase in the number and activity of macrophages. They have
a larger number of mucous producing cells, an increase of lung compliance, and an increase of airways resistance. Smokers may develop hy-
perreactive airways (Zuskin et al. 1974; Barter and Campell, 1975).
Lung function in some smokers deteriorates at an accelerated rate as
compared to non-smokers.

Chronic effects of air pollution have been described in section IV.
They include increases of the production of phlegm, coughing, and
slight decreases of lung function in certain individuals. These symptoms and signs are similar to those found in smokers; their frequency,
however, and severity is less.

An interesting aspect of the comparison of smoking with pollution
is to study of the effect of air pollution has on smokers. Numerous
studies have shown that individuals who smoke and live in a polluted
environment have an increased incidence of persistant cough and phlegm
production, chest discomfort, and cancer with non-smokers living in the
same area (Ferris and Shepard, 1978). Lambert and Reid
(1970) showed that smokers living in a polluted area have an 18 to 34%
incidence of persistant cough and phlegm, while smokers of the same
age groups living in a low pollution area had an incidence of 12 to 28%.

Non smokers had an incidence of 4-6% of the same symptoms dependent on their exposure to pollution. Air pollution may act differently in smokers because of the selected deposition of particulates in areas of the lungs which already have an impaired function, and because chronic smoking inhibits mucociliary clearance (Stuart 1973; Cohen et al. 1979).

Another point for comparison is cancer. Lung, throat, and lip cancer are known to result from smoking. Air pollution has also been invoked as a possible cause. In one study conducted in Japan (Hirayama, 1976) the relative risks of lung cancer were as follows:

nonsmokers, light pollution	1
nonsmokers, heavy pollution	2.5
heavy smokers, light pollution	6.1
heavy smokers, heavy pollution	9.6

Thus a smoker compromises his or her health to a great extent by smoking; and living in a polluted environment furhter endangers personal health.

VI. CONCLUSION

We have documented in the previous sections of this chapter that the results of air pollution research are inconclusive. Different authors have sometimes found opposing results even while studying the very same topic. When issues are controversial, it is difficult to convince disbelievers or to discourage believers.

The lungs are exposed daily to about 12,000 liters or 3000 gallons of air. Even if relatively low levels of pollution prevail, the insidence and accumulation of pollutants on the internal surfaces of the respiratory tract must be appreciable. We would like to summarize this process and the present evidence of its health effect as follows; emphasizing air pollution resulting from coal combustion.

Coal combustion emits small-sized particulates, (< 2 μm), NO_x, SO_2, and sulfates and, to a lesser degree, causes the formation of secondary products of the interaction of pollution with air such as O_3 and NO_2. The particles contain a large number of various elements, some in trace amounts. Most of these pollutants are chemically highly reactive. They

all have been found to elicit physiological responses in animals and
in man at concentrations that are only slightly higher than the present
National Air Quality Standards. Some act in combination to produce
potentiated effects, a fact not recognized by present regulations. In
the southeastern part of this country the large scale burning of coal
might result in high levels of secondary products because of high solar
irradiation and high humidity. The site of action of most pollutants
in the respiratory tract is fairly well understood and is related to the
physicochemical properties of each pollutant. The deposition and retention fraction, i.e., the fractional amount of a pollutant retained by
the lungs during a brief exposure, is high: 100% for SO_2 and 60 to 80%
for particulates smaller than 2 μm. Defense mechanisms exist that buffer
acids and clear the lungs of particulates. But these defense mechanisms
are partially inhibited by chronic exposure to pollutants. For all pollutants physiological response are dose dependent. When doses are very
high, extremely high as in certain animal experiments, the response e-
licited may not be reversible.

Human studies have demonstrated that some individuals are particularly susceptible to the effects of air pollution. Asthmatics be-
long in that group. Nitrogen dioxide seems to make asthmatics more
susceptible to some drugs. Ozone may induce an increased sensitivity
to histamine in normals. Other groups of subject such as children,
elderly people, and chronic smokers may also react differently than
"normal" middle aged adults to pollution. It has also been shown that
subjects can develop a diminished susceptibility to air pollution, which
indicates some kind of tolerance. Only some of these reactions are
well documented. But because of their multiplicity and diversity,
these reactions add confusion to the controversy of the subject.

The result of large epidemiological studies that compare past and
present levels of pollution with health effects are, in part, in contra-
diction to the information derived from the acute animal and human experiments mentioned. The pollutants known to be most toxic (O_3 and NO_2)
have not been found to correlate with an increased morbidity and/or
mortality. There is evidence, however, that SO_2, sulfates, and particulates do have some detrimental effect on the health of children and

adults. But these effects do not allow the relationships between pollutants and health to be quantified in some simple way. It may be that
the paucity of strong epidemiological evidence exists, because the average age levels of many pollutants are low and have been declining even
in large cities. If so, epidemiological studies would document the effectiveness of past pollution control and encourage us not to deviate
from that trend.

A medically and perhaps legally interesting aspect of research is
the observation that certain subjects, who can be identified, are more
susceptible to air pollution than other healthy individuals. This observation allows one to narrow down predictions in quite a different
fashion than if a random distribution of individual responses was observed. It may be that the subjects or patients who accounted for
some of the epidemiologist's findings belonged in large part to the
selected few who react abnormally, i.e., are most susceptible to air
pollution.

Smoking is a special form of self-induced air pollution. Its
effects can be compared to those of ambient air pollution. While effects of ambient air pollution can, apparently, not be quantified epi-
demiologically, there is no problem indicating the numbers of deaths
that result from smoking because of lung cancer and because of chronic
obstructive respiratory disease. Another special form of air pollution
is related to occupational exposures. Here, too, the known health effects
far exceed the effects of ambient air pollution in severity and frequency.

References

Amdur, M. O., J. Bayles, V. Ugro, and D. W. Underhill, Comparative irritant potency of sulfate
salts, Envir. Res., 16:1-8, 1978.
Amdur, M. O., L. Silverman and P. Drinker, Inhalation of sulfuric acid mist by human subjects,
Arch. Ind. Hyg. Occup. Med., 6:305, 1952.
American Thoracic Society, Health Effect of Air Pollution. Authors and committee members: C.M.
Shy, J. R. Goldsmith, J. D. Hackney, M. D. Lebowitz and D. B. Menzel, 1978.
Alarie, Y., W. M. Busey, A. A. Krumm, C. E. Ulrich, Long-term continuous exposure to sulfuric acid
mist in Cynomolgers monkeys and guinea pigs. Arch. Envir. Health, 27:16-24, 1973.
Alarie, Y. C., A. A. Urumm, W. M. Busey, C. E. Ulrich and R. J. Kantz II, Long-term exposure to
sulfur-dioxide, sulfuric acid mist, flyash, and their mixtures, Arch. Env. Health, 30:254-62,1975.
Anderson, D. O., and B. G. Ferris, Jr., Air pollution levels and chronic respiratory disease,
Arch. Environ. Health, 10:307, 1965.
Angel, J. H., C. M. Fletcher, I. D. Hill et al., Respiratory illness in factory and office workers.
A study of minor respiratory illnesses in relation to changes in ventilatory capacity, sputum
characteristics, and atmospheric pollution, Br. J. Dis. Chest. 59:66, 1965.
Aronow, W. S., and M. W. Isbell, Carbon moxoxide effect on exercise-induced angina pectoris, Ann.
Intern. Med., 79:392, 1973.

Astrup, P., K. Kjeldsen, and J. Wanstrup, Enhancing influence of carbon monoxide on the development of atherosclerosis in cholesterol fed rabbits, J. Atherosclerosis. Res., 7:343, 1967.
Balchum, O. J., J. Dybicki and G. R. Meneely, Pulmonary resistance and compliance with concurrent readioactive sulfur distribution in dogs breathing $S^{35}O_2$, J. Appl. Physiol., 15:62-66, 1960.
Barter, C. E. and A. H. Campbell, Relationship of constitutional factors and cigarette smoking to decrease in 1-second forced expiratory volume, Amer. Rev. Resp. Dis., 113:305-314, 1976.
Bartlett, D., Jr. Effects of carbon monoxide on human physiological processes. In: Proceedings of the Conference on Health Effects of Air Pollutants, National Academy of Sciences, prepared for the Committee on Public Works, U.S. Senate, Committee Print, Serial No. 93-15, U.S. Government Printing Office, Washington, D.C., 103-126, 1973.
Bates, D. V., Air pollution and chronic bronchitis, Arch. Environ. Health, 14:220, 1967.
Bates, D. V., The fate of the chronic bronchitic: A report of the ten-year follow-up in the Canadian Dept. of Veteran's Affairs Coordinated Study of Chronic Bronchitis, Am. Rev. Respir. Dis., 108:1043, 1973.
Bates, D. V., G. M. Bell, C. D. Burnhave, M. Hazucha, J. Mantha, L. D. Pengelly and F. Silverman, Short-term effects of ozone on the lungs, J. Appl. Physiol. 32:176-181, 1972.
Bates, D. V., C. A. Gordon, G. I. Paul, R. E. G. Place, D. P. Snidal, and C. R. Woolf (with special sections contributed by M. Katz, R. G. Fraser, and B. B. Hale): Chronic bronchitis. Report on the third and fourth stages of the Coordinated Study of Chronic Bronchitis in the Dept. of Veterans Affairs, Candada, Med. Serv. J. Can., 22:5, 1966.
Bates, D. V., C. R. Woolf, and G. I Paul, Chronic bronchitis: A report on the first two stages of the Coordinated Study of Chronic Bronchitis in the Dept. of Veterans Affairs, Canada, Med. Serv. J. Can., 18:211, 1962.
Becklake, M. R., F. Aubiy, et al., Health effects of air pollution in the greater Montreal region, Dept. of Epidemiology, McGill University, Montreal, 1975.
Boyd, J. T., Climate, air pollution, and mortality, Br. J. Prev. Soc. Med., 14:123, 1960.
Brain, J. D., D. F. Proctor and L. M. Reid, Respiratory defense mechanisms, Parts I and II. Marcel Dekker, Inc., New York and Basel, 1977.
Brasser, L. J., P. E. Joosting, and D. von Zuilen, Sulfur oxide-to what level is it acceptable? Report G-300, Research Institute for Public Health Engineering, Delft, Netherlands (originally published in Dutch, 1966), 1967.
Buchley, Ramon, J. D. Hackney, K. Clark, C. Posin, Ozone and human blood, Arch. Environ. Health, 30, 1975.
Buechley, R. W., W. B. Riggan, V. Hasselblad, and J. B. Van Bruggen, SO_2 levels and perturbations in mortality. A study in the New York-New Jersey metropolis, Arch. Environ. Health, 27:134, 1973.
Buell, P., and J. E. Dunn, Relative impact of smoking and air pollution on lung cancer, Arch. Environ. Health, 15:291, 1967.
Burrows, B., R. J. Knudson, M. G. Cline and M. D. Lebowitz, Quantitative relationships between cigarette smoking and ventilatory function, Amer. Rev. Resp. Dis. 115:195-205, 1977.
Burrows, B., R. J. Knudson, M. D. Lebowitz, The relationship of childhood respiratory theory to adult obstructive airway disease, Am. Rev. Resp. Dis., 115:751, 1977.
Calvert, J. G., Interaction of air pollutants. In: Proceedings of the conference on health effects of air pollutants. Serial No. 93-15. Com. on Pub. Works, U.S. Senate, p. 19-102, 1973.
Cederlof, R., Urban factor and the prevalence of respiratory symptoms and angina pectoris. A study on 9168 twin pairs with the aid of mailed questionnaires. Arch. Env. Health, 13:743, 1966.
Chapman, R. S., C. M. Shy, J. F. Finkles, D. E. House, H. E. Goldberg, and C. G. Hayes, Chronic respiratory disease in military inductees and parents of school children. Arch. Environ. Health, 27:138, 1973.
Chiaramonte, L. T., J. R. Bongiorno, R. Brown and M. E. Laano, Air pollution and obstructure respiratory disease in children, N.Y. State J. Med., 70:394, 1970.
Cohen, A. A., S. Bromberg, R. W. Buechley, L. T. Heiderscheit, and C. M. Shy, Asthma and air pollution from a coal-fueled power plant, Am J. Public Health, 62:1182, 1972.
Cohen, C. A., A. R. Hudson, J. L. Clausen, J. H. Knelson, Respiratory symptoms, spirometry and oxidant air pollution in non-smoking adults, Am. Rev. Respir. Dis., 105:251, 1972.
Cohen, D., S. F. Arai and J. D. Brain, Smoking impairs long-term dust clearance from the lungs, Science, 204:514-516, 1979.
Cohen, S. I., M. Deane, and J. R. Goldsmith, Carbon monoxide and survival from myocardial infarction, Arch. Environ. Health, 19:510, 1969.
Cole, P., and M. B. Goldman, Occupation in Persons at High Risk of Cancer. An Approach to Cancer Etiology and Control, J. F. Fraumeni, Jr., ed., Academic Press, New York, 1975, pp. 167-184.
Colley, J. R. T., and D. D. Reid, Urban and social origins of childhood bronchitis in England and Wales, Br. Med. J., 3:213, 1970.
Comstock, G. W., R. W. Stone et al., Respiratory findings and urban living, Arch. Environ. Health, 27:143, 1973.
Da Silva, A. M. T., and P. Hamosh, Effect of smoking one cigarette or the "small airways", J. Appl. Physiol., 34(3):361-365, 1973.
Dohan, F. C., and E. W. Taylor, Air pollution and respiratory disease. A preliminary report, Am. J. Med. Sci., 240:337, 1960.
Douglas, R., Effect of nitrogen dioxide on resistance to respiratory infection, Bacteriol. Rev. 30:604, 1966.

Ehrlich, R, Effect of nitrogen dioxide on respiratory infection, Bacteriol. Rev. 30:604, 1966.
Fairbairn, A. S. and D. D. Reid, Air pollution and other local factors in respiratory disease, Br. J. Prev. Soc. Med., 12:94, 1958.
Fairchild, G. A., J. Roan, and J. McCarrol, Atmospheric pollutants and the patholgenesis of viral respiratory infection. Sulfur dioxide and influenza infection in mice. Arch. Environ Health, 25:174, 1972.
Fenters, J. D., J. N. Bradof, C. Aranyi, K. Ketels, R. Ehrilich, and D. E. Gardner, Health effects of long term inhalation of sulfuric acid mist-carbon particle mixture. Environ. Res., 19:244-257, 1979.
Ferris, Benjamin, Epidemiology Standardization Project, Am. Rev. Resp. Dis., 118:1, 1978.
Ferris, B. G., Jr., and D. O. Anderson, The prevalence of chronic respiratory disease in a New Hampshire town, Am. Rev. Respir. Dis., 86:165, 1962.
Ferris, B. G., Jr., I. T. T. Higgins, M. W. Higgins, J. M. Peters, W. F. Van Gause, and M. D. Goldman, Chronic non-specific respiratory disease, Berlin, New Hampshire 1961-1967: a cross-sectional study. Am. Rev. Resp. Dis., 104:232, 1971.
Ferris, B. G., Jr., I. T. T. Higgins, M. W. Higgins, and J. M. Peters, Chronic non-specific respiratory disease in Berlin, New Hampshire, 1961-1967, a follow-up study. Am. Rev. Respir. Dis., 107:110, 1973.
Ferris, B. G., Jr., H. Chem, S. Puleo, and R. L. H. Murphy, Jr. Chronic non-specific respiratory disease in Berlin, New Hampshire, 1967-1973. A further follow-up study. Am. Rev. Respir. Dis., 113:457, 1976.
Finklea, J. F., J. G. French, G. R. Lowrimore, J. Goldberg, C. M. Shy, and W. C. Nelson, Prospective surveys of acute respiratory disease in volunteer families. EPA, 1974a.
Folinsbee, L. J., S. M. Horvath, P. B. Raven, J. F. Bedi, A. R. Morton, B. L. Drinkwater, N. W. Boldman, and J. A. Gliner, Influence of exercise and heat stress on pulmonary function during ozone exposure, J. Appl. Physiol: Respirat. Environ. Exercise. Physiol. 43:409-413, 1977.
Frank, R., The effects of inhaled pollutants on nasal and pulmonary flow-resistance, Am. Otolo. Rhin. Laryn. 79:545:546.
Frank, N. R., M. O. Amdur, J. Worcester, and J. L. Whittenberger, Effects of acute controlled exposure to SO_2 on respiratory mechanics in healthy male adults, J. Appl. Physiol., 17:252,1962.
Frank, R., J. D. Brain, D. E. Knudson and R. A. Kronmal, Ozone exposure, adoptation, and changes in lung elasticity, Environ. Res. 19:449-459, 1979.
Freeman, G., S. C. Crane, R. J. Stephens, and N. H. Furiosol, Pathegenesis of the nitrogen dioxide induced lesion in the rat lung. A review and presentation of new observations, Am. Rev. Respir. Dis., 98:929, 1968.
Freeman, G., S. C. Crane, R. J. Stephens, and M. J. Furiosi, Environmental factors in emphysema and a model system with NO_2, Yale. J. Biol. Med., 40:566, 1968a).
Fujita, S., T. Motoichi, K. Shoji, Y. Ichiro, F. Takashi, S. Seigo, K. Tatsuo, and M. Michiko, Studies on chronic bronchitis-epidemiological survey (2nd report) Teishin Igaku, 21:13, 1969. English translation no. 1734, APTIC No. 28558, EPA Air Pollution Technical Information Service.
Glasser, M., L. Greenburg, and F. Field, Mortality and morbidity during a period of high levels of air pollution, New York, Nov. 23-25, 1965, Arch. Environ. Health, 15:684, 1967.
Goldberg, I. S., and R. V. Lourenco, Deposition of aerosols in pulmonary disease, Arch. Intern. Med. 31:88-91, 1973.
Golden, J. A., J. A. Nadel, and H. A. Boushey, Bronchial hyperirritability in health subjects after exposure to ozone, Amer. Rev. Resp. Dis., 118:287, 1978.
Goldsmith, J. R., and W. S. Arnow, Carbon monoxide and coronary heart disease: A review. Environ. Res., 10:236-248, 1975.
Goldstein, I. F., and G. Block, Asthma and air pollution in two city areas in New York City, J. Air Pollut. Control Assoc., 24:665, 1974.
Green, S. M., G. J. Jakab, R. B. Low and G. S. Davis, Defense mechanisms of the respiratory membrane, Amer. Rev. Resp. Dis., 115:479-514, 1977.
Greenberg, L., F. Field, J. I. Reed, and C. L. Erhardt, Asthma and temperature change. An epidemiological study of emergency clinic visits for asthma in three large New York Hospitals, Arch. Environ. Health, 8:642, 1964.
Haenszel, W., D. B. Loveland, and M. G. Sirken, Lung cancer mortality as related to resistance and smoking histories. II. White females, J. Natl. Cancer Inst., 32:803, 1964.
Hammond, E. C., Smoking habits and air pollution in relation to lung cancer, In: Environmental Factors in Respiratory Disease, D. H. K. Lee (eds.), Academic Press, N.Y., pp. 177-198, 1972.
Hammond, E. C., and D. Horn, Smoking and death rates - report on 44 months of follow-up of 187, 783 men. I. Total mortality II. Death Rates by cause, JAMA, 166, 1159, 1294, 1958.
Hexter, A. C., and J. R. Goldsmith, Carbon monoxide: association of community air pollution with mortality, Science, 172:265, 1971.
Hirayama, T., Cancer and air pollution and smoking, Chapter 5, In: Clinical Implications of Air Pollution Research, A. J. Finkel and W. C. Duel (Eds.), Acton, Mass., 1976.
Hackney, J. D., W. S. Linn et al., Effects of ozone exposure in Canadians and Southern California, Arch. Environ. Health, 32:110-116, 1977.

Hazucha, B., and D. V. Bates, Combined effects of ozone and sulfur dioxide on human pulmonary function, Nature, 257:50-51, 1975.
Hazucha, M., F. Silverman, C. Parent, S. Field and D. Bates, Pulmonary function in man after short-term exposure to ozone, Arch. Environ. Health, 27:183-188, 1973.
Hoffman, E. F., and A. G. Gilliam, Lung cancer mortality. Geographic distribution in the United States for 1948-1949, Public Health Rep., 69:1033, 1954.
Horvath, S. M., T. E. Dahms, J. F. O'Hanlon, Carbon monoxide and human vigilance. A deleterious effect of present urban concentrations, Arch. Environ. Health, 23:342, 1971.
Hrubec, Z., R. Cederlog, L. Friberg, and R. Horton, Respiratory symptoms in twins: Effects of residence-associated air pollution, tobacco and alcohol use, and other factors, Arch. Environ. Health, 27:189, 1973.
Ipsen, J., M. Deane, and F. E. Ingenito, Relationships of acute respiratory disease to atmospheric pollution and meterological conditions, Arch. Environ. Health, 18:463, 1969.
Irwig, L., D. G. Altman, R. J. W. Gibson, and C. Du V. Florey, Air pollution: methods to study its relationship to respiratory disease in British schoolchildren, Proceedings of the Internat'l. Symp. on Recent Advances in the Assessment of the Health Effects of Environmental Pollution, Luxembourg: Commission of the European Communities, vol. 1, pp. 289-300, 1975.
Jaeger, M. J., D. Tribble and H. J. Wittig, Effect of 0.5 ppm sulfur dioxide on respiratory function of normal and asthmatic subjects, Lung, 156:119-127, 1979.
Kagawa, J. and T. Toyama, Effects of ozone and brief exercise on specific airway conductance in man, Arch. Environ. Health, 30:36-39, 1975.
Kerr, H. D., T. J. Kulle, M. L. McIlhany and P. Swidersky, Effects of ozone on pulmonary function in normal subjects, Amer. Rev. Respir. Dis., 111:763-773, 1975.
Kreisman, H., C. A. Mitchell, H. R. Hosin and A. Bouhuys, Effect of low concentrations of sulfur dioxide on respiratory function in man, Lung, 154:25-34, 1976.
Lambert, P. M., and D. D. Reid, Smoking, air pollution and bronchitis in Britain, Lancet, 1:853, 1970.
Larson, T. V., D. S. Covert, R. Frank and R. J. Charlson, Ammonia in the human airways: Neutralization of inspired acid sulfate aerosols, Science, 197:161, 1977.
Lawther, P. J., A. J. MacFarlane, R. E. Waller and A. G. F. Brooks, Pulmonary function and sulfur dioxide, some preliminary findings, Environ. Res., 10:355-367, 1975.
Lebowitz, M. D., E. J. Cassell and J. D. McCarroll, Health and urban environment. XV. Acute respiratory episodes are reactions by sensitive individuals to air pollution and weather. Environ. Res., 5:135, 1972.
Lebowitz, M. D., A comparative analysis of the stimulus-response relationship between mortality and air pollution-weather, Environ. Res., 6:106, 1973.
Lindeberg, W., Air pollution in Norway. III. Correlations between air pollutant concentrations and death rates in Oslo, Smoke Damage Council, Oslo, Norway, 1968.
Linn, W. S., R. D. Buckley, C. E. Spier, R. L. Blessey, M. P. Jones, D. A. Fischer and J. D. Hackney, Health effects of ozone exposures in asthmatics, Amer. Rev. Resp. Dis., 117:835-843,1978.
Longo, L. D., The biological effects of carbon monoxide on the pregnant woman, fetus, and newborn infant, Amer. J. Obstet. Gyn., 129:69-103, 1977.
Love, G. J., A. A. Cohen, J. F. Finklea, J. G. Frence, G. R. Lowrimore, W. C. Nelson, and P. B. Ramsey, Prospective surveys of acute respiratory disease in volunteer families: 1970-1971, New York studies, in Health Consequences of Sulfur Oxides: A report from CHESS, 1970-1971, EPS-650/1-74-004, U.S. EPA, Off. of Res. & Dev., U.S. Gov't. Print. Off., Washington, 5:49-69,1974.
Lunn, J. E., J. Knowelden and A. J. Handyside, Patterns of respiratory illness in Sheffield infant school children, Br. J. Prev. Med., 21:7, 1967.
Lunn, J. E., J. Knowelden and J. W. Roe, Patterns of respiratory illness in Sheffield junior schoolchildren. A follow-up study. Br. J. Prev. Soc. Med., 24:223, 1970.
McJilton, C., R. Frank and R. Charlson, Role of relative humidity in the synergistic effect of a sulfur dioxide-aerosol mixture on the lung, Science, 182:503, 1973.
Martin, A. E., Mortality and morbidity statistics and air pollution, Proc. R. Soc. Med, 57:969,1964.
Martin, A. L. and W. Bradley, Mortality, fog and atmospheric pollution, Monthly Bull. Ministry Health., 19:56, 1960.
Medical Research Council, Committee on the Aetiology of Chronic Bronchitis: Standardized questionnaire on respiratory symptoms, Br. Med. J, 3:1665, 1960.
Miller, F. J., A mathematical model of transport and removal of ozone in mammalian lungs. Ph.D. Thesis, North Carolina State University, Raleigh, N.C., 1977.
Mork, T., A comparative study of respiratory disease in England and Wales and Norway, Norwegian University Press, Oslo, 1962.
Morrow, P. E., An evaluation of recent NO_x toxicity data and an attempt to derive an ambient air standard for NO_x by established toxicological procedures, Environ. Res., 10:92-112, 1975.
Mustafa, M. G., and D. F. Tierney, Biochemical and metabolic changes in the lung with O_2, O_3 and NO_2 toxicity, Amer. Rev. Resp. Dis., 118:1061-1078, 1978.
Nadel, J. A. and J. H. Comroe, Jr., Acute effects of inhalation of cigarette smoke or airway conductance, J. Appl. Physiol., 16:713-716, 1961.

Nadel, J. A., H. Salem, B. Tamplin and Y. Tokiwa, Mechanism of bronchoconstriction during inhala-
    tion of sulfur dioxide, J. Appl. Physiol., 20:164-167, 1965.
Newhouse, M. T., M. Dolovich, G. Obiminsk and R. K. Wolff, Effect of TLV levels of SO_2 and H_2SO_4
    on bronchial clearance in exercising man, Arch. Environ. Health, 33:24-31, 1978.
Norris, R. M. and J. M. Bishop, The effect of calcium carbonate dust on ventilation and respira-
    tory gas exchange in normal subjects and in patients with asthma and chronic bronchitis,
    Clin. Sci., 30:103, 1966.
Office of Technology Assessment, Direct use of coal, Congress of U.S.A., Washington, 20510, 1979.
Orehek, J., J. P. Massori, P. Gayrard, C. Grimarid and J. Charpin, Effect of short-term, low-
    level nitrogen dioxide exposure on bronchial sensitivity of asthmatic patients, J. Clin. In-
    vest., 57:301, 1976.
Petrilli, F. L., G. Angese, and S. Kanitz, Epidemiologic studies of air pollution effects in
    Genoa, Italy, Arch. Environ. Health, 12:733, 1966.
Rao, M., P. Steiner, Q. Qazi, R. Padre, J. E. Allen, and M. Steiner, Relationship of air pollu-
    tion to attack rate of asthma in children, J. Asthma Res., 11:23, 1973.
Reichel, G., Effect of air pollution on the prevalence of respiratory diseases in West Germany,
    In: Proceedings of the Second Internal Clean Air Congress, Washington, D.C., 1970.
Reid, D. D., The beginnings of bronchitis, Proc. R. Soc. Med, 62:311, 1969.
Roehm, J. N., S. G. Hadley and D. B. Menzel, Oxidation of unsaturated fatty acids by O_3 and NO_2:
    A common mechanism of action, Arch. Environ. Health, 23:142, 1971.
Sackner, M. A., D. Ford, R. Fernandez, et al., Effects of sulfuric acid aerosol on cardiopulmonary
    function of dogs, sheep and humans, Amer. Rev. Resp. Dis., 118:497-510, 1978.
Said, S. I., Environmental injury of the lung: role of humoral mediators, Fed. Proceed., 37:
    2504-2506, 1978.
Schimmel, H., and L. Greenburg, A study of the relation of pollution to mortality. New York City,
    1963-1968, J. Air Pollut. Control Assoc., 22:607, 1972.
Schrenk, H. H., H. Heinmann, G. D. Clayton, W. Gafafer, and H. Wexler, Air pollution in Donora,
    Pennsylvania. Epidemiology of the unusual smog episode of October 1948, Public Health Bulletin
    306, U.S. Government Printing Office, Washington D.C., 1949.
Schwartz, L. W., D. L. Dungworth, M. G. Mustafa, B. K. Tarkington and W. S. Tyler, Pulmonary re-
    sponses of rats to ambient levels of ozone, Lab. Invest. 34:565-578, 1976.
Shepard, R. J., Cigarette smoking and reactions to air pollutants, CMA Journal, 118:379-383, 1978.
Sim, V. M., and R. E. Pattle, Effect of possible smog irritants on human subjects, J. Amer. Med.
    Assoc., 165:1903-1913, 1957.
Stuart, B. O., Deposition of inhaled aerosols, Arch. Intern. Med., 131:60-73, 1973.
Tsenetoshi, Y., T. Shimizu, H. Takahasi, A. Ichinosawa, M. Ueda, N. Nakayama, Y. Yamagata, Epi-
    demiological study of chronic bronchitis with special reference to effect of air pollution,
    Int. Arch. Arbeitsmed., 29:1, 1971.
Ulmer, W. T., G. Reichel, A. Czeike, and A. Leuschner, Regional incidence of non-specific
    respiratory diseases, IV. Communication, Int. Arch. Arebeitsmed., 27:73, 1970.
United States Dept. of Energy, April 1979, Fuel Use Act-Final Environmental Impact Statement,
    DOE-EIS-0038, Washington, D.C., 30461.
Verma, M. P., F. J. Schilling, and W. H. Becker, Epidemiological study of illness absences in re-
    lation to air pollution, Arch. Environ. Health, 18:536, 1969.
Waldbott, G. L., Health Effects of Environmental Pollutants, 2nd Ed., The C.V. Mosby Co., St.
    Louis, 1978.
Watanabe, H., and F. Kaneko, Excess death study of air pollution, In: Proceedings of the Second
    International Clean Air Congress, H. M. Englund and W. T. Beery (Eds.), Academic Press,
    pp. 199-200, 1971.
Wehner, A. P., O. R. Moss, E. M. Milliman, G. E. Dagle and R. E. Schirmer, Acute and sub-chronic
    inhalation exposures of hamsters to nickel-enriched flyash, Environ. Res. 19:355-370, 1979.
Weill, H., M. M. Ziskind, V. Derbes, R. Lewis, R. J. Horton, and R. O. McCaldin, Further observa-
    tion on New Orleans asthma, Arch. Environ. Health, 8:184, 1964.
Woolcock, A. J., and N. Berend, The effects of smoking on the lungs, Aust. N.Z.J. Med., 7:649-
    662, 1977.
Zuskin, E., C. A. Mitchell, and A. Bouhuys, Interaction between effects of beta blockade and
    cigarette smoke on airways, J. Appl. Physiol., 36:449-452, 1974.

CHAPTER 15

QUANTITATIVE PUBLIC POLICY ASSESSMENTS

Alex E. S. Green and Daniel E. Rio

I. INTRODUCTION TO QUANTITATIVE PUBLIC POLICY ASSESSMENT

In this chapter we will report the overall approach and principal results of a study (ICAAS, 1978) which attempted to carry out an integrated interdisciplinary assessment of air pollution abatement alternatives in the Tampa area of Florida, a region where coal is a major source of electric power. The study was carried out under the auspices of the Florida Sulfur Oxide Study, Inc. (FSOS-1978), a non-profit corporation funded by the Electric Utilities of Florida and supervised by a board consisting of representatives from the Electric Utilities, environmental groups and the Florida Department of Environmental Regulation. The study provided realistic tests of public policy decision methodologies (PPDM) applicable to coal burning issues.

ICAAS's initial conceptualization of models for public policy decisions on regulating air pollution began in connection with a proposed Air Quality Index (AQI) Project (ICAAS, 1970). This AQI project was designed to be a broad socio-technical research program leading to the establishment of a quantitative scale (or scales) for air quality, to validation of this scale (or scales), and to development of laws to implement air quality control, particularly in Florida and the Southeast. From this AQI program plan, we developed our first practical cost/benefit analysis approach. Figure 1 is an adaptation of a flow diagram which originated in a proposal entitled, "Preservation and Enhancement of Air Quality (PEAQ)," (ICAAS, 1971).

303

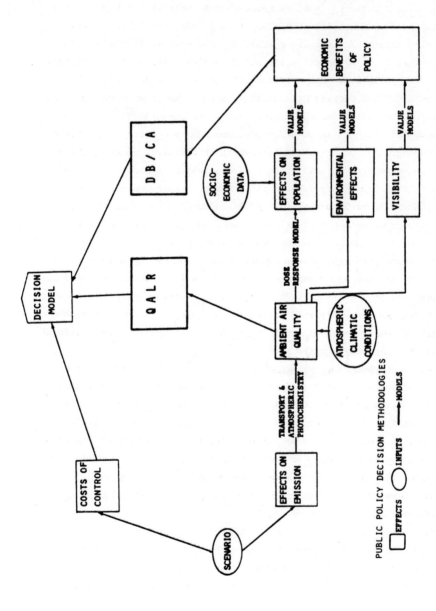

Fig. 1. Systems flow diagram of ICAAS-FSOS public policy decision methodologies (PPDM).

These two overall systems approaches illustrated in Figure 1 were implemented in the ICAAS-FSOS study (ICAAS-FSOS, 1978, 1979, 1980). The flow diagram illustrates a chain of component studies within the framework of two types of public policy decision methodologies on sulfur oxide pollution. One methodology, the Disaggregated Benefit/Cost Analysis (DB/CA), is essentially an advanced form of economic analysis in which the distributional aspects of B/C are considered (i.e. the question of who gets the benefits and who pays the costs). The second methodology, the Quantitative Assessment of the Level of Risk (QALR), is a non-economic analysis that by-passes many of the difficult problems of translating important decision factors into monetary terms. In particular, it by-passes the difficulties which arise because costs for pollution abatement are usually expressed in current dollars, whereas many benefits in human health, aesthetics, social good, etc. are expected to manifest themselves decades later. Hence, to the extent that benefits can be translated into dollar values, the currency of the future will be involved. Thus, the long term discount rate must play a major role in B/C analyses. Unfortunately, all of the world's problems influence the long term discount rate, not only the particular regional pollution abatement problem.

In this chapter we will concentrate on a few specific but vital facets of the flow diagrams in Figure 1. In particular, in Section II we discuss quantitative characterizations of ambient air quality from air pollution health effects viewpoint. In Section III we describe an approach to quantitative dose response relations based on the use of an air quality index (AQI). In Section IV we describe the application of the QALR-PPDM used in the ICAAS-FSOS study which also uses an AQI. finally, in Section V we describe the essence of the DB/CA-PPDM used in the ICAAS-FSOS study.

II. AIR QUALITY INDICES AND FACTORS OF SAFETY

The present state of knowledge concerning health effects of low levels of pollutants is still at a very primitive stage. Indeed, ex-

tensive medical testimony at Congressional hearings held in November,
1975, (Brown, 1975) on "The Costs and Effects of Chronic Exposure to
Low Level Pollutants in the Environment" indicate that we know almost
nothing about the health effects of long-term exposure to low levels
of pollutants. The discussion by Jaeger and Schlenker (Chap. 14)
confirms this assessment. Since quantitative approaches are needed in
most decision methodologies for air pollution control, we have examined
the use of a variety of air quality indices based on various algebraic
functions of the concentrations of major air pollutants. Using these
air quality indices, we have adapted for air pollution the "Factor of
Safety" concept (Green et al., 1978, 1979) used in engineering fields.

To define the Factor of Safety (FS) in engineering practice, the
best technical judgment first must be used to estimate the load limit
(i.e., the load level at which the system would be expected to fail).
The factor of safety is then defined as this load limit divided by the
maximum applied load reported or foreseen for that structure or system.

Our present lack of knowledge about effects of low pollutant levels makes it impossible to specify a precise load limit for the human
pulmonary system. However, for now, we identify as the pulmonary load
limit an air quality index corresponding to air in which all five major
pollutants are present simultaneously at the National Ambient Air Quality Standards (NAAQS). The air quality factor of safety based upon any
index I_j is then defined as the ratio

$$FS_j = \frac{I_j(C_1^*, C_2^*, C_3^*, C_4^*, C_5^*)}{I_j(C_1, C_2, C_3, C_4, C_5)} \tag{1}$$

where C_1^*, C_2^*, C_3^*, C_4^*, C_5^* denote the NAAQS concentration for the five
major pollutants TSP, SO_2, NO_2, Ox, and CO and C_1, C_2, C_3, C_4, and C_5
denote the actual reported concentrations. Thus, larger values (>1)
indicate cleaner air. Green et al. (1979) (to be referred to as GBRMM)
have examined six air quality indices for the purposes of providing factor of safety ratings for selected American cities. These indices en-

compass a variety of linear, quadratic, biquadratic, and nonlinear mathematical forms.

The PSI Index (Ott and Thom, 1976; Ott and Hunt, 1976) has been recommended (EPA, 1976) for use by the Council on Environmental Quality (1976) and by the Environmental Protection Agency as a common air pollution reporting system. This index consists of an array of sub-indices, each based on concentrations of five major pollutants. Algebraic prescriptions are given for calculating these sub-indices and the PSI is defined as the largest of all these sub-indices. With one exception, the sub-indices are all based upon single pollutants. The exception, in recognition of the synergistic involvement of total suspended particulates and sulfur dioxide, utilizes a sixth sub-index based on products of the concentration of SO_2 and TSP. Table 1 gives the critical levels used in the PSI index system and various descriptions characterizing the impacts of these levels.

The remaining air quality indices examined are continuous mathematical functions. They are the ICAAS index (Green et al., 1978), the Ottawa index (Inhaber, 1975), the Oak Ridge National Laboratory Air

Table 1: *Pollution Standard Index Levels and Health Descriptor*

Index value	TSP (24-hr), $\mu g/m^3$	SO_2 (24-hr), $\mu g/m^3$	CO (8-hr), mg/m^3	O_3 (1-hr), $\mu g/m^3$	NO_2 (1-hr), $\mu g/m^3$	Air Quality level	Health descriptor
500 (5)	1000 (3.85)	2620 (7.18)	57.5 (5.75)	1200 (7.5)	3750 (7.5)	Significant harm	Hazardous
400 (4)	875 (3.37)	2100 (5.75)	46.0 (4.6)	1000 (6.25)	3000 (6.0)	Emergency	
300 (3)	625 (2.4)	1600 (4.38)	34.0 (3.4)	800 (5)	2260 (4.38)	Warning	Very unhealthful
200 (2)	375 (1.44)	800 (2.10)	17.0 (1.7)	400 (2.5)	1130 (2.19)	Alert	Unhealthful
100 (1)	260 (1)	365 (1)	10.0^a (1)	240 (1)	500^a (1)	NAAQS	Moderate
50 (½)	75^b (½)	80^b (½)	5.0 (½)	120 (½)	100^b (½)		Good

a - 40 mg/m^3 used as 1-hr. max.   b - annual average NAAQS

Quality Index (ORNL) (Thomas et al., 1971), the Toronto index (Shen-
feld, 1970), and an index based on the American Council of Governmental
Industrial Hygienist (ACGIH) threshold pollution values (ACGIH, 1977).
These are presented in Table 2 in terms of the symbols C_i - concentration of pollutants; S_i, scale factors and T_i, threshold limit values.

Table 2: *Air Quality Index Formulas*

$$I(ICAAS)=\left[\sum_i (C_i/S_i)^2 + b_{ij}(C_i/S_i)(C_j/S_j)\right]^{\frac{1}{2}} \quad I(OTTAWA)=\left[\sum_i (C_i/S_i)^2\right]^{\frac{1}{2}}$$

$$b_{ij} = 0 \text{ except } b_{12} = 1.5 \quad I(ORNL)=\left[\sum_i 5.7\ C_i/S_i\right]^{1.37}$$

$$I(TORONTO)=2(30.5[COH]+126.0[SO_2])^{1.35} \quad I(ACGIH) = \sum_i C_i/T_i$$

To illustrate the calculation of FS, Table 3 gives the levels of
pollutants for 1976 for various regions as reported by the Florida Department of Environmental Regulation (FDER). The row labeled NAAQS
gives the National Ambient Air Quality Standards for these time periods.
The row labeled FAAQS shows the corresponding Florida standards.

Table 3: *Pollution levels at various locations in Florida in 1976
reported by the Florida Department of Environmental
Regulation.*

Loc.	POLLUTION LEVELS 1976	C_{TSP} $\mu g/m^3$	C_{SO_2} $\mu g/m^3$	C_{NO_2} $\mu g/m^3$	C_{O_x} (a)	C_{CO} (a)
1.	Tampa-Hills	69.3	39.6	58.0	.13	18.5
2.	Miami-Dade	64.4	8.7	31.3	.07	22.0
3.	Duval-	59.7	22.3	59.0	.18	17.5
4.	St. Pete-Pinellas	59.7	18.3	43.1	.132	12.0
5.	Gainesville-Alachua	47.1	5.62	26.4	.110	5.0
6.	Tallahassee-Leon	37.3	6.0	26.4	.114	2.4
7.	Orange-Orlando	49.4	6.3	2.9	.130	5.0
8.	Palm Beach	56.4	26.2	23.4	.148	10.5
9.	Ft. Laud.-Broward	59.0	5.24	35.6	.105	23.0
NAAQS		75.0	80.0	100.0	.08	35
FAAQS		60.0	60.0	100.0	.08	35

(a) 1 hr. max. (ppm)

Table 4 gives the safety factors for the regions in Florida for various indices where column (2) is based upon the PSI system, (3) the ICAAS index, (4) the ORNL index, (5) the Toronto index, and (6) an index based upon the ACGIH rule for threshold limit values. Clearly the nine areas of Florida have significantly differing air qualities. The relative assignments of the air quality FS do not differ substantially for any of the indices.

Table 4: *Safety factors for 1976 concentrations for various indices*

(1) LOC	(2) PSI	(3) ICAAS	(4) ORNL	(5) TOR	(6) ACGIH
1	0.87(4)	1.14	1.29	1.60	1.06
2	1.36(1)	1.74	2.23	2.71	1.57
3	0.67(4)	0.98	1.18	2.32	0.90
4	0.86(4)	1.27	1.66	2.51	1.19
5	0.98(4)	1.62	2.61	4.23	1.60
6	0.96(4)	1.64	2.84	5.54	1.65
7	0.87(4)	1.43	2.54	3.93	1.99
8	0.78(4)	1.18	1.65	2.32	1.16
9	0.01(4)	1.48	1.85	3.22	1.27
FAAQS	1.23(6)	1.13	1.14	1.42	1.05
L.L.	100	2.55	98.6	28.1	0.60

(1) TSP, (2) SO_2, (3) NO_2, (4) O_x, (5) CO, (6) $TSPxSO_2$

The factors of safety for the ICAAS, OTTAWA, ORNL, TOR, ACGIH, PSI and ICAAS RISK systems have been calculated by GBRMM on a nationwide data base. Some results are given in Table 5. The 82 cities listed all have populations greater than 100,000. The only cities considered were those for which all five pollutant readings were present in the Air Quality Data - 1976 Annual Statistics (EPA. 1978) report.

The FS factors in Table 5 reflect, in part, varying practices for monitoring pollution. They illustrate the potential usefulness of the FS technique, should uniform standards of reporting be adopted, and suggest some interesting trends. The column to the right of the PSI column gives the pollutant responsible for the PSI-FS value. This number varies

Table 5: *Air Pollution in Major U.S. Cities in 1976*

ST. CITY	ICAAS	OTTAWA	ORNL	TOR	ACGIH	PSI		RISK
ACGIH STDS.								
STANDARDS								
AL BIRMINGHAM	1.29	1.16	1.48	1.80	1.11	1.37	4	1.28
HUNTSVILLE	1.84	1.63	3.00	4.19	1.61	1.46	4	1.88
AZ PHEONIX	1.15	1.03	1.30	1.20	1.16	0.97	1	1.07
TUSCON	1.57	1.41	1.91	2.19	1.40	1.79	4	1.55
CA ANAHEIM	0.81	0.72	0.80	1.53	0.63	0.87	4	0.87
FRESNO	0.96	0.85	1.01	1.12	0.86	0.91	1	0.92
LOS ANGELES	0.75	0.68	0.73	1.20	0.62	0.85	4	0.81
OAKLAND	1.54	1.36	1.90	3.61	1.23	1.38	4	1.56
PASADENA	0.75	0.68	0.78	1.62	0.60	0.81	4	0.84
SACRAMENTO	1.60	1.42	1.99	3.37	1.28	1.46	4	1.62
SAN BERNADIN	0.86	0.76	1.03	1.68	0.71	0.86	4	0.95
SAN DIEGO	1.38	1.24	1.68	2.88	1.14	1.27	4	1.41
SAN FRANCISC	1.66	1.49	1.95	3.56	1.35	1.69	4	1.65
SAN JOSE	1.19	1.05	1.28	2.46	0.94	1.20	4	1.21
CO DENVER	1.02	0.92	1.03	1.57	0.83	1.11	4	1.03
CT BRIDGEPORT	1.05	0.98	1.18	1.67	0.86	1.01	4	1.10
HARTFORD	1.04	0.96	1.19	1.91	0.84	0.99	4	1.11
NEW HAVEN	0.92	0.85	0.99	1.60	0.73	0.92	4	0.99
DC WASHINGTON	1.18	1.10	1.37	1.86	0.96	1.09	4	1.22
FL JACKSCNVILLE	2.06	1.94	2.73	3.53	1.72	1.90	4	2.04
MIAMI	1.96	1.75	2.55	3.79	1.61	1.92	4	1.95
ST PETERSBUR	2.09	1.89	2.96	4.21	1.72	1.76	4	2.10
TAMPA	1.89	1.78	2.47	2.69	1.69	1.91	4	1.88
HI HONOLULU	1.71	1.54	2.26	3.64	1.35	1.50	4	1.73
IL CHICAGO	1.28	1.22	1.47	1.55	1.19	1.46	4	1.27
PEORIA	1.22	1.27	1.48	1.10	1.35	1.04	6	1.20
IN EVANSVILLE	1.75	1.73	2.36	2.01	1.74	1.96	4	1.74
INDIANAPOLIS	1.44	1.40	1.70	1.68	1.35	1.74	4	1.43
IA CEDAR RAPIDS	1.18	1.19	1.45	1.11	1.20	1.05	6	1.16
DES MOINES	1.49	1.34	1.89	1.75	1.45	1.37	1	1.44
KS KANSAS CITY	1.58	1.50	2.19	1.59	1.77	1.45	1	1.53
TOPEKA	2.32	2.09	3.54	2.97	2.52	2.40	1	2.31
WICHITA	1.86	1.68	2.48	3.18	1.55	1.75	4	1.86
KY LEXINGTON	2.03	1.95	2.72	3.35	1.79	2.13	4	2.06
LOUISVILLE	1.32	1.32	1.54	1.44	1.25	1.30	6	1.31
MD BALTIMORE	1.29	1.22	1.49	1.67	1.14	1.36	4	1.29
MI DETROIT	1.09	1.01	1.22	1.55	0.91	1.06	4	1.12
GRAND RAPIDS	1.53	1.39	2.01	3.14	1.25	1.31	4	1.57
MN MINNEAPOLIS	1.61	1.57	1.95	1.92	1.55	2.05	4	1.59
ST. PAUL	1.69	1.59	2.13	2.15	1.60	1.93	4	1.67
MO ST. LOUIS	0.75	0.70	0.79	1.01	0.63	0.82	4	0.82
NB OMAHA	1.46	1.33	1.74	2.13	1.24	1.50	4	1.45
NV LAS VAGAS	1.29	1.17	1.60	1.40	1.26	1.08	1	1.23
NM ALBUQUERQUE	1.39	1.29	1.66	1.65	1.31	1.51	1	1.36
NY BUFFALO	1.49	1.53	1.87	1.53	1.59	1.42	6	1.47
N.Y. CITY	1.54	1.60	2.09	1.53	2.06	1.40	6	1.53
ROCHESTER	1.94	1.95	2.56	2.41	1.87	2.22	4	1.92
SYRACUSE	1.80	1.85	2.51	1.85	2.21	1.98	6	1.78
NC CHAROLETTE	1.35	1.23	1.69	3.23	1.04	1.15	4	1.40
DURHAM	1.63	1.50	1.99	2.63	1.36	1.64	4	1.63
OH AKRON	1.04	0.99	1.08	1.43	0.90	1.25	5	1.06
CANTON	1.55	1.50	2.05	1.89	1.48	1.61	4	1.55
CINCINNATI	1.24	1.17	1.46	1.70	1.07	1.21	4	1.26
CLEVELAND	0.98	0.96	1.00	1.03	0.96	1.03	6	0.98
COLUMBUS	1.62	1.57	2.04	1.92	1.55	1.86	4	1.60
DAYTON	1.23	1.16	1.45	1.67	1.06	1.21	4	1.25
TOLEDO	1.39	1.35	1.67	1.70	1.26	1.48	4	1.39
OK TULSA	1.20	1.06	1.38	3.23	0.95	1.11	4	1.25
OR PORTLAND	1.62	1.52	1.88	2.56	1.39	1.82	4	1.61
PA ALLENTOWN	1.58	1.61	2.07	1.69	1.66	1.62	6	1.57
ERIE	1.96	1.97	3.22	2.03	2.47	2.33	1	1.95
PHILADELPHIA	1.08	1.07	1.20	1.16	1.00	1.06	6	1.09
RI PROVIDENCE	1.17	1.11	1.33	1.61	0.98	1.14	4	1.20
TN CHATTANOOGA	1.74	1.62	2.26	2.50	1.53	1.71	4	1.73
KNOXVILLE	1.63	1.49	2.06	2.62	1.38	1.54	4	1.63
MEMPHIS	1.36	1.29	1.55	1.81	1.21	1.50	4	1.36
NASHVILLE	1.62	1.47	1.94	2.88	1.33	1.64	4	1.62
TX AUSTIN	1.98	1.77	2.82	3.10	1.75	1.82	4	1.98
CORPUS CHRIS	1.65	1.46	2.38	2.75	1.42	1.44	4	1.67
DALLAS	1.55	1.37	2.28	3.61	1.29	1.26	4	1.61
EL PASO	1.25	1.13	1.57	1.56	1.12	1.27	1	1.23
FORT WORTH	1.51	1.33	2.08	3.44	1.22	1.25	4	1.56
HOUSTIN	0.95	0.84	1.27	2.56	0.78	0.88	4	1.07
SAN ANTONIO	1.36	1.20	1.79	3.44	1.06	1.15	4	1.42
UT SALT LAKE CI	1.28	1.21	1.37	1.65	1.16	1.72	4	1.27
VA ALEXANDRIA	1.23	1.14	1.45	2.28	1.00	1.12	4	1.28
NORFOLK	1.28	1.19	1.43	2.14	1.03	1.35	4	1.30
RICHMOND	1.54	1.48	1.82	2.01	1.37	1.72	4	1.53
WA SEATTLE	1.75	1.67	2.14	2.34	1.74	2.50	1	1.73
SPOKANE	1.39	1.35	1.76	1.40	1.64	1.39	1	1.35
TACOMA	2.15	2.15	3.10	2.56	2.61	2.74	1	2.14
WI MILWAUKEE	1.14	1.07	1.36	1.73	0.96	1.06	4	1.19

from 1 to 6 corresponding to TSP, SO_2, NO_2, O_x, CO, and the TSP x SO_2 synergism, respectively. By examining this list for the 82 cities, we see that 12 of the PSI values are due to TSP, 56 to oxidants, and 9 to the TSP x SO_2 synergism.

The ordering of air quality of cities by the FS method is relatively insensitive to the index formulation used. To verify this, GBRMM calculated correlation coefficients for the values of the FS, using each index against the rest from the results. The four combinative indices (ICAAS, OTTAWA, ORNL, and ACGIH) all correlate very well with each other (with correlation coefficients near .9). Even the PSI index correlates reasonably well with the combinative formulas. The only index which does not correlate with any of the others is the Toronto index, a result undoubtedly due to the lack of consideration of oxidants. Since the ICAAS index has the highest correlation with the PSI index (0.89), it can be used as a continuous combinative formula giving air quality values similar to those assigned by the PSI. Finally, since a factor of safety less than 1 exceeds the defined pulmonary load limit, Table 5 identifies the more marginal cities in the U.S. from a standpoint of overall air pollution.

As an alternative to the approach based upon AQI's, one might develop a risk assessment for each pollutant (i.e., a response which in effect is a dose-response formula or dose-risk formula). Then, one could use combinatorial rules to develop a composite risk. The health risk levels associated with various pollutant levels as given in Table 1, which are in the pollution standard index scheme and are sanctioned by the Council on Environmental Quality (1976), can serve in this regard. The index value for various pollution concentration levels constitutes a risk level on a judgmental scale. The points in Figure 2 illustrate these risk levels for scaled TSP, O_3, and SO_2 concentrations. The curves are sigmoidal logistic curves constrained to pass through the 0,0 and 1,1 points and to achieve an optimum fit in a least sense to the remaining points. The other curves show smooth risk-concentration curves obtained in a similar way from the data in Table 1. Table 6

gives the equations and parameter values for two and one parameter
logistic curves. The fits of the one parameter functions are not
significantly different from the corresponding two parameter fits shown
in Figure 2.

Table 6:   *Logistic Function and Parameters*

	2 Parameters $y = p \dfrac{a^x - 1}{Pa-P-a + a^x}$		1 Parameter $y = P \dfrac{x}{P-1 + x}$
	P	a	P
TSP	72.4	1.94	
SO_2	11.4	0.99	11.3
CO	21.0	1.02	27.9
O_3	34.4	1.04	∞
NO_2	10.7	0.97	10.4

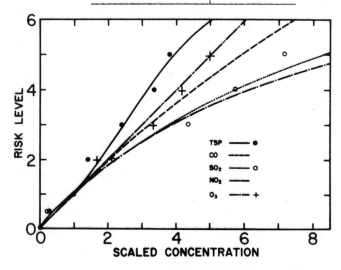

*Fig. 2.   Relative risk levels vs concentrations scaled by NAAQS
for the five major atmospheric pollutants. The points correspond to
the critical PSI index assignment for TSO, O_3 and SO_2. The smooth
curves correspond to two parameter logistic function fits to the PSI
index critical level for all five pollutants (see Table 1).*

Having defined reasonable risk levels on a common scale for various pollutant concentrations, we may seek out various combinatorial rules in terms of the risks rather than in terms of the scaled concentrations used previously. One could then calculate factors of safety in an analogous fashion to that used previously with scaled concentrations.

An evaluation of this Risk-Factor of Safety for the cities listed is given in the last column of Table 2. Essentially the 2 parameter logistic was used along with the ICAAS combination rule for the RISK. The TSP x SO_2 synergism term was taken as a measure of risk itself and was combined according to the ICAAS formula. These risk factors of safety obtained are very close to the values for the ICAAS index, based upon scaled concentrations.

III. DOSE RESPONSE RELATIONSHIPS

One of the great problems in all epidemiological studies is to define what constitutes a dose, dose-rate, or other quantitative measure of the hazardous environmental agent. The factor of safety concept, as defined in previous sections, provides a number of way to define a composite air pollution dose. The reciprocal of the factor of safety which increases as air pollution gets worse might be used as a measure of a normalized dose. Thus we might use as a basic definition

$$d = 1/FS .$$

For human variability we could assume that failure (illness) would occur when $d > 1$. In point of fact, we know that the probability of induced illness in a population is expected to be a sigmoidal function of dose.

Such a dose response curve has been noted in investigations of ultraviolet (UV) dose and skin cancer response (Green and Hedinger, 1978), Green, 1977), Green, 1978), (Fears et al., 1977), (Green et al., 1976), Blum, 1959), which is now one of the best quantified environmental carcinogens (Setlow, 1978). The same dose response curve is also suggested in studies of chemical-and radiation-induced carcinogenesis (Druckery,

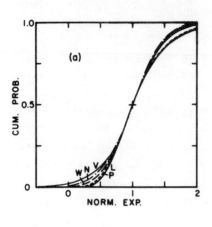
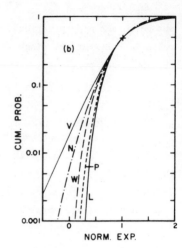

Fig. 3. Two graphs of cumulative probability (CUM. PROB.) vs normalized exposure (NORM. EXP.). The distributions are: L, lognormal; P, Photographic; W, Weibull; N, Normal; and V, Verhulst. The types of paper are (a) linear, and (b) semi-log. The parameters of the five distributions have been chosen to have the same value and slope at the 50th percentile (see +).

Table 7: *Analytic sigmoid type cumulative distribution functions $F(x)$ where $x = E/E_O$ where $E = \phi DT$ and the corresponding differential distributions $R = dF/dT$. For air pollution studies E_O serves as a scaling dose, ϕ as a relative exposure factor, D as an annual dose and T as the number of years.*

Symbol	Model	Cumulative Dist.	Differential Dist.
L	Log Normal	$\frac{1}{2}\{1 + \mathrm{erf}\,[\zeta \ln x]\}$	$\frac{\zeta\phi}{\sqrt{\pi}\,E}\exp-(\zeta \ln x)^2$
P	Photographic	$\frac{1}{\eta}\ln\frac{1 + x^\gamma}{1 + x^\gamma e^{-\eta}}$	$\frac{\gamma\phi D}{E_0}\frac{x^{\gamma-1}}{[1 + x^\gamma][1 + x^\gamma e^{-\eta}]}$
W	Weibull	$1 - e^{-x^\gamma}$	$\frac{\phi D}{E_0}x^{\gamma-1}e^{-x^\gamma}$
N	Normal	$\frac{1}{2}[1 + \mathrm{erf}\,(x - \eta)]$	$\frac{\phi D}{E_0\sqrt{\pi}}e^{-(x-\eta)^2}$
V	Verhulst	$[1 + \exp-(x - \eta)]^{-1}$	$\frac{\phi D}{E_0}\frac{\exp-(x - \eta)}{(1 + \exp-(x - \eta))^2}$

1966), (Albert and Altshuler, 1972), in studies of aging (Burch, 1969), pharmacology (Garrett, 1973), and in stress failure studies in relia- bility theory (Shooman, 1968).

Figure 3 illustrates a number of cumulative sigmoidal functions such as those listed in Table 7 which have been used in epidemiological studies. Preliminary applications of the sigmoidal dose response model to air pollution have been initiated. Toward this end, we present two models:

(1) A model for the number of acute asthma attacks a person will suffer per year as a function of age and air pollution normalized dose (1/FS). Data on the functional age dependence was obtained from a study (Broder et al., 1974) which measured the prevalence of asthma by age in a large population. In modeling the dose dependency of asth- matic people, we used the Weibull cumulative distribution model given in Table 8. The parameters of this model were adjusted to fit data from two studies (Yoshida et al., 1966). (See also Figure 4.)

(2) A model for the prevalence of chronic bronchitis as a function of age, rate of cigarette smoking, and air pollution normalized dose. Chronic bronchitis, as the name implies, is an illness with a very long recovery time and we will, therefore, use a dose response model that predicts prevalence as a function of dose. The model is similar to the Weibull cumulative model (Table 7).

Having examined many of the epidemiological studies on chronic bronchitis prevalence, we have used two studies (Tsunetoshi et al., 1967; Lambert and Reid, 1970) with reasonable sample sizes to obtain data for detailed modeling. In both studies, bronchitis rate data were presented for various levels of air pollution levels and smoking rates. In addition, Lambert and Reid gave age specific data. Tsunetoshi gives an adult age-adjusted rate. It was found in both studies that smoking habits were a major predictor of chronic bron- chitis prevalence and we have included it in our dose term as being additive to d (see Figures 5a and b, and Table 9).

Table 8: *Asthma Functions and Parameters*

	Parameters	Male	Female
A - age	β	4.35	5.15
d = 1/FS - normalized dose	ρ_1	2.68×10^{-4}	7.13×10^{-5}
$R(Ad) = F(A)K(d) - \dfrac{\text{\# asthma attacks}}{\text{person @ age A/year}}$	ρ_2	6.56×10^{-2}	4.75×10^{-2}
$F(A) = A^{\beta}\rho_1 e^{-A/A_1} + \rho_2 e^{A/A_2} - \dfrac{\text{\# asthmatics}}{\text{person @ age A}}$	A_1	2.298	1.946
$F(d) = 365\left[1 - e^{-(d/14.7)^{1.33}}\right]\dfrac{\text{\# asthma attacks}}{\text{asthmatics/year}}$	A_2	1.34×10^{2}	$.57 \times 10^{2}$

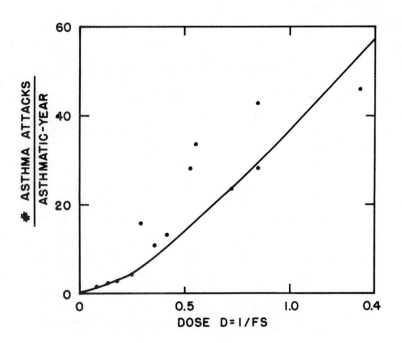

*Fig. 4. Plot of K(d) - # asthma attacks/asthmatic year from
Table 8. Data is from Yoshida, et al., 1966.*

Table 9: *Chronic Bronchitis Functions and Parameters*

A – age, R – smoking rate (packs/day), $S = Re^{-Au/A}$ – smoking index

I_j – index (i.e., ICAAS), $I_j' = (I_j^2 + 2I_j S + S^2)^{\frac{1}{2}}$ – generalized index

$d' = 1/FS'$ – modified dose where $FS' = \dfrac{I_j'}{I_j} \cdot \dfrac{(C_1^*, C_2^*, C_3^*, C_4^*, C_5^*, S=0)}{(C_1, C_2, C_3, C_4, C_5, S)}$

$P(A,d') = 1 - e^{-\{\phi_1 A + [\phi_2 A(d'-d_o)]^{\alpha}\}}$ – prevalence @ age A

Parameters	
A_o	37.7
β	5.84
d_o	-.0711
γ	1.8
ϕ_1	1.62×10^{-4}
ϕ_2	3.82×10^{-3}

Pts.	ICAAS Dose	Data Smoking Rate	Fitted Line
○	.5	0	———
△	.5	.75	– – –
□	1.25	0	—·—·—
◇	1.25	.75	··—··—

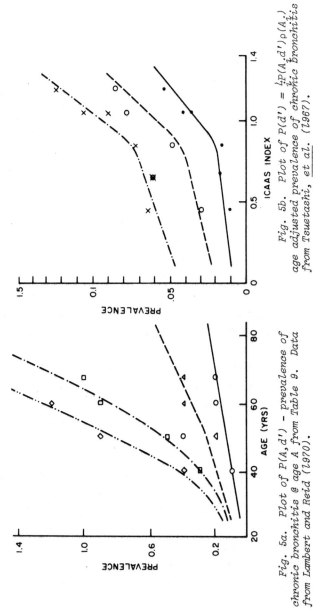

Fig. 5a. Plot of $P(A,d')$ – prevalence of chronic bronchitis @ age A from Table 9. Data from Lambert and Reid (1970).

Fig. 5b. Plot of $P(d') = \int P(A,d')\rho(A,)$ age adjusted prevalence of chronic bronchitis from Tsuetashi, et al. (1967).

IV. QUANTIFIED ACCEPTABLE LEVEL OF RISK (QALR)

The Clean Air Act of 1977 contains Prevention of Significant Deterioration (PSD) provisions that impose rigid constraints on possible
public policy decisions. At the regional and state levels this clause,
together with an air quality index (AQI) or factor of safety (FS) des-
ignation, provides a simple public policy decision methodology (PPDM)
for use in air pollution control issues. For example, many regions of
Florida and the Southeast are still Class 2 regions. (A Class 2 is a
region with ambient pollution levels below the National Ambient Air
Quality Standards.) For such regions, Congress has specified that the
maximum allowable annual average increases over the present baseline
shall not exceed 19 $\mu g/m^3$ of TSP and 20 $\mu g/m^3$ of SO_2. These regulations
are compromise attempts at a middle course between strict PSD and the
realities of our energy and economic crises. By specifying these incremental limits, Congress has, in effect, given its quantitation of a
maximum acceptable level of risk for the nation as a whole. States are
given only the option to place more stringent incremental limits in
areas with air quality better than national standards. Accordingly,
the range of decision options has been narrowed greatly. States can
either conform to these limits or can adopt lower increments (Stern,
1977).

This reduction in the range of options provides a quantification
of acceptable level of risk (QALR) PPDM adequate for public policy decisions. To illustrate this approach, Table 10 gives the safety factors
in each of the Florida geographic regions characterized in Tables 3 and
4 if the PSD increments were allowed. With this simple quantitative
transformation of air pollution concentrations, a knowledgeable control
board can quickly assess problem areas (e.g. location 3) in order to
utilize their option to restrict the SO_2 and TSP increments to less
than those permitted by Federal law.

Thus, by providing a reasonable relative rating system one easily
identifies the "hot cities" from an air pollution viewpoint. This initial identification immediately suggests where environmental stance must
be given greater consideration. On the other hand - again from a

Table 10: *Safety factors with PSD increments of 19 µg/m³ TSP and 20 µg/m³ SO₂ (up to NAAQS).*

(1) LOC	(2) PSI-1	(3) ICAAS	(4) ORNL	(5) TOR	(6) ACGIH
1	0.61(6)	1.07	1.18	1.28	1.03
2	0.65(6)	1.35	1.68	1.35	1.43
3	0.67(4)	0.91	1.03	1.42	0.86
4	0.71(6)	1.13	1.38	1.42	1.13
5	0.89(6)	1.33	1.88	1.66	1.45
6	0.96(4)	1.36	2.00	1.91	1.50
7	0.85(6)	1.22	1.84	1.62	1.36
8	0.74(6)	1.08	1.41	1.47	1.11
9	0.71(6)	1.22	1.42	1.43	1.17
FAAQS	1.00(6)	1.00	1.00	1.00	1.00

(1) TSP, (2) SO_2, (3) NO_2, (4) O_x, (5) CO, (6) $TSPxSO_2$

pragmatic point of view - cities with high factor of safety ratings can more confidently proceed with additional development.

V. THE ICAAS-FSOS REGIONAL DISAGGREGATED BENEFIT/COST ANALYSIS

The Regional DB/CA component of the ICAAS-FSOS study is an analysis of benefits and costs of various air pollution control strategies in the Tampa Bay area, the major coal burning region of Florida. The fact that regional B/C analyses must be carried out with detailed consideration of the spatial distribution of sources, geographic features, population groups, etc. opened up the possibility for consideration of the distri-butional aspects of benefits and cost--one of the timely problems in B/C analyses in general. Traditional benefit-cost analysis adds benefits and costs from a policy change according to "to whomsoever they may ac-crue." The benefit-cost decision criterion for whether or not to adopt

a policy depends on whether or not total benefits exceed total costs
(or the ratio B/C is greater than one). This criterion is closely related to notions of economic efficiency but ignores distributional issues. Thus, the benefit-cost criterion might be satisfied under conditions where costs of a policy are borne by a majority while beneifts
accrue to a minority.

An alternative criterion to the benefit-cost criterion in our society is one based on majority voting. In that case, costs to a majority and benefits to a minority would be unacceptable considerations. A
majority voting criterion may result in an inefficient choice since a
majority may receive benefits while a minority bear costs greater than
the benefits.

In a disaggregated B/C analysis DB/CA, one seeks acceptable policies that satisfy both efficiency and equity criteria. To carry out
such an analysis, in addition to information on total benefits and costs,
information on disaggregation of benefits and costs is needed for a complete evaluation of air quality control alternatives. The ICAAS regional B/C analysis provides an opportunity for attempting such a complete
evaluation. The flow diagram used in the DB/CA analysis was essentially
without the QALR channel and without the environmental and visibility
effects channels.

The distributed effects of emissions on various geographic locations
are predicted by using an atmospheric dispersion model (HEW, 1969; EPA,
1973; EPA, 1975). Dose-response models are then used to predict health
effects on populations from local ambient air quality levels. Finally,
an economic model is used to value effects according to demographic characteristics. These models and demographic data were computer-integrated
and linked with demographic data. The computer model developed for the
Florida study gives a framework for distributional issues for any region
by specifying appropriate regional demographic and climatic inputs. Benefits and costs are computed by both location and age group, and then may
be aggregated to be comparable to a traditional benefit-cost analysis.

In the FSOS study, ICAAS examined only direct household effects, inasmuch as the household is the unit most closely tied to voting issues. If persons in the household perceive that their benefits from control exceed household costs, they will favor controls; otherwise, the household will oppose control.

Within the household, we modeled only morbidity effects because they affect the most people and are most likely to influence household voting behavior. To value health effects, "willingness-to-pay" was the criterion used. There are three alternative approaches to measure willingness-to-pay (Zeckhauser, 1976). The first involves directly asking individuals their willingness to pay for pollution reduction. The second uses market transactions (waves or property values) to infer values of pollution reduction; and the third determines what individuals should be willing to pay for pollution reduction when the cost is based on direct cost accounting of damage due to pollution. Each method has its problems and weaknesses (EPRI, 1977). The ICAAS-FSOS study used the direct questionnaire approach.

Costs of pollution control go either into the rate base for capital items or into fuel adjustment charges; in either case, costs of abatement are distributed according to consumption of electricity, with the amount dependent on the rate structure. The policies examined result in changes in fuel use, the cost of which would be passed on through a fuel adjustment charge. In the ICAAS study, we assumed this charge would have no effects on the rate structure.

Economic models of energy consumption (Halvorsen, 1976) show that electricity consumption by a household is dependent on household income. A study of Florida electricity usage and income showed that the share of abatement costs for different household income groups was allocated based on the energy consumption of these households by income.

A. Dose Response Models

A dose response approach of the nature of that described in Section III was used in the DB/CA-PPCM. However, to arrive at a response, the

approach used in the ICAAS-FSOS study was based upon the prevailing
practice in air pollution of using a "hockey-stick" threshold type dose-
response model (EPA, Chess, 1974). This response was, however,
weighted by the expected number of times per year that the ambient air
quality in a region exceeds the threshold. This process yields a curvilinear dose-response model relating annual average doses to annual averages responses quite similar to the sigmoidal dose response curves
shown in Section III.

One key assumption of the dispersion model involves the time period of analysis. Although it is recognized that "worst conditions and
epidodes" account for the most serious health effects, health data as
related to pollution levels is available mainly on an annual average
basis. Thus, as is usual in epidemiological studies, the ICAAS-FSOS
study used only annual averages for emissions and meteorological conditions. The contours in Figure 2 (Chap. 8), show ambient levels for
the baseline scenario predicted from the model that uses the Climatological Dispersion Model (CDM).

B.   Integration of Models

A computer program combined the submodels illustrated in Figure 1
together with census tract socioeconomic data (number in each age and
household income group and ambient air quality by tract as computed by
the CDM model. Health effects by tract (t) in days of each symptom (s)
for each age group were computed as the difference in symptom days in
groups with and without control. The bid per person by tract age and
symptom was then computed. Finally, the aggregate bid by tract was obtained by multiplying the bid for a group by the number of persons in
each group for the tract and adding over symptoms and groups. Total
household benefits for the area are obtained by summing over the tracts.
To estimate costs by tract, the cost per household by income group is
multiplied by the number of households in an income group in tract t and
added over income groups.

In the ICAAS-FSOS study the above procedures were followed for six-

teen scenarios. Table 11 shows the aggregated benefits and costs found
for these scenarios. Table 12 illustrates the disaggregated benefits
and costs for the scenario "urban plants only controlled." Here the
total morbidity benefits exceed direct household by $2 million.

To determine whether control of emissions beyond federal primary
standards is justified, benefits and costs for households in affected
areas besides those for the 88 tracts of this study also were considered.
To do this, three residential customer groupings were defined, based
upon whether they experienced pollution reduction (get benefits) or buy
electricity (pay costs).

Some rough magnitudes obtained for these groupings are shown in
Table 12. A strong assumption was made that Group 2 obtained the same
benefit per person (about $10) as Group 1 because of adjacent locations.
It was also assumed that Group 3 obtained a morbidity benefit of $5 per
person (about half that for Group 1). Thus, if one counts only total
direct household costs and benefits, the control of emissions in urban
areas seems justified. On the basis of electricity consumers only,
there would be more persons against control (Group 3) than for (Group 1)
though the sums of benefits and costs for Groups 1 and 3 are roughly
equal. The decision for a control policy still requires some ethical
judgment as to who should bear costs and receive benefits. One possible way to gain acceptance from all groups would be to give Group 3 a
smaller share of abatement costs in their rate structure and Group 1 a
larger share.

The overall result of the DB/CA analyses as applied to the Tampa
region is that the requirement to use 0.5% sulfur fuel (equivalent to
0.8 lbs. SO_2 per 10^6 BTU) in urban plants gives net household benefits
to the majority (both by age and location). Thus both efficiency and
equity tests appear to be satisfied by this policy.

The results of the ICAAS-FSOS study are based upon the use of a
standard CDM air quality model, a curvilinear dose-response model which
incorporates all pollutants, a value model derived from a "willingness-
to-pay" survey and upon disaggregation methods which permit each house-

Table 11: *Scenario Results Summary*

Number	Test Area Cost ($)	Modeled Health Benefit ($)	HB/C Ratio	Remarks
1	0	4,378,264	--	
"1976 Post-B.B. #3 Construction Comparisons"				
2	5,307,708	1,378,418	.30	NSPS @ B.B #3
3	18,693,856	4,285,681	.23	NSPS @ all plants
4	4,333,079	3,57o,472	.82	NSPS @ G. & H.P.
5	4,520,327	4,171,077	.92	NSPS with Stack Height
6	110,023	2,307,583	20.97	Stack Height Increase
"1976-1975 During-B.B. #3 Construction Comparisons"				
7	5,307,708	4,600,451	.87	NSPS @ B.B. #3
8	18,693,856	4,826,598	.31	NSPS @ all plants
9	4,333,079	5,380,697	1.24	NSPS downtown
10	4,520,327	5,751,488	1.27	NSPS with Stack Height
11	110,034	4,848,663	44.07	Stack Height Increase
"1975 Pre-B.B. #3 Construction Comparisons"				
12	5,307,708	2,136,855	.40	NSPS @ B.B. #1-2
13	13,918,030	5,696,913	.41	NSPS @ all plants
14	4,333,079	5,424,630	1.25	NSPS downtown
15	4,520,327	5,792,550	1.28	NSPS with Stack Height
16	110,034	2,409,095	21.89	Stack Height Increase

Table 12: *Benefits and Costs for Households in Three Groups for control "downtown," Scenario 14*

	Group	Population[a]	Morbidity Benefits	Abatement Costs
Tampa Costs and benefits	1	369,518	5,125,555	3,098,435
S. Hillsborough, Manatee, Pinellas No costs, benefits	2	740,182	7,401,820	0
Polk, Pasco, Plant Cost reduced benefits	3	432,000	2,160,000	4,461,746
		1,541,700	14,686,375	7,560,181

[a] est. based on 1970 census

hold to be considered separately. Each of these models has uncertainties characteristic of the various frontiers of knowledge involved.
Furthermore, there are also uncertainties in the meteorological and
socioeconomic inputs, as well as in the calculated effects based on
these inputs and models which also involve various assumptions. Despite
these caveats the overall policy machinery is quite tractable, efficient
and should certainly be useful in examining the relative merits of various possible policy scenarios.

VI. OTHER STUDIES

For a complete study of benefits due to decreases in pollution
we must include not only human morbidity but human mortality, plant
damage, materials damage, ecological changes and aesthetic values.
In the studies so far carried out by ICAAS, morbidity, especially with
respect to respiratory illness has been the major response modeled
(ICAAS, 1978). Also important is the change in the mortality rate
caused by changes in pollutant concentrations. This response has been
examined in various studies (Lave and Seskin, 1973; Sprey et al.)
1973; Buechley et al., 1973) and functional relationships (dose
response functions) have been inferred (Liu and Yu, 1976). Furthermore, with increasing incidences of cancer and the relationship seen
between cancer and air pollution in general (Winkelstein
and Kantor, 1969) and certain specific chemicals released or produced
by coal burning (Sparrow and Schaner, 1974; Peacock and Spence, 1967),
dose response functions modeling this relationship are becoming important.

The uncertainties in the measurement of pollutants, the spatial
variation of the pollutant concentration and the interference of co-
factors such as meterological and social economic variables present
formidable problems in defining the relationship between air pollutants and health effects (Ferris and Anderson, 1962; Lambert and Reid,
1970; Biller, 1978; Feagans, 1978). The three principle techniques
usually used to infer dose response functions are cross tabulation,
multivariate regression and non-parametric or distribution-free

analysis (Hershaft, 1978). The ICAAS study represented a limited
application of the regression method, in which a priori judgment has
been made that the response is sigmoidal in nature. This method has
been used to varying degrees in other studies.

Ferris (1978) has recently carried out a "Critical review of
our knowledge of exposure to low levels of the regulated air pollutants. He concludes that the primary standards for the six regulated
pollutants seem adequate to protect the health of the public. From
the discussion papers (First et al., 1978) following this critical
review, one might conclude that strong differences of opinion among
recognized public health experts still exist on this topic and that
the overall state of knowledge has not changed greatly from that
summarized in the Brown report (1975).

In connection with other environmental impacts it might be noted
that plant damage has been discussed in Chapter 10, although no
specific dose-response functions were presented. Most of the problems
in modeling health effects in humans are also incurred in developing
plant damage functions. In addition, we are faced with the large
variations seen in the response of various species of plants to air
pollution. Thus, we are faced with analyzing plant damage on a dis-
aggregate basis. Examples of some dose damage functions for plants
due to SO_x can be seen in Table 13. Materials damage can be precisely measured. The largest drawback which exists with these
established dose damage functions is in their application. This occurs
because of the almost infinite variety in material composition, each
variation reacting quite differently. Examples of dose materials
damage functions are given in Table 13.

Returning to the use of dose response relations for benefit
cost analysis, even if health effects were understood it would still
be difficult to establish the dollar value of benefits of air pollution abatement. This is due to the diverse nature of the effects
produced by air pollution, and the latitude in establishing the cost
of human illness and death, and the dollar damage to plants and
materials. There are three approaches to measuring benefits

Table 13: *Dose Response Functions*

FUNCTION	PARAMETERS	VARIABLES	REFERENCES
Plants:			
$T(C-C_o)=K_R$ @ P% leaf destruction	C_o, K_R	T-duration exposure $C-[SO_2]$	Thomas, 1956
$T=T_R \exp K_2 P_2\left(\dfrac{S-S_R}{\sqrt{10C}}\right)$	K_2, P_2, S_R, T_R	S-degree folior damage $(.5 \leq S \leq 5.), C, T$ as above	Zahn, 1963
Materials:			
$Y=.131x+.0182+.787$ -steel corrosion rate $(mg/dm^2/wet\ day)$		$X-SO_2$ pollution rate Z-avg. monthly temp.	Sereda, 1960
$CR=.001028(RH-48.8)C$ -corrosion rate zinc $(\mu m/yr)$		RH-relative humidity $C-[SO_2](\mu g/m^3)$	Haynie and Upham, 1970

(Zeckhauser, 1976; ICAAS, 1978). (1) One approach is to calculate the damages averted and income loss. This has been used extensively (Liu and Yu, 1976). (2) The second approach is to look at the values of products or services which will be affected by air quality. The most widely used studies have involved correlation of air pollution with property values (Crocker, 1971; Waddell, 1974; Rubinfeld, 1978; Freeman, 1979; Milliman and Sipes, 1979). (3) The third approach involves a survey of public willingness to pay to avoid certain health effects (ICAAS, 1978),which while subject to various criticisms, remains one of the very few consumated studies.

Mention should also be made with respect to the establishment of aesthetic quality values (or indexes), and furthermore, on the monetary values placed on such intangibles. These effects may include odor, visibility and vegetation covering of polluted area. Dose response functions are usually developed using an instrument reading and a sub-jective scale. Odor, for example, is judged as being acceptable, critically acceptable or not acceptable in parallel with the measure-ments of concentration of the pollutant in question. This is then transformed to a scale reading by dividing the observed concentration c by the maximum (critically) acceptable concentration (Summers, 1971).

Monetary values are assessed by the survey method of willingness to pay
to avoid the offensive condition (ICAAS, 1978).

VII. THE INTERNATIONAL CONTEXT

Let us suppose that in time the possible atmospheric modifications
discussed in Chapter 11 gain general scientific acceptance. Then it will
be necessary to develop an international framework for controlling coal
burning emissions to forestall global disasters occasioned, say, by a
runaway greenhouse effect. Even when concerned with domestic issues it
would be helpful to envisage reasonable future international frameworks.

In view of the differences in the political and economic structures
of the various nations which eventually would have to participate in an
international treaty on coal utilization it would probably be fruitless
to use cost/benefit decision methodologies such as the techniques discussed in Section V or VI. However, a risk assessment technique analogous
to those described in the earlier sections of this chapter might be useful.

For example, let us suppose that there is some carbon dioxide level
(e.g., 1000 ppm as compared to present day 330 ppm) at which the vast
majority of knowledgeable atmospheric and optical scientists would agree
could trigger a runaway greenhouse effect. Then we must seek agreement
among scientists and engineers knowledgeable in coal and fossil fuel combustion, in atmospheric dynamics and chemistry and in oceanography as to
the CO_2 emission levels which would lead to the 1000 ppm ambient level.
Suppose, for example, this is 40 Gigatons per year (1GT is 10^{15} gr). Obviously we must next develop some international policy procedure for
allocating CO_2 emissions to various countries. For example, if projections are made on current use ratios the world might then reasonably
allocate, say, 10GT CO_2 for USA emissions. Such a national emission
limit could then be taken as a base case for the examination of internal
U.S. regulatory policies which could be undertaken using the most refined public policy decision methodologies.

Clearly the world is not now ready to discuss the sophisticated
international measures which might be needed to prevent global climate
change. When it is, the world might also be ready to discuss how to

defuse the 100 or so Gigatons TNT equivalent locked in the world supply of nuclear bombs (1 GT TNT $\approx$ 4 x 10^{18} joules $\approx$ 4 quads). Perhaps the fissionable materials could be placed in stockpiles for contingency use as breeding stock should fusion reactors not work out.

References

Albert, R. E., and B. Altshuler, Proceedings of the Twelfth Annual Hanford Biology Symposium, CONF-73050, U.S. Atomic Energy Commission, 1972.

ACGIH TLVS: Threshold limit values for chemical substances and physical agents in the workroom environment with intended changes for 1977. American Council of Governmental Industrial Hygients, c/o Secretary, P. O. Box 1937, Cincinnati, Ohio 45201, 1977.

Biller, W. F., Estimation of expected number of people in a population experiencing adverse health effects at a given level of air quality, Environmental Protection Agency, Research Triangle Park, N.C., unpublished, 1979.

Broden, I., M. Higgins, K. Mathews and J. Keller, Epidemiology of asthma and allergic rhinitis in a total community, Tecumseh, Michigan, J. of Allergy and Clinical Immunology, 54:100, 1974.

Brown, G. E., Jr., The costs and effects of chronic exposure to low-level pollutants in the environment. Subcommittee on the Environment and the Atmosphere of the Committee on Science and Technology, U.S. House of Representatives, 94th Congress. No. 49, November 17, 1975.

Blum, M. F., Carcinogenesis by ultraviolet light, Princeton University Press, Princeton, NJ,1959.

Buechley, R.W., Riggan, W.B., Hasselblad, V., and Van Bruggen, J.B. SO$_2$ levels and perturbations in mortality, A study in the New York-New Jersey metropolis, Arch. Environ. Health, 27 (1973).

Burch, P. R. J., Growth, disease and aging, University of Toronto Press, Toronto, Canada, 1969.

Busse, A. D. and J. R. Zimmerman, User's guide for the climatological dispersion mode, EPA-R4-73-024, EPA, Research Triangle Park, N.C., 1973.

Druckery, H., Naturliob vorkomende carcionogene, Med. Mschr., 20, 154-157, 1966.

EPA, Guideline for public reporting of daily air quality -- Pollutant Standards Index (PSI). Guideline Ser. OAQPS No. 1.2-044, EPA-450/2-76-013, Washington, D.C., 1976.

EPA, Air quality data - 1976 Annual Statistics, U.S. Environmental Protection Agency, EPA-450/2-78-009, March, 1978.

Feagans, T., A method for assessing the health risks associated with alternative air quality standards for ozone, Environmental Protection Agency, Research Triangle Park, N.C., unpublished,1978.

Fears, T., J. Scotto, and M. A. Schneiderman, Mathematical models of age and ultraviolet effects on the incidence of skin cancer among whites in the United States, Am. J. Epidemiol. 105:420-427, 1977.

Ferris, B. G., Health effects of exposure to low levels of regulated air pollutants: A critical review, J. Air Poll. Contr. Assoc., 28:481-497, 1978.

Ferris, B. G., Jr., and D. O. Anderson, The prevalence of chronic respiratory disease in a New Hampshire town, Ann. Rev. Resp. Dis. 86:165-185, 1962.

First, M. W., and H. H. Hovey, Health effects of exposure to low levels of regulated air pollutants, discussion papers, J. Air Poll. Contr. Assoc., 28:883-894, 1978.

Freeman, A. M. III, The Benefits of Environmental Improvement: Theory and Practice, John Hopkins University Press, Baltimore, Md., 1979

FSOS, Wilson, S. U., D. S. Anthony, E. R. Hendrickson, C. L. Jordon and P. Urone, Final report of Florida Sulfur Oxide Study, Post, Buckley, Schuh and Jernigan, Inc., Orlando, Fl., 1978.

Garrett, E. R., Classical pharmacokinetics to the frontier, J. of Pharmacokinetics and Biopharmaceutics, 1:341-361, 1973.

Green, A. E. S., Ultraviolet exposure and skin cancer response, Am. J. of Epidemiol., 107:277-380, 1978.

Green, A. E. S., G. B. Findley, Jr., K. F. Klenk, W. M. Wilson and T. Mo, The ultraviolet dose dependence of nonmelanoma skin cancer incidence, Photochem. and Photobiol., 24:353-362, 1976.

Green, A. E. S., and R. A. Hedinger, Models relating ultraviolet light and non-melanoma skin cancer incidence, Photochem. and Photobiol., 28:283-292, 1978.

Green, A. E. S., D. E. Rio and R. A. Hedinger, Florida's air quality, present and future, Florida Sci., 41:183-190, 1978.

Green, A. E. S., T. J. Buckley, A. MacEachern, R. Makarewicz and D. E. Rio, Factor of safety method. application to air and noise pollution, Atmospheric Environment (to be published 1980).

Halvorsen, R., Demand for electric energy in the U.S., Southern Economic Journal, 610-625, 4/1976.

Haynie, F. H., and J. B. Upham, Effects of atmospheric sulfur dioxide on the corrosion of zinc, Mater. Prot. Perform. 9:35-40, 1970.

Hershaft, A., J. Morton and G. Shea, Critical reviews of air pollution dose effect functions, Final Report EQ5-ACC12, Council on Environmental Quality, 722 Jackson Place, N.W., Washington, D.C., March 1976.

Hrubec, 1973 - used in Chapter 11.

ICAAS, An Air Quality Index (AQI) Project, a proposal by the Interdisciplinary Center for Aeronomy and (other) Atmospheric Sciences (ICAAS), University of Florida, Gainesville, FL., 1970.

ICAAS, Preservation and Enhancement of Air Quality (PEAQ) in Florida and the Southeast, 1971.

ICAAS-FSOS, Green, A. E. S., S. V. Berg, E. T. Loehman, M. E. Shaw, R. W. Fahien, R. A. Hedinger, A. A. Arroyo, V. H. De, R. P. Fishe, L. L. Gibbs, G. P. Penland, D. E. Rio, W. F. Rossley, W.A. Wallace, and J. M. Schwartz, An interdisciplinary study of the health, social and environmental economics of sulfur oxide pollution in Florida, ICAAS report, February 1978.

ICAAS-FSOS, Loehman, E. T., S. V. Berg, A. A. Arroyo, R. A. Hedinger, J. M. Schwartz, M. E. Shaw, R. W. Fahien, V. H. De, R. P. Fishe, D. E. Rio, W. F. Rossley and A. E. S. Green, Distributional analysis of regional benefits and cost of air quality control, accepted by J. Environmental Economics and Management, 1979.

Inhaber, H., Environmental quality: National index for Canada, Science, 186:798-805, 1975.

Lambert, P. M. and D. D. Reid, Smoking, air pollution and bronchitis in Britain, The Lancet, 853-857, 1970.

Lave, L. B., and E. P. Seskin, An analysis of the association between U.S. mortality and air pollution, J. Amer. Stat. Assoc., 68:284-290, 1973.

Liu, B. C., and E. S. Yu, Physical and economic damage functions for air pollutions by receptor, Midwest Research Institute Project for U.S. E.P.A., 600-5-76-011, Corvallis, Oregon, 1976.

Milliman, J. W. and N. G. Sipe, Benefit measures of air quality regulation in Florida, Star Grant 78-104, Bureau of Economics and Business Resources, University of Florida, 1979.

NAS, Sulfur Oxides, Office of Publications NAS, 2101 Constitution Avenue, N.W., Wash. D.C.,1978.

Ott, W. R., and G. C. Thom, A critical review of air pollution index systems in the United States and Canada, J. Air Poll. Contr. Assoc., 26:460-470, 1976.

Ott, W. R., and W. F. Hunt, Jr., A quantitative evaluation of the pollutant standards index, J. Air Poll. Contr. Assoc., 26:1050-1054, 1976.

Peacock, P. R., and J. B. Spence, Incidence of lung tumours in LX mice exposed to (1) free radicals; (2) SO_2. Br. J. Cancer, 21:606-618, 1967.

Rubinfeld, D. L., Market approaches to measurement of benefits of air pollution abatement, in Friedlaender, editor, Approaches to Controlling Air Pollution, MIT Press, Cambridge, MA, 1978.

Sereda, P. J., Ind. Eng. Chem., 52:157, 1960.

Setlow, R. B., Review-repair deficient human disorders and cancer, Nature, 271:713-717, 1978.

Shenfield, L., Ontario's air pollution index and alert systems, J.A.P.C.A., 20, 614-614, 1970.

Shooman, M. L., Probabilistic Reliability: An Engineering Approach, McGraw-Hill, New York,1968.

Sigma Research, Proceedings of a workshop on the EPRI Report, EQ-405-SR, Palo Alto, CA, 1977.

Sparrow, A. H., and L. A. Schaner, Mutagenic response of Tradescantia to treatment with X-rays, EMS, DBE, ozone, SO_2, N_2O and several insecticides, Mutat. Res., 26:445, 1974. Abstract

Sprey, P., et al., Health effects of air pollutants and their interrelationship, U.S. Environmental Protection Agency, 1974.

Stern, A. C., Prevention of significant deterioration: A critical review, J. Air Poll. Contr. Assoc., 27:440-453, 1977.

Summer, W., Odor Pollution of Air, CRC Press, Cleveland, Ohio, 1971.

Thomas, M. D., The invisible injury theory of plant damage, J.A.P.C.A., 205-208, 1956.

Thomas, M. D., and R. H. Hendricks, Effect of air pollution on plants, 9-1 -- 9-44. In P. L. Magill, F. R. Holden, and C. Ackley, Eds., Air Pollution Handbook, McGraw-Hill, N.Y.. 1956.

Thomas, W. A., L. R. Babcock and W. B. Schults, Oak Ridge air quality index. ORNL-NSF-EP-8, Oak Ridge National Laboratory, Oak Ridge, TN, 1971.

Tsunetoshi, Y., et al., Epidemiological study of chronic bronchitis with special reference to effect of air pollution, Intern. Arch. Arbeitsmed., 29:1, 1967.

Turner, D. B., User's guide to UNAMAP, environment applications branch, meteorology laboratory, EPA, Research Triangle Park, N.C., 1975.

U.S. Dept. of Health, Education and Welfare, Air quality display model, PB-189-194, National Air Pollution Control Administration, Washington, D.C., 1969.

U.S. EPA, Health consequences of sulfur oxides: A report from CHESS, 1970-71, EPA-640/1-74-004, EPA, Research Triangle Park, N.C., 1974.

Waddell, T. E., The economic damages of air pollution, Environmental Protection Agency, EPA 600-5-74-012, Washington, D.C., 1974.

Williams, J. W., and N. G. Sipe, Benefit measures of air quality regulation in Florida: A pilot study, Star Grant, 78-104, Bureau of Economic and Business Research, Univ. of Florida, 1979.

Winkelstein, W., Jr., and S. Kantor, Prostatic cancer: Relationship to suspended particulate air pollution, Amer. J. Pub. Health, 59:1134, 1969a.

Yoshida, K. M. D., et al., Air pollution and asthma in Yokkaichi, Arch. Env. Health, 13:763,1966. 1966.

Zahn, R., Untersuchunger über die Bedeutung kontinuierlicher und intermittierender Schwefeldi-oxideinwirkung für die Pflanzenreaktion. Staub 23:343-352, 1963, Summary in English.

Zeckhauser, R., Willingness to pay as an efficiency guide for the regulation of noise disturbance, Economic Welfare Impacts of Urban Noise, EPA-600/S-76-002, EPA, Res. Tri. Park, N.C. 1976.

CHAPTER 16

FINANCING CAPACITY GROWTH AND COAL CONVERSIONS
IN THE ELECTRIC UTILITY INDUSTRY

Eugene F. Brigham and Louis C. Gapenski

Demand for electric power is expected to triple from 1979 to the
year 2000. To meet this projection, the electric power industry must
undertake a truly huge construction program which will require many
billions of dollars. At the same time, many utilities will be converting oil- and gas-fired plants to coal, both on economic grounds and
to meet national policy goals. To raise the capital needed to finance these monumental construction programs will be an exceedingly
difficult task, perhaps an impossible one without major alterations in
the industry's organization and regulation. The purposes of this chapter are (1) to explore the magnitude of capital requirements of the industry; (2) to discuss the ability of the industry to meet these capital
requirements; and (3) to consider some possible remedies to what appears
to be a very serious financing problem.

I. INDUSTRY GROWTH: 1961-1977

Table 1 gives a sketch of several dimensions of the utility industry's recent growth. From 1961 through 1977, the average growth
rate in output of electric power was 6.4%. The growth rate was faster
before the oil embargo and the subsequent huge increases in oil prices
and utility service rates, with the slow-down resulting from a combination of price elasticity, the 1974-75 recession, and conservation efforts. Future growth in output, assuming that capacity is available to
meet demand, will doubtless be much higher than the 3.4% 1973-1977 rate,
but probably it will not be as high as the 1961-1973 rate.

331

Table 1: *Historical Growth:* 1961-1977 [1]

Year	Output (billion kwh)	Output (mwh/ person)	Revenues (billion dollars)	Cost per kwh (cents)	Generating Capacity (million kw)	Capital Expenditures (billions of 1978 dollars)
	(1)	(2)	(3)	(4)	(5)	(6)
1961	722	4.8	$12.2	1.69	198	$ 5.1
1962	778	5.1	13.0	1.67	209	4.3
1963	833	5.4	13.7	1.64	229	4.9
1964	895	5.7	14.4	1.61	240	5.7
1965	957	6.0	15.2	1.59	255	5.8
1966	1,042	6.4	16.2	1.55	267	7.0
1967	1,111	6.7	17.2	1.55	288	8.1
1968	1,207	7.2	18.6	1.54	310	10.1
1969	1,311	7.7	20.1	1.53	333	11.0
1970	1,396	8.0	22.1	1.58	360	13.1
1971	1,470	8.3	24.7	1.68	388	16.0
1972	1,580	8.9	27.9	1.71	418	17.6
1973	1,706	9.4	31.7	1.86	459	18.7
1974	1,703	9.4	39.1	2.30	496	21.7
1975	1,738	9.4	46.9	2.70	528	23.4
1976	1,852	9.9	53.5	2.89	550	25.7
1977	1,954	10.3	62.6	3.20	576	32.5
Growth rates:						
1961- 1977	6.4%	4.9%	10.8%	4.1%	6.9%	12.3%
1961- 1973	7.4%	5.8%	8.3%	0.8%	7.3%	11.4%
1973- 1977	3.4%	2.3%	18.5%	14.5%	5.8%	14.8%

Column 2 shows that about two-thirds of the total growth came from increased usage by individual customers, with the other one-third coming from a larger number of customers. This suggests that electric demand will continue to grow even if population growth slows. However, higher prices and conservation efforts will tend to reduce usage per customer, although a large-scale substitution of electricity for space heating and auto fuel could increase per capita usage of electricity.

Column 3 shows that the utilities' revenues -- which are equal to the bills paid by customers -- have grown much faster than kilowatt hour (kwh) usage, while column 4 shows that the cost per kwh has increased at an average rate of 4.1%. Note, however, that the cost per kwh actually declined all during the 1960's, and first turned up in 1970 as a result of the higher U.S. monetary inflation rate in the late 1960's. The cost per kwh has risen quite rapidly during the 1970's as a result of the general inflation rate and especially because of rising fuel costs.

[1] US DOE Report to Congress (1978) and Edison Electric Institute Yearbook (1977).

Generating capacity as shown in column 5 matched growth in output (column 1) closely, prior to 1974. However, in 1974 and thereafter,
higher prices, recession, and conservation efforts stimulated by higher
costs caused a slowdown in growth, but construction "in the pipeline"
could not be stopped; so, in 1974 and since, capacity has grown faster
than output. Construction program cutbacks suggest that this situation
has been reversed, and that demand will grow faster than capacity in the
next few years, reducing reserve generating capacity to the point where
some utility systems may experience shortages in capacity.

Capital expenditures (in 1978 dollars) as shown in column 6 have
grown faster than production capacity, indicating an increasing real
cost per unit of capacity added. Had construction expenditures in nominal dollars been reported, the growth rate would have been much higher,
to reflect the effects of inflation on construction costs. Since inflation is not reflected in column 6, the increases in cost per unit of
capacity reflect primarily (1) an increasing shift from oil and gas to
more capital-intensive coal and nuclear units and (2) increased environmental control requirements.

II. INDUSTRY GROWTH: 1980-1995

Growth in the demand for electric power is influenced by population increases, by the level of economic activity, by the prices of
electricity and alternative forms of energy, and by efforts to conserve
energy.[1] As noted above, the price of electricity has risen rapidly
during the last few years, and continued increases can be expected in
the future. This will lead to conservation and more efficient usage
of electricity, reducing the growth in demand. Forecasts of economic
activity during the 1980's and 1990's are varied and uncertain, to say

---

[1] Most energy conservation efforts are stimulated by prices. On a short-
run basis, "patriotic" appeals to conserve may be effective, but beyond
a few days or weeks, what motivates residential and industrial customers
is price. If the price of electricity rises, all users, industrial and
residential alike, will insulate better, use load management and be careful.

the least, but if the economy is to provide jobs for the people who will
be entering the work force, the economic activity will have to be at
least as high as during the 1970's. Demand for electricity was restrained during the 1960's and 1970's by the artificially low cost of
natural gas. As gas supplies decline, and as artificial controls are
lifted, the demand for electricity for space heating will doubtless
rise. Similarly, in the years ahead electric automobiles will probably
replace gasoline autos to at least some extent, further increasing the
demand for electricity.

With so many factors at work to both increase and decrease demand,
forecasts of actual demand can at best be taken only as rough indicators of future demand levels. Still, for planning purposes, forecasts
must be made: Table 2 gives the latest U.S. Department of Energy projections for both output and generating capacity. Both output and generating capacity are expected to grow at a rate of just over 4% per
year from 1980 to 1995.

Table 2: *Forecast of Growth:* 1980-1995

Year	Output (billion kwh)	Generating Capacity (million kilowatts)
1980	2,171	586
1985	2,757	700
1990	3,358	879
1995	4,017	1,064
Growth rate, 1980-1995	4%	4.1%

Source: U.S. DOE Energy Information Administration Annual
Report to Congress, 1978.

Table 3: *Forecast of Capital Requirements,* 1980-1995
(billions of 1978 *dollars)*

Cost per kw	New Capacity Required (million kw)	Fuel Mix Scenarios				
		Only oil	Only coal	Nuclear only	One Third each fuel	0% oil 80% coal 20% nucl.[a]
Oil $435		$207.9	–	–	$ 69.3	$ 0.0
Coal $650	478	–	$310.7	–	103.6	248.6
Nuclear $1,000		–	–	$478.0	159.3	95.6
Cost		$207.9	$310.7	$478.0	$332.2	$344.2
Transmission and distribution costs		150.0	150.0	150.0	150.0	150.0
Total cost		$357.9	$460.7	$628.0	$482.2	$492.2

[a]averaged for high and low sulfur coal and with and without scrubber

III.   CAPITAL REQUIREMENTS

A vast amount of new-money capital will be required to meet the
projected expansion in generating capacity, to replace aged facilities,
to convert from oil and gas to coal, to meet environmental protection
requirements, and to provide the transmission and distribution facilities needed to service the projected load.   The capital needed to meet
generation requirements depends upon the type of fuel used,(see Table 3
and U.S. DOE, 1979).

Based on these figures and the projected capacity increases, we
may estimate capital requirements for new generating equipment from
1980 through 1995, depending on the type of fuel used, as shown in the
top section of Table 3.   Transmission and distribution costs have been
averaging about 44% of generating expenditures, with the percentage
being higher for oil, lower for nuclear.   A reasonable estimate of T&D
expenditures from 1980 through 1995 is about $150 billion (1978 dollars).
Projected cost per KW for new generation and T&D range from $357.9 for
oil only to $628 for nuclear only; and our estimate of the most likely
situation, with 00% oil, 80% coal, and 20% nuclear would call for expenditures of about $490 billion of 1978 dollars.

Environmental protection expenditures for new capacity are built
into the estimates given above, but added equipment for existing plant
is not included.   Also, no allowance is made for conversions from oil/
gas plants to coal, or for the replacement of aged and obsolete plant.
Accordingly, our expenditure estimates are probably on the low side.

The utility industry's construction program from 1961 to 1977 was
only $231 billion in 1978 dollars, or less than half the projected
level for the next 15 years.   Thus, the utility industry must raise
about twice the amount of capital in real terms during the next decade
and a half that it did during the last.

The magnitude of the utilities' financing problem can also be seen
by comparing net plant at year-end 1978 with the projected capital expenditures from 1980 to 1995.   On December 31, 1978, the net plant of
the 100 largest U. S. utilities, which have about 80% of U. S. generating
capacity, was $183 billion (Value Line, 1979).   A small part (about 3%)

of this plant represents gas distribution plant, not electric plant, so
the net electric plant of the 100 companies was about $177 billion, and
for the entire U.S., net plant as carried on the balance sheets was
about $220 billion. With projected expenditures of $492 billion, we
see that the U.S. electric power industry must raise new capital equal
to 224% of its current amount of capital.

IV. THE FINANCIAL CONDITION OF THE PUBLIC UTILITY INDUSTRY

At the start of the 1960's, the electric power industry was the
epitome of financial strength. Most of the companies carried bond ratings of AAA or AA, indicating virtually unlimited borrowing capacity,
and their stocks were highly regarded and in great demand by the investment community.

Today, the situation has been completely reversed. Only one U.S.
utility company (Texas Utilities) has retained its AAA rating; income
available for interest payments has declined from over 6 times in the
1960's to just over 3 times in 1978 (versus over 8 times for the industrial companies); rates of return on equity have declined from an approximate parity with industrial companies in the 1960's to only 72% of
the industrials' ROE in 1978 (11.2% vs. 15.5%). Investors were willing
to pay over $20 in market value for every $1 of utility earnings in the
early 1960's; at the end of 1978, the price/earnings ratio was only 7.8
times. Similarly, the utilities' market value-book value ratio declined from a high of 2.8 in 1964 to about 0.85 today. Finally, if an
investor had put $1,000 into a typical utility stock in 1965, his port-
folio would be worth only about $600 today versus about $1,100 had he
put his money into an average industrial stock.

This deterioration of the utilities' financial position, both in
absolute terms and vis-a-vis the industrial companies, was caused by a
combination of political and economic factors. The first is inflation.
As the rate of inflation rose in the late 1960's and early 1970's, utilities' costs also increased. Cost increases were held down by productivity gains, but productivity could not match the inflation rate. Net
cost increases rose dramatically after 1974, when oil prices quadrupled

and the general rate of inflation exceeded 10%.

Utility prices are set by regulatory commissions. Therefore, if costs rise, a utility must apply to its commission for a rate increase. Time lags are inherent in the process. First, the company usually must wait until costs have actually risen, and profits have been squeezed, before applying for a rate hike. Then further delays are encountered before hearings can be scheduled, arguments can be presented, and a decision reached. At a minimum, these regulatory delays, or regulatory lags, take six months; on average, the lag between applying for a rate increase and being allowed to put new rates into effect is about one year. (However, most electric utilities do have fuel-adjustment clauses which allow them to pass along higher fuel costs with lags of about three months.) If costs rise, but prices can be increased only after a long regulatory delay, then obviously profit margins are squeezed, rates of return on invested capital decline, and the company's financial position suffers.

Increasing operating costs in the face of regulatory lag is not the only problem caused by inflation. As we have seen, the cost of building new plant also rises. Suppose a utility had net assets of $2,000 per residential customer in 1979. If the cost of new plant is twice that of current plant in service, this means that $4,000 of investment is required to serve each new customer. Now, suppose the utility had been earning its cost of capital -- say 10% -- on the old plant. With the same service rates, the rate of return on the $4,000 investment required to serve the new customer will fall. Although the new plant may be somewhat more efficient, efficiency gains cannot offset the higher depreciation charges and return requirements on the new plant.

In theory, utility commissions are supposed to operate without unduly long lags; and when new service rates are set, these rates cover all operating costs and provide a fair rate of return on capital invested in the companies. In fact, in most states, the lags are quite long, and service rates are not adequate to cover costs. By the time the new rates are in effect, costs have already risen again, so profits do not

attain the level prescribed by the commissions. To illustrate this
point, in 1978 commissions across the country authorized utilities to
earn approximately 13.5% on common equity capital, but the actual rate
of return was only 11.2% (versus 15.5% for the average industrial company). This same situation has existed continuously since the early
1970's.

Even the depressed utility profits are questionable on two counts.
First, about half of the utility industry "flows through" tax savings
from liberalized depreciation rather than setting up reserves for future tax payments ("normalizing"). The accounting profession
and the Securities and Exchange Commission (SEC) refuse to permit industrial firms to state profits on a flow-through basis, on the
grounds that this overstates true profits. The second questionable
factor in utility profits relates to the "Allowance for Funds Used
During Construction" or AFUDC. Utilities obtain profits from operations -- selling electricity for more than it costs to produce the
power -- and also by a simple stroke of the accountants' pen. The accountants simply multiply the amount of the utility's investment in
construction times an allowed rate of return to create AFUDC income,
which is then added to operating profit, to produce reported net income. Because of their very large construction programs, the long time
required to complete generating plants and transmission lines, and very
high interest rates and equity capital costs, AFUDC income has become
quite large. As a percentage of total income, AFUDC averaged about 5%
until the late 1960's. Since then, it has risen steadily and hit a
record 40.9% in 1978.

AFUDC income is noncash income, so it cannot be used for interest
or dividend payments, nor can it be plowed back into the business to
help support the construction program. Investors recognize the low
"quality" of AFUDC income, and the existence of such large amounts of
AFUDC credits has increased the perceived riskiness of the industry,
has increased the cost of capital to the companies, and has reduced
the utilities' ability to attract new capital. Firms in the industrial
sector generally do not calculate and report AFUDC income--thus,

the quality of profits reported by nonutilities is higher than that of
utility profits.

V.   CAN THE UTILITY INDUSTRY FINANCE THE NECESSARY EXPANSION?

We have seen that the electric power industry must undertake a truly massive investment program in order to meet projected demand. Capital investment, in 1978 dollars, must be over twice the level of the
1961-1977 period, and the net investment from 1980 through 1995 must
total over twice the level of 1978 assets. Can the industry raise the
capital necessary to finance an investment program of this magnitude?
The answer is "maybe."

In its currently very poor financial condition, the average utility
simply could not raise the necessary capital. Bonds could not be sold
without massive stock sales, and such stock sales, at today's prices
(about 15% below book value) would greatly dilute book value and earnings per share. If earnings fell, dividend cuts would soon follow.
Investors recognize all this, of course, so the utilities would be cut
off from the equity market, hence also from the debt market, early in
the game.[1]

The situation could change fairly rapidly if utility commissions
would allow the companies to set rates which actually cover their cost
of capital. If the companies were earning their cost of capital, then
by definition they could be able to pay the going price for new capital
and hence attract it. Also, the higher rates would help reduce the rate
of growth in demand.

------

[1]This very situation has been threatening the utilities for some time
now, but they have not had to sell large quantities of securities in recent years because of capacity buildups that inadvertantly resulted from
the post-1974 slowdown in growth. The utilities had been projecting a
continuation of the 7% pre-1974 growth. With the much higher prices
that have prevailed since 1974, demand has slowed, This meant that
construction programs could be reduced. However, the excess capacity
of most systems will have been eliminated by the early- to mid-1980's.
In any event, the combination of a large amount of plant coming on
line during a period of slow growth has kept most utilities away from
the capital markets, except for relatively small sums, in recent years.

Thus, the key question is this: Will utility commissions across the country permit the companies to raise prices to a level which fully covers the cost of producing electricity, including the cost of capital to finance the plants required to generate and distribute power? It is not possible, without being clairvoyant, to answer this question. The alternatives, though, are somewhat more predictable.

(1) If commissions do grant adequate rate relief, in the near-term future, then the utilities will be able to raise the capital necessary to complete existing construction programs and to embark upon and carry through new programs needed to provide power in the late 1980's and 1990's.

(2) The utilities can survive with their current rate levels for at least several years. The companies in some states (for example, Texas and Florida) are in relatively good financial shape at present, so they can raise capital without too much strain. Companies in certain other states (for example, Alabama and Georgia) are currently unable to raise much, if any, capital. However, many plants started before 1973 have either just come into service or are scheduled to go on line soon, and because demand forecast at the time these plants were started has not materialized, many individual systems, and the industry as a whole, has excess generating capacity. For several years growth can be met out of this excess capacity. Thus, the industry as a whole will be able to meet demand over the next few years, even though the inability to attract capital prevents companies from starting new plants that they think will be necessary in the mid- to late-1980's and the 1990's.

(3) The situation noted in the preceding paragraph can delude people -- "the public" and perhaps also some utility commissions -- into thinking that the utilities' financial situation, and its long-run production capabilities, are really satisfactory, and that the utilities are merely "crying wolf." If this occurs on a large scale, then power shortages will start to occur, and the scenario depicted in Hailey's book, Overload will materialize. At this point, however, it will be too late to avoid at least some economic hardships, because of the long lead-time required to build generating capacity.

(4) If shortages do develop, they will be spotty, for some companies are in better financial shape than others, and some have more current excess capacity than others. Thus, the appearance of problems in certain areas can lead to corrective actions in other areas.

(5) If "overload" problems do start to occur on a wide scale, then governmental intervention will undoubtedly be necessary. Electricity is too important for governments to permit long-term power shortages. Thus, one might anticipate various types of federal and state loans, loan guarantees, joint ventures, and the like. In any event, the possibility of at least a partial shift from private ownership to governmental ownership and control of the electric power industry might result from a continuation of the utility companies' present financial weakness.

VI. CONCLUSION

The picture painted in this chapter is bleak. The utility industry as a whole faces huge future demands for power and, to meet these demands, the companies must raise and invest unprecedentedly large sums of money. However, the average company today is so weak financially that it simply cannot meet its capital requirements. The situation is masked by the fact that the companies now have excess capacity that arose from the sudden, sharp reduction in growth after 1973 -- this excess capacity has permitted the companies to survive and to meet current power demands, so the public has not suffered to any significant extent.[1]

The current situation cannot continue: excess reserves will soon be used up and, if construction programs have not been started well in advance, power shortages accompanied by severe economic problems will follow. It is possible, however, to prevent all this. What is needed is for utility commissions across the country to realistically analyze the situation and then to allow the utility companies to charge prices that cover the cost of providing service, including the cost of the capital invested in order to provide that service.

---

[1]Of course, power costs have gone up recently, but not nearly to the same extent as the prices of gasoline and a host of other products.

In this chapter we have not estimated the capital needed to meet requirements of synthetic fuel plants. Such capital costs cannot be realistically forecast since general engineering cost estimates are not available. Each specific conversion proposal would require extensive analysis to determine costs, and these costs would vary widely from plant to plant. In addition, the total number of plants to be converted cannot be easily estimated. However, regardless of the magnitude of capital required for conversion, it is in addition to that needed to meet new generation requirements.

References

Brigham, E. F., The Changing Investment Risk of Public Utilities, Working Paper 2-79, Public Utilities Research Center, College of Business Administration, University of Florida, Gainesville, Florida, 1979.

EEI (Edison Electric Institute), Statistical Year Book of the Electric Utility Industry for 1977, EEI, 90 Park Avenue, New York, N.Y., 1978.

Lerner, E. M., "On Utility Financing," Public Utilities Fortnightly, 95, 10, 30, 1975.

Thompson, W. R., "Preparing for the Future in Electric Power," Public Utilities Fortnightly, 103, 8, 19, 1979.

U.S. DOE Energy Information Administration, Annual Report to Congress 1978, DOE/EIA - 0173, Vol. 3, Government Printing Office, Washington, D.C., 1979.

Value Line Investment Survey, "Electric Utility (East) Industry," Ratings and Reports, Edition 1, Oct. 5, 1979.

CHAPTER 17

COAL AND THE STATES: A PUBLIC CHOICE PERSPECTIVE

Walter A. Rosenbaum

Twice within his Administration's first three years, President
Jimmy Carter exhorted Americans to reduce their menacing dependence
upon imported petroleum by consuming, instead, prodigious new quantities of domestic coal. In 1977, the President's first Energy Message
proposed that the United States increase domestic coal combustion by
two-thirds before 1985 -- a strategy requiring the U.S. to consume
yearly 1.2 billion tons of coal by the mid-1980's. In 1979, the President advocated a national investment of $88 billion to produce enough
synthetic fuel from coal to replace 2.5 million barrels of imported oil
daily while, simultaneously, converting to coal as many industrial and
utility boilers as possible. Coal will be irresistable to national energy planners, for the American earth covers perhaps half the world's
known coal reserves. Consequently, virtually all new federal energy
programs will depend heavily upon massive new coal consumption.

The most immediate, tangible costs and benefits of increased coal
combustion will be felt in the states. Moreover, a national coal policy
will depend in the future, as in the past, upon state implementation
during which states will enjoy considerable discretionary authority.
Thus, how the States define their interest in future coal development
and what institutional arrangements they develop to implement national
coal policy, will profoundly affect the nature of the nation's future
coal management plan.

I. THE STATES AND COAL UTILIZATION

State governments participate directly and continually in fashioning national coal policies. Historically, Washington has delegated to
the states generous discretion in implementing virtually all aspects of
federal coal programs. Also, the states retain independent authority
over many other aspects of coal utilization besides. This combination
of powers makes the role of state governments as important as that of
Washington in shaping the operational character of any coal program.

The State's strategic position relevant to existing coal utilization policy can be appreciated by examining, briefly, the range of the
State's authority. States now possess authority to deal with some, or
all, of the following coal issues:

1. Surface and underground coal mine siting. Many federal laws
delegate to the states the authority to determine when coal mines may
be sited on public or private lands. The Strip Mining Control and Reclamation Act of 1977 (Pub. L. 98-87), for instance, permits the states
to determine which lands are "unsuitable for surface mining" and which
coal reserves west of the 100th meridian are inappropriate for development because agricultural production on alluvial valley floors would be
endangered.[1] The same legislation requires Washington to work collab-
oratively with the states in creating arrangements for deciding when
federal coal reserves will be mined. The states also possess authority,
both delegated and inherent, to control underground coal mine siting,
safety conditions, and environmental impacts.

2. Siting, size and design of power generating plants. More than
70% of the nation's current coal production is consumed by electric
utilities. Historically, the states have licensed power plants and, in
the process, have regulated virtually every aspect of plant development
in conjunction with federal agencies. This traditional exercise of
state police powers has a direct affect upon the volume of national

---

[1]The discretionary authority delegated to the states in the Surface
Mining Control and Reclamation Act is examined in Walter A. Rosenbaum,
Coal and Crisis (New York: Praeger Publishing Co., 1978, pp. 58-62.

coal consumption by influencing the rate and location of utility coal demand.

3. Location of logistical facilities for coal utilization. The states share with the federal government authority over the location, design and rate schedules for railroads, waterways and slurries necessary to transport coal from mine to consumer. Any significant future enlargement of national coal consumption will require a proliferation of railroad lines and probably the construction of several major slurry lines crossing thousands of miles. State determinations affecting the location, size and rate structure for these logistical facilities will affect the rate of future coal utilization.[1]

4. Air and water standards applicable to coal burning facilities. Pollution standards, together with market price, are the most significant constraints upon future coal utilization. Federal air and water quality legislation permits the states considerable latitude in implementing federal air and water quality standards and, sometimes, in setting these standards. The Federal Water Pollution Control Act Amendments of 1972 (Pub. L. 92-500) and the Clean Air Act Amendment of 1970 (Pub. L. 91-604) both permit the states to set air and water quality standards more stringent than federal ones and, under some circumstances, to grant to specific pollution sources variations or exemptions from these standards. The states may also permit relatively unpolluted bodies of air and water to be "degraded" under specified conditions. The State Implementation Plans for enforcing federal air and water quality standards allow local officials discretion in deciding when, where and under what conditions a polluting activity, such as coal combustion, may occur. All these decisions directly affect national coal demand by determining the circumstances under which future utility and industrial coal-fired boilers may operate.[2]

---

[1] A useful review of state authority over all aspects of electric power plant development can be found in Association of the Bar, City of New York, Electricity and the Environment: The Reform of Legal Institutions (St. Paul, Minn.: West Publishing Co., 1972), Part IV.

[2] Both major federal programs are exhaustively described in Erica L. Dolgin and Thomas G. P. Guilbert, eds., Federal Environmental Law (St. Paul, Minn.: West Publishing Co., 1974), Chapters 10 and 15.

5. <u>The magnitude of State government coal utilization</u>. The
states themselves are major coal consumers. Approximately 10% of all
electric power generated in the U.S. is produced by state or local government authorities. Decisions made by state institutions about future
power generating facilities will directly influence the volume of prospective utility coal demand generated by governmental entities.

6. <u>Utility rate structures and the amount of pollution control
costs "passed through" from utilities to customers</u>. Coal demand is
responsive to unit pricing for electric power. The states largely set
utility rates and, together with federal agencies, determine how much
of pollution abatement costs may be passed to utility consumers by the
power companies. These market determinations, in turn, effect electric
power demand and, ultimately, contribute to the rate of future utility
coal demand.

These disparate, sometimes overlaping, powers collectively constitute an impressive array of state authority with obvious relevance to
coal utilization. However, such powers seldom are exercised simultaneously or coherently by one, or a few, state agencies. Rather, the delegation and exercise of these powers usually is fragmented. In most
states, fashioning a coordinated approach to coal utilization will require a degree of integrated planning and policy implementation which
may be beyond the capability of many state governments.

II. COMPETING POLICY OBJECTIVES

Virtually none of the multitude of different state powers affecting
coal utilization can be exercised without public officials having to
choose among competing, conflicting and sometimes irreconcilable policy
objectives. Dissonance created by these different objectives intrudes
constantly upon coal policy formulation. Still, to govern is to choose.
To exercise authority in any specific domain of coal utilization, state
officials will have to establish, if only implicitly, the relative priorities among four policy objectives: (A) environmental protection and
enhancement; (B) economic growth; (C) distributive equity; and (D) state
vs. national concerns. The relevance of these issues to state coal policy

can be appreciated by examining how they arise in coal policy development.

A. The Environmental Issue: Its Political Geography

The environmental issue stalks all discussions of coal utilization. Increased national coal consumption will sharply increase the magnitude of surface mining across the nation. In 1975, about 54% of the nation's coal was produced by surface mining; by 1985, three of every four tons of domestic coal will be scoured from the earth through strip mining. Stripping inflicts a violent, pervasive and frequently devastating impact upon the exposed lands. Uncontrolled, it is usually an ecological catastrophe whose risks have been documented amply.[1] Also, coal is the dirtiest of all fossil fuels to burn without pollution constraints. Unregulated coal combustion in industrial and utility boilers pours large volumes of carbon monoxide, sulfur oxides, hydrocarbons, nitrogen oxides, and particulates into the atmosphere, together with lesser amounts of ratioactive material, heavy metals, and trace elements. Thus, coal usually inflicts severe environmental impacts in both production and consumption. States likely to experience an increase in either activity, or in both, must formulate a calculus for setting future coal policy in which the magnitude of environmental risks, the acceptable levels of environmental damage, the costs of environmental protection, and the prospective benefits of coal utilization all are reckoned.

These environmental impacts will be experienced with different severity among the states and, thus, the salience of the environmental issue in deciding future coal policy will also vary. Table 1 provides a rough comparison of the anticipated environmental impact of future coal utilization among the states. The states have been categorized according to the concurrent impact upon the states of new national coal production and consumption. Generally, states in Category I will ex-

[1]See, for example, U.S. Congress, House, Committee on Interior and Insular Affairs, Surface Mining Control and Reclamation Act of 1976: A Report (Washington, D.C.: Government Printing Office, 1976), H. Rept. 94-1445; and William Ramsey, Unpaid Costs of Electric Energy (Baltimore: Johns Hopkins University Press, 1979).

Table 1: *The Environmental Impact of Increased Coal Utilization*
          on the States: A Comparative Summary Classification

---

I.   Low to Moderate Impact:[a] Alabama, Arizona, California, Connec-
     ticut, Delaware, District of Columbia, Florida, Georgia, Idaho,
     Illinois, Indiana, Kentucky, Maine, Maryland, Massachusetts,
     Michigan, Mississippi, Nevada, New Hampshire, New Jersey, New
     York, North Carolina, Ohio, Oregon, Pennsylvania, Rhode Island,
     South Carolina, Tennessee, Utah, Vermont, Virginia, Washington,
     West Virginia, Wisconsin. (34)

II. High Impact:[b] Arkansas, Iowa, Kansas, Louisiana, Minnesota,
     Missouri, Nebraska, Oklahoma, South Dakota. (9)

III. Extremely High Impact:[c] Colorado, Montana, New Mexico, North
     Dakota, Texas, Wyoming. (6)

Note: Estimates for future coal consumption and production upon which
the table is based are found in U.S. Congress, Office of Technology
Assessment, The Direct Use of Coal (U.S. Government Printing Office,
1979, p. 48).

[a]States in this category are in the low to moderate range in both estimated coal production and utilization. This placed them between 3.2 -
8.6 million tons in anticipated increased coal utilization by 1985 and
from -13 to +10 million tons per year in new coal production by 1985.

[b]States in this category exceeded 15 million tons per year in anticipated new coal consumption and 11 million tons per year in anticipated
new coal production.

[c]States in this category ranged above 30 million tons per year in anticipated new coal consumption and above 30 million tons per year of
new coal production by 1985.

---

perience little to moderate impact from either activity. Those states
in Category II can expect moderate environmental spillovers from coal
consumption, coal production or both. In Category III are the most severely affected states: those likely to experience concurrently grave
environmental risks from both coal production and coal utilization. A
politically significant aspect of Table 1 is the disparity between the
distribution of the U.S. population and the distribution of environmental risks from coal utilization. The states vulnerable to the greatest
ecological risks from new coal utilization are generally the least pop-

ulous -- at times a serious strategic disadvantage (for example, in
Congress where the legislative delegations from coal-using states
heavily outnumber representatives from those states exposed to the
greatest potential environmental damages). Further, lands now most exposed to ecological degradation from increased coal utilization are
among the nation's most biologically fragile and, thus, the potential
for "restoring" strip mined lands is most clouded where the magnitude
of stripping is likely to be greatest.

B. Coal Utilization as a Growth Stimulant

There is a more benign, even benevolent, aspect to coal utilization which may figure prominently in future state policies affecting
coal mining or combustion. Coal utilization is an economic growth
stimulant. Especially in Appalachia and the Plains States where coal
seams abound, escalating coal production could become those states'
most important economic stimulus within the next two decades. Thus,
the environmental impacts of coal utilization will almost inevitably be
considered as a possible "trade-off" with invigorated state economy.

Generally, new coal mining creates jobs, generates mining royalties for affected state governments, enhances the tax base for local
government, spurs rapid urbanization with all its associated service industries, and sometimes leads to the appearance of ancillary service industries taking advantage of the proximity of coal supplies. Mine development, particularly when high-volume and sustained, can lead to the
appearance of mammouth mine-mouth generating plants and the growth of
transportation systems linking coal to its markets. A rough estimate
of the potential for mining development within the states is suggested
in Table 2 which identifies the demonstrated coal reserves in the U.S.
in the mid-1970's.[1] The volume of mining at these sites will vary according to numerous factors including federal and state environmental
regulations, the market price for coal and the rapidity of electric util-

---

[1]See also U.S. Congress, Office of Technology Assessment, The Direct Use
of Coal (Washington, D.C.: Government Printing Office, 1979), Chapters
IV-VI.

Table 2: *Demonstrated Coal Reserves in the United States, 1974*
(millions of tons)

Region	Estimated Reserves Strip Mine	Underground	Percent of Total Reserves
Appalachia			
Alabama	1,184	1,798	
Kentucky	7,354	18,185	
Maryland	146	902	
Ohio	3,654	17,423	
Pennsylvania	1,181	29,819	
Tennessee	319	667	
Virginia	679	2,971	
West Virginia	5,212	34,378	
Subtotal	19,729	106,143	29.0
Midwest			
Arkansas	263	402	
Illinois	12,223	53,442	
Indiana	1,674	8,949	
Iowa	--	2,885	
Kansas	1,388	--	
Michigan	1	118	
Missouri	3,414	6,074	
Oklahoma	434	860	
Subtotal	19,397	72,730	21.3
Rocky Mountain-Pacific Coast			
Alaska	7,399	4,246	
Arizona	350	--	
Colorado	870	13,999	
Montana	42,561	65,834	
New Mexico	2,258	2,137	
North Dakota	16,003	--	
South Dakota	428	--	
Utah	262	3,781	
Washington	508	1,446	
Wyoming	23,845	29,491	
Subtotal	94,484	120,934	49.7

Source: U.S. Department of the Interior, Bureau of Mines, <u>Minerals Yearbook</u>, 1975, p. 353.

ity boiler conversions. At least 20 states face the prospect of major
economic growth attributable to mining. In the Western states, especially, this growth can be extremely rapid and traumatic. A 1975 federal
study of six Western States most likely to experience new coal production suggests that the growth impacts will touch 99 communities. More
than a third of these are settlements of less than 1,500 individuals;
only 11 exceed populations of 5000.[1] Among all ten Western states with
significant coal reserves, the study estimates that population growth by
1985 would approach 300,000 people under circumstances of moderate new
coal demand. This figure excludes any population growth associated with
new generating plants, transportation systems or other industries ar-
riving in the wake of new mine production.

In recent years, the federal government has undertaken a number of
programs which mitigate many adverse impacts from such rapid growth and
offer states an incentive for accepting massive new mining programs.
Many of these actions relate to coal in deposits on federal lands.
About 50% of all Western coal reserves are directly controlled by Washington as part of the public domain; consequently, federal conditions
and constraints attached to this coal utilization will profoundly affect the general nature of future Western coal mining. The Federal Coal
Leasing Amendments Act (1975) increased state royalties from new mineral
leases on federal lands from 37.5% to 50%; also, royalties from surface-
minded land rose from 5¢ per ton to at least 12.5% of the selling price.
In 1976 Congress passed the Federal Land Policy and Management Act,
giving state legislatures greater discretion in awarding portions of state mining revenues to those subdivisions (such as cities and
counties) affected by mine development on federal lands.[2] It is very
likely that other federal legislation will enable Washington to make
available even more attractive subsidies to communities heavily impacted

---

[1] U.S. Comptroller General, Rocky Mountain Energy Resource Development:
Status, Potential and Socioeconomic Issues (Washington, D.C.: Government Printing Office, 1977), pp. 31-57.

[2] The use of subsidies as incentives for coal development and risks are
discussed in Walter Rosenbaum, op. cit., Chapter 5.

by new mining or related facilities, in an effort to diminish resistance to such mining.

C. The Distributive Equities in Coal Utilization

The benefits and costs of future coal utilization are unevenly distributed across the nation's geographic regions. The burden and rewards
of coal combustion are also dispersed inequitably across the nation's
major social groups. There is no simple answer to the question: "Who
benefits from new coal utilization?" The mix of social groups aligned
on the differing sides of the combustion issue will depend upon what
government policies are used to encourage more coal combustion and upon
what pollution control strategies are retained or modified. For instance, a federal coal utilization program leaning heavily on the crea-
tion of a new economic intra-structure to produce synthetic fuel would
probably bestow particular economic rewards on the large petroleum corporations and their stockholders. In contrast, a strategy that encour-
ages more coal utilization by decreasing industrial and utility demand
for oil and natural gas is likely to impose short-term costs upon those
corporations and their stockholders.

There are, however, a number of social groups whose interests will
unquestionably be affected by accelerated coal utilization; the nature
and magnitude of the effects will depend upon which coal policies are
pursued. Among these groups are the following:

1. Coal miners and allied workers. Almost a quarter million Amer-
cans currently work in surface and underground coal mines. Increased
coal utilization will probably enlarge this work force from 45 to 110%
in the next twenty years. Increased coal production means more jobs but
also greatly increased risks of disabling injuries. The U.S. Office of
Technology Assessment has estimated, for instance, that injuries from
major new mining activities could increase from 86 to 167% during the
same period. This is likely to create further difficulties in one of
the most turbulent U.S. labor markets and to disturb further the already
troubled working relations between labor and management in the mining

industry.[1]

2. Utility rate payers. Consumers of electric power in the United
States will also be affected. Slightly less than half the nation's
present electric power is coal generated. Conversion of existing and
future electric facilities from petroleum and natural gas to coal could
increase this to 50% of all electric power by 1985. If present environmental regulations affecting coal burning prevail, the effect will be to
drive up the price of electric power, as utilities "pass through" their
pollution abatement costs to their customers. Cost increases would also be experienced by customers of utilities switching from petroleum
and natural gas to coal. Thus, the environmental context in which coal
policy is developed will economically affect a large portion of the U.S.
population. Relaxing environmental standards may inhibit the rate of
which electric power costs mount for customers. Maintaining or increasing ecological safeguards on coal combustion may drive up the cost of
electric power quickly and, perhaps, dramatically.

3. Construction trades, technical workers and allied crafts associated with synthetic fuel facility construction and maintenance. A
massive U.S. drive to develop a synthetic fuel production capability,
such as President Carter's 1979 Energy Message proposed, would create a
new, and potentially large, market for both skilled and unskilled labor.
A major reason is that the construction and operation of synthetic fuel
facilities is likely to be labor-intensive. One study by the Department
of Energy suggests, for instance, that constructing a small coal liquefaction plant would probably require 2500 workers, the subsequent operation would involve 725 individuals, and the cost would exceed $15 bil-
lion.[2]

---

[1]U.S. Office of Technology Assessment, op. cit., Chapter VI.

[2]The author is indebted to officials in the U.S. Department of Energy
for this information.

4. <u>The economically disadvantaged</u>. Any increase in the cost of electricity, or any other product such as petrochemical derivatives based on coal, is felt with particular acuteness among the poor. The economically disadvantaged have less total income from which to pay the costs of energy utilization or any other forms of coal products. Further, a far larger portion of their disposable income must be allocated to such necessities as housing, food and utilities. Thus, groups and individuals representing the economic interests of the poor must regard any upward movement in the absolute or relative costs of energy to the disadvantaged as a political matter of great importance.

5. <u>Shareholders and management in major holding corporations for the mining industry</u>. Coal mining is rapidly becoming a captive industry. In 1977, 13 of the top 15 coal producers were part of corporate conglomerates. Five of these controlling companies were petroleum corporations; four more were steel companies. The corporate stake in future coal policy is obvious and makes these corporate interests active participants at whatever governmental levels where coal policy will be formulated.[1]

This belief inventory only suggests some of the major political influences on the process of coal policy formulation but does identify some of the salient social and economic sectors to have a voice in shaping such policy.

D. National vs. State Interests

Implicit in any state policies concerning coal combustion is a judgment about the relative importance to be accorded national interests in state coal perspectives. This is a potentially divisive issue because state and federal interests often seem inconsistent if not incompatible. Indeed, the issue of future coal utilization has <u>created</u> some of these cleavages while exacerbating others.

---

[1]The implications of this corporate concentration are usefully summarized in the U.S. Comptroller General, <u>The State of Competition in the Coal Industry</u>, Document No. EMD 78-22 (December 30, 1977).

Perhaps the most intense expression of this cleavage is the so-called "East-West" controversy -- essentially, a conflict arising from the different perspectives of coal-producing and coal-utilizing states. Presently, four Western states -- Montana, New Mexico, North Dakota and Wyoming -- overlie almost two-thirds of U.S. strippable control reserves yet constitute only 4% of the U.S. population. The largest coal consuming states, entirely in the East, Midwest and South Atlantic, consume almost two-thirds of all U.S. coal production and contain about a third of the nation's population. Spokesmen of the Western coal states have frequently asserted that the nation -- and particularly the Eastern U.S. -- want to turn the West into an "energy colony" bearing the risks of grave ecological devastation in order to provide the energy-hungry population East of the Mississippi with needed coal supplies. Spokesmen for Eastern interests and, more generally, for increased coal production, have asserted that the national interest lies in new coal to replace dependence on imported petroleum to feed Eastern furnaces. In effect, it is an admonition to Western interests to elevate the national interest-- or at least one version of it -- above parochialism.

In formulating a national coal utilization strategy, precisely this kind of balancing -- state vs. state, region vs. region, nation vs. state -- must be done by Congress and the President. However, the Congress is, among other things, a convocation of ambassadors from local constituencies. Thus, when a national coal policy is formulated, it becomes, in large part, the product of decisions made by legislators who must singly and collectively come to some verdict about the relative importance assignable to local vs. national coal utilization impacts.

III. THE FEDERAL ROLE: OPTIONS AND OPPORTUNITIES

The competing policy objectives we have examined are implicit in the formulation of any current state coal combustion policies. The obligation to choose between them, and to reconcile them when possible, is difficult yet inevitable. However, the Federal government -- which will establish broad national coal policies to constitute the context in which state decisions are made -- can simplify state options and reduce the risks inherent in new coal burning policies.

It is essential that the federal government should avoid the temptation to extricate the U.S. from its current energy dilemmas through a massive, long-term, open-ended commitment to increased coal combustion. Clearly, the nation will burn more coal, perhaps twice the present quantity in the next decade. But the ecological risks of a massive, continuing raid on coal resources are so grave that they must be avoided. Instead, the nation should seek to inhibit coal production to a short-term surge while energy conservation and alternative, more efficient new energy systems are gradually developed. In short, any state coal combustion policies should evolve within the context of a restrained, short-term national commitment to new coal combustion.

There are several specific actions Federal planners can take to reduce the dangers of an unrestrained new raid on coal. First, projections of future energy demand, both nationally and with individual states, should be formulated more critically and cautiously than in the past. Projections of future energy demand are often based upon presumptions about future levels of economic activity; indeed, anticipated energy demand is often an expression of growth rates considered desirable. Federal planners seldom challenge the ideas that future U.S. economic growth should be predicated on past trends; neither do most Congressional proposals for coal development. In effect, governmental planning of future coal needs is largely captive to the assumptions (often unstated) about likely or desirable economic growth rates.

It is possible to manipulate energy demand by moderating the stimulants to economic growth, or by applying them selectively as to time and place. This applies to federal coal planning, of course, but to state planning also. The federal government can take measures in several ways to moderate future coal demand. In the most literal sense, coal demand is electric power driven. The federal government can discourage the Federal Power Commission (and even state utility regulators) from the continuing promotion of more electric power use by means of preferential rates to large users, "block rates," and other pricing policies catering to increased utility growth. More importantly, the federal

government needs to explicitly formulate a target national economic
growth rate adequately reflecting a balanced concern for economic expansion and resource conservation.

The federal government, working in harmony with Western states,
should encourage institutional arrangements in formulation of national
coal policy which allow for a vigorous, continuing expression of Western
state viewpoints. This is important because, as we have already observed, the Western states that might experience particularly severe
environmental impact from new coal production are thinly populated and
meagerly represented in Congress. For example, the four Western States
with the largest remaining coal reserves (Montana, New Mexico, North
Dakota and Wyoming) contain less than 1.5% of the U.S. population; the
total Congressional delegation from these four states (15) is less than
the Congressional representation from Ohio alone, the nation's largest
coal consuming state. It is readily imaginable that this enormous disparity between resource reserves and political representation can work
to the extreme disadvantage of the Western environment and energy conservation. The great political weight carried by the Congressional
delegations of coal consuming states can be placed behind ambitious new
coal combustion programs with the extensive new strip mining they will
inflict on Western lands. The Western states, however, inadvertently
have become a major conservative force in coal utilization, out of an
acute concern with their own regional interests. Thus, arrangements
which encourage a vigorous expression of Western regionalism in national
coal policy formulation are likely to work to the advantage of energy
conservation as well. It is especially important, in this respect, that
the ambitious provisions for public involvement in the regulation of
strip mining, as provided in the Strip Mining Control and Reclamation
Act (1977), be zealously enforced by the U.S. Department of the Interior
and by the states assuming responsibility for implementing the regulatory programs under the Act. Further, the federal government should
provide the statutory authorization to federal and state agencies with
strip-mine regulatory authority to allocate funds to public interest
groups for purposes of educating the state publics on such programs.

Finally, it is important that Federal development of new energy technologies -- especially through Federal "R & D programs" -- maintain a balanced commitment to alternative technologies rather than continuing past, almost exclusive preoccupation with coal and nuclear technologies. Especially, the federal government should not be permitted largely to underwrite the development of new coal technologies such as coal gasification and liquefaction through massive R & D funding. Once such a federal commitment has been made, it is very difficult for Washington to curtail or substantially redesign the developing systems. Beneficiaries of the existing investment become politically organized into powerful constituencies to defend the existing policies or to enlarge their scope. As "sunk costs" in existing R & D programs rise, Congress and the bureaucratic stewards of the programs are also increasingly reluctant to abandon the programs. Further, the massive, continuing investment of Federal dollars in synthetic coal technologies will feed the demand for new coal combustion and will appear to justify -- if not to demand -- enlarged projections of future national energy use. In brief, a major federal commitment ot R & D funding of synthetic fuel technologies is likely to accelerate the growth of coal demand and advance escallating use by a great many decades. All these circumstances work against a short-term, limited commitment to extensive new coal combustion. President Carter's recent allocation of approximately $500 million in the FY-1980 federal budget for solar-related R & D is a promising start to a more balanced future investment in energy technologies. However, his more recent proposal that the federal government invest $88 billion in the development of new synthetic fuel technologies indicates a continuing federal preoccupation with increasing coal demand

As these proposals clearly suggest, the states' large role in future coal policy formulation should not obscure the increasingly substantial part played by Washington in defining the nature of future coal policy and the range of alternatives available to the states in implementing it.

CHAPTER 18

FEDERAL REGULATORY AND LEGAL ASPECTS

Joseph W. Little and Lynne C. Capehart

I.   INTRODUCTION

Federal laws affect the use of coal in a variety of ways. Laws mandating coal conversion and prohibiting the use of natural gas and petroleum encourage coal use; whereas laws that protect the environment impose conditions on coal production and combustion that impede coal use.
Mine safety laws and transportation regulations may also discourage coal
use by imposing higher costs.

This chapter is intended to provide the layman with an overview of
federal laws that may have an important influence on coal utilization in
the United States. The accompanying table lists and briefly summarizes
the federal laws that are discussed more fully below. Although lawyers
and legal researchers should find this treatment to be a good place to
start in acquainting themselves broadly with this field, it does not
purport to be a research resource.

The reader should be aware that state and local laws can also have
a powerful influence on coal use in specific areas. Apart from a general overview of certain water resource issues, however, the brevity of
this work precludes examination of them.

II.   FEDERAL LEGISLATION ENCOURAGING COAL USE

The 1973 Arab oil embargo stimulated federal interest in the use of
coal as an alternative fuel. Federal legislation and policy since then
have been largely directed toward providing both incentives and mandates

359

TABLE OF FEDERAL STATUTES
FEDERAL LEGISLATION AFFECTING COAL UTILIZATION

STATUTE	MAJOR AREAS OF IMPACT

I. INCENTIVES FOR COAL USE

A.	Energy Supply and Environ-mental Coordination Act of 1974 15 U.S.C. §791-798 (P.L. 93-319, June 22, 1974) Energy Policy Conservation Act of 1975 42 U.S.C. §6201-6422 (P.L. 94-163, Dec. 22, 1975)	May require electric powerplants and other major fuel burning in-stallations to burn coal.
B.	Powerplant and Industrial Fuel Use Act of 1978 42 U.S.C. §8301 et seq. (P.L. 95-620, Nov. 9, 1978)	Prohibits new electric powerplants and other major fuel burning in-stallations from burning natural gas or petroleum.

II. ENVIRONMENTAL LAWS

A.	National Environmental Policy Act of 1969 42 U.S.C. §4321 et seq. (P.L. 91-190, Jan. 1, 1970)	Affects "any major action signi-ficantly affecting the quality of the human environment."
B.	Air Quality	
	Clean Air Act Amendments of 1970, P.L. 91-604 (Dec. 31, 1970) Amendments of 1977, 42 U.S.C. §1857 et seq. (P.L. 95-95, Aug. 7, 1979)	Requires air emission controls for coal burning facilities to meet air quality standards.
C.	Water Quality	
	Clean Water Act of 1977 Pollution Control, 33 U.S.C. §1251 et seq. (P.L. 95-217, Dec. 27, 1977)	Requires NPDES permits for: coal mine drainage, coal storage pile runoff, discharges from electric powerplants.
	Safe Drinking Water Act of 1977 42 U.S.C. §300(f)-(j)(9) (P.L. 94-580, Oct. 21, 1976)	Regulates waste disposal by in-jection wells.

TABLE OF FEDERAL STATUTES (con't.)

D. Strip Mining

Surface Mining Control and Reclamation Act of 1977 30 U.S.C. §1201 et seq. (P.L. 95-87, Aug. 3, 1977)	Specifies areas unsuitable for surface mining. Requires reclamation.

III. OTHER REGULATORY LAWS

A. Transportation

Interstate Commerce Act 49 U.S.C. §1 et seq.	Regulates commodities moving in interstate commerce. Especially affects railroads.
Railroad Revitalization and Regulatory Reform Act of 1976 45 U.S.C. §801 et seq. (P.L. 94-210, Feb. 5, 1976)	Provides some economic and regulatory assistance to railroads.
Federal-Aid Highway Act of 1978 23 U.S.C. §101 et seq. (P.L. 95-599, Nov. 6, 1978)	Requires states to enforce weight limits. Could affect coal trucks.

B. Mine Safety

Mine Safety and Health Act of 1977 30 U.S.C. §801 et seq. (P.L. 95-164, Nov. 9, 1977)	Sets operational standards for coal mines. Requires training for miners.

C. Mineral Leasing

Coal Leasing Amendments Act of 1976 30 U.S.C. §201-209 (P.L. 94-377, Aug. 4, 1976)	Intended to improve federal leasing procedures.

IV. ANTITRUST LAWS

Sherman Act 15 U.S.C. § et seq.	Outlaws restraints of trade and commerce.
Clayton Act 15 U.S.C. §12 et seq.	Outlaws anti-competitive mergers and acquisitions.
Federal Trade Commission Act 15 U.S.C. §41-58	Outlaws unfair or deceptive acts or practices.

for increasing the use of coal as a replacement for petroleum and natural gas.

A.   Initial Legislation

The Energy Supply and Environmental Coordination Act of 1974 (Energy
Supply Act) was the initial Congressional effort at encouraging coal utilization. The Energy Supply Act required the Federal Energy Administration (FEA), the forerunner of the Department of Energy, to prohibit any
powerplant from using oil or gas as a primary fuel if (1) burning coal
was practicable, (2) coal and coal transportation facilities were available, and (3) reliability of electric service would not be impaired.
Conversion to coal could be required for any other major fuel burning
facility if the installation had the capability and the equipment available to burn coal.

These mandates were strengthened by the Energy Policy and Conservation Act of 1975 (Energy Policy Act). In addition, these Acts grant to
FEA the authority to establish a coal allocation program which would include allocating low-sulfur coal on a priority basis to avoid or minimize adverse public health impacts and authorize loan guarantees for the
development of new underground coal mines.

B.   The Powerplant and Industrial Fuel Use Act of 1978

Because the process under the Energy Supply Act was cumbersome and
time-consuming, the rate of coal-conversion was disappointingly slow.
To remedy this, Congress enacted the Powerplant and Industrial Fuel Use
Act of 1978 (Fuel Use Act).

1.   *Conversion to Coal or Alternate Fuels*

The Fuel Use Act contains a general prohibition on the use of natural gas or petroleum as a primary energy source in any new electric powerplant or major boiler-fired fuel-burning installation. Restrictions on
non-boiler fired installations are authorized but not mandated.

Both temporary and permanent exemptions are authorized. Three kinds
of temporary exemptions are permitted: (1) for a period of 5 years when
an applicant can show either that issuance of an exemption would be in

the public interest or that the installation will meet certain design
criteria, and will use coal or an alternate fuel for at least 75% of its
fuel needs after the exemption expires; (2) for up to 10 years if the
installation will be using a synthetic fuel when the exemption expires
but is unable to do so at an earlier time; (3) for up to 10 years if (a)
the cost of using imported petroleum will be substantially less than the
cost, including transportation, of obtaining an adequate, reliable supply
of coal; (b) site limitations prevent the location or operation of a
coal-fired installation; or (c) a violation of applicable environmental
laws would occur.

A permanent exemption will be granted if: (1) the cost of coal or
alternate energy sources will substantially exceed the cost of imported
petroleum over the useful life of the installation; (2) site limitations
prevent the use of coal; (3) using coal or alternate fuels would cause a
violation of environmental laws; or (4) adequate capital would not be
available to finance an installation using coal or alternate fuels.
Other grounds for permanent exemptions are: (1) state or local requirements make construction infeasible and the exemption is in the public
interest; (2) the installation is a cogeneration facility and only use of
petroleum or natural gas will yield cogeneration benefits; (3) the installation uses an approved mixture of petroleum/gas and coal/alternate
fuel; (4) the installation will be used only for emergency purposes; (5)
the powerplant must use natural gas/petroleum to maintain reliability of
service; (6) the installation is a peakload or intermediate load powerplant; (7) the installation requires natural gas/petroleum to maintain
product quality control; or (8) the installation requires natural gas/
petroleum in order to meet scheduled equipment outages.

The Act permits exemptions for existing installations similar to
those for new installations. In addition, temporary exemptions of up to
10 years will be granted to installations planning to adopt an innovative
technology for coal use upon expiration of the exemption.

Whether the conversion goals of the Fuel Use Act will be fulfilled
remains to be seen. Because of the breadth of the exemptions, performance may be strongly influenced by the administrative criteria applied

by the Department of Energy.

2. *Other Provisions*

The Fuel Use Act prohibits the use of natural gas in boilers that provide space heating or in decorative outdoor lighting. Electricity converted from coal seems a probable substitute energy source.

The Act also authorizes the President to allocate coal in emergencies; authorizes federal assistance to areas of the United States impacted by increased coal production; and establishes a loan program to assist utilities in acquiring equipment that will enable converted powerplants to meet air quality standards.

The Fuel Use Act requires: (1) a national coal policy study of the alternative national uses of coal available in the United States; (2) a performance and competition study of the coal industry that must examine the extent of the control of commodity transportation systems owned by coal companies; (3) a socio-economic study of the impact of increased coal production; and (4) a study of the problems peculiar to the operation of small utility systems that will be posed by compliance with the Act.

C.   Energy Mobilization Board

The federal government has established a policy of reducing this country's dependence on foreign oil without an accompanying policy of reducing the nation's energy consumption proportionately. Cutting oil imports will, therefore, necessarily require the construction of a large number of new energy-related facilities such as electric powerplants, synthetic fuel plants and coal slurry pipelines. Environmental laws (discussed in the next section) will be an impediment to expanded use of coal burning facilities by either increasing the costs, slowing the rate of conversion to coal, or both. Nevertheless, Congress is aware of the fact that the weakening of the environmental laws would reverse hard won air-and-water quality improvements throughout the nation, which would be resisted strongly by some sectors.

So far, Congress has not directly weakened the environmental laws. Instead, it is considering an approach that would result in an overriding

of environmental controls under certain conditions. One proposal is to
create an Energy Mobilization Board to oversee certain priority energy
projects. The Board could be empowered to steer critical projects
through regulatory red tape and might even be authorized to override
federal, state or local laws that stand in the way of an approved facility. The power of the Board could be made even greater by precluding
judicial review of its decisions. If Congress adopts the Board approach,
the extent of the authority given it will be a measure of how Congress
views the relative importance of coal use versus clean environment.

III. FEDERAL ENVIRONMENTAL LEGISLATION REGULATING COAL USE

Coal is abundant, but its use poses severe threats to air and water
quality. Laws regulating air emissions, water discharges, surface mining and waste disposal are all likely to have direct impacts on the use
of coal.

A.   National Environmental Policy Act

The most sweeping federal statute pertaining to environmental protection is the National Environmental Policy Act of 1969 (NEPA). NEPA
prescribes procedures to assure the deliberation of environmental questions in planning processes but does not impose substantive regulations.
All federal agencies must prepare Environmental Impact Statements (EIS)
to accompany every proposal for legislation or other major federal action "significantly affecting the quality of the human environment" before
it is adopted. For example, NEPA required the Department of Energy to
prepare a programmatic EIS on the nationwide environmental impact of the
coal conversion promoted by the Fuel Use Act. In addition, proposals
pertaining to a site specific project, such as a single powerplant, must
be accompanied by an EIS if they are of the magnitude to meet the "major
federal action significantly affecting the human environment" criterion.
EIS's for site specific proposals must also consider regional impacts if
other pollution sources are nearby.

The massive synthetic fuels program proposed by President Carter in
July, 1979, would require a programmatic EIS and each synthetic fuel

plant would require an individual EIS.

Congress may grant exemptions to NEPA requirements. The 1978 Fuel
Use Act excludes EIS preparation for applications for certain permanent
exemptions from the coal use requirements. Furthermore, if an EIS is
already required for some other reason for the project for which an exemption is sought and the environmental effect of the exemption is described adequately, then a separate EIS for the exemption is not required.
Apart from these exemptions, the Fuel Use Act specifically requires compliance with all applicable environmental requirements. This may change
if Congress determines that environmental degradation must be accepted
to enhance coal use.

B.   Air Quality Laws

The Air Quality Act of 1967, the Clean Air Act Amendments of 1970,
and the Clean Air Amendments of 1977 set the national policy for clean
air. The 1970 Clean Air Act Amendments mandated the setting of uniform
federal air quality standards and granted to the states the power to enforce them through the operation of State Implementation Plans. Under
the 1970 Amendments, the Environmental Protection Agency promulgated two
sets of national ambient air quality standards: primary standards to
protect public health; and secondary standards to protect the public welfare from non-health related pollution damage. Standards of both classes
have been set for the primary by-products of coal burning; namely, sulfur oxides, nitrogen dioxide, and particulate matter. When it became
apparent that the states would not attain air of the quality required by
the primary standards by the established deadline, Congress enacted the
1977 Amendments, extending that deadline and requiring the states to revise their Implementation Plans by July, 1979, to assure attainment of
primary standards by 1982.

1.   *New Source Performance Standards*

In addition to mandating ambient air quality standards, the 1970
Amendments also required EPA to set performance standards for 19 categories of large new stationary sources of air pollution, such as powerplants. The goal was to avoid making air pollution worse or creating

new pollution sources. The 1977 Amendments added nine more categories of regulated new sources and required that all these sources apply the best technological system to reduce emissions continuously. This provision may have the effect of requiring that sulfur oxide scrubbers be installed on all new coal-fired powerplants regardless of the sulfur content of the coal, but the EPA may waive the use of scrubbers on any source that will utilize an innovative emission reduction system that has a substantial likelihood of success.

2.   *Non-Attainment Areas*

The 1970 Amendments prohibited further construction of polluting sources in any region where the national primary ambient air quality standard is exceeded for any pollutant until the area is "in attainment." The 1977 Amendments moderated this restriction allowing construction of new sources in non-attainment areas where: (1) the state adopts a revised State Implementation Plan that provides for attainment of primary air standards in non-attainment areas by December 31, 1982; (2) the area is making "reasonable further progress" toward attainment of primary standards; and (3) a permit is issued stating the ability of the source to meet the applicable standards.

The "reasonable further progress" criterion is a more stringent regulation when applied to new sources, such as coal-fired powerplants, than are the new source performance standards. This is true because the economic consequences to an industry are one of the criteria that affect the application of new source standards; whereas, "reasonable further progress" requires the source to install equipment that meets the "lowest achievable emission rate" attainable by that source regardless of cost. It may be lower than but can never exceed new source performance standards.

The stringencies of the non-attainment regulations are loosened to some extent by the EPA's "emission offset" policy that permits a construction of a major new source if its sponsors can obtain a reduction of emissions from existing sources by an amount that more than offsets the expected emissions from the proposed new source. The 1977 Amendments further loosen the non-attainment regulations by providing a new growth allowance that may be used conjunctively with the emission offset.

To obtain the benefits of an emissions offset, the owner of a proposed new source must certify that all existing sources within the state and under his ownership or control either comply with all applicable air pollution laws or have had compliance schedules approved. Thus, a utility must clean up all of its non-complying powerplants in a state before it may build a new plant in a non-attainment area of the state. It could also be required to reduce emissions on existing complying plants within the non-attainment area and to install the lowest achievable emission technology on the new plant. This could require installing scrubbers and burning low-sulfur coal.

3.   *Prevention of Significant Deterioration*

The Clean Air Act has been judicially interpreted to require maintaining the existing state of air purity in areas where it is cleaner than required by national ambient air quality standards. This policy was incorporated into EPA regulations and later enacted into statutory law by the 1977 Amendments. Prevention of Significant Deterioration (PSD) regulations also affect the siting of coal burning facilities within a PSD region by virtue of the fact that the regulations pertain to effects on ambient air quality not only at the proposed site but also in areas within a downwind range of 50km of the facility.

The PSD program creates three clean air area categories and specifies a maximum allowable "PSD increment" for each pollutant in each category. Class I areas, such as national parks and wilderness areas, allow such slight degradation of air quality that little development can be permitted. Class II areas allow moderate growth, and Class III areas permit degradation of air quality up to the ceiling set by the ambient quality standards.

Any major emitting facility must obtain a PSD permit before beginning construction in a PSD area, or in a non-attainment area if its emissions will have significant impact on any PSD area. Not only must the applicant show that new emissions will be within the permitted increment, but it must also show that the permitted increment has not been used up by growth in minor sources that require no permit. This poses a threat to large facilities if growth in the number of small users of coal, such as residential coal

stoves and small industrial boilers, should happen to consume the PSD
increments for sulfur oxides, particulate matter and nitrogen dioxide,
thereby precluding larger facilities with lower total emissions levels
from receiving permits.

4.   *Delayed Compliance Orders*

Installations converted to coal burning under the Energy Supply Act
may be relieved of some of the stringent emission controls required in
non-attainment areas. The compliance deadline has been extended to
December 31, 1985 for these installations, but they must be fitted with
the "best practicable system of emission reduction" during the interim
period.

5.   *Conclusion*

Moving from gas and oil to coal as the nation's chief energy source
will cost more money to keep the air clean. The federal clean air laws
lay down standards that will force industry to make these expenditures,
but the coal industry charges that they will merely increase the cost of
energy without adding appreciably to the protection of the public. Congress was aware of the conflicts between environmental and energy policies when it enacted the 1977 Amendments and appears to be standing by
the choices it has made.

C.   Water Quality

Three major stages in the processes in which coal is used that
threaten the purity of water are: (1) mining operations, (2) coal preparation facilities, and (3) coal burning installations. Federal water
quality legislation imposes regulations on all three stages.

1.   *Clean Water Act of 1977*

The Clean Water Act of 1977, formerly the Federal Water Pollution
Control Act, establishes a comprehensive federal program to clean up the
nation's waters and to obtain "zero discharge" of pollutants by 1987.
Two kinds of regulations are imposed. Each state must set water quality
standards based on the use made of the waters receiving the polluting
wastes, and the federal government must set effluent limitations on pollutants discharged by point sources. In an arrangement similar to the

Clean Air Act permit program, the Clean Water Act authorizes the states
to manage a federal water quality permit program, the National Pollution
Discharge Elimination System (NPDES), that employs the state and federal
water quality standards noted above. Every activity, including construction and operation of facilities, that may discharge pollutants into
navigable waters must receive an NPDES permit. Specific coal-use applications include the regulation of discharges from the generation process,
runoff from storage and construction sites at powerplants, and drainage
from active coal mines and secondary coal recovery and preparation facilities.

2.   *Safe Drinking Water Act*

The Safe Drinking Water Act of 1974 sets national drinking water
standards to assure ample supplies of safe drinking water throughout the
United States. To protect ground water, the Act requires states to establish a permit program for regulating the injection of wastes into
wells. Thus, coal burning installations must be permitted by the state
(or the EPA if the state has not developed an EPA-approved control program) before using injection wells for waste disposal.

If the wastes are hazardous, the wells will be "hazardous waste
management facilities" subject to further regulation under the Resource
Conservation and Recovery Act.

D.   Water Quantity

Congress has not adopted a comprehensive federal policy pertaining
to the consumption and allocation of water supplies. These matters have
historically been governed by state law. Nevertheless, because energy
facilities require massive quantities of water, it is appropriate to
present a national overview of basic American water law.

There are two common law systems of water allocation in the United
States. The system of law prevailing east of the Mississippi River is
known as the riparian system, and usually rests upon the reasonable use
doctrine. Riparian land abuts a body of water. The riparian owner is
entitled to use the water for any beneficial purpose so long as the use
is reasonable and does not interfere with reasonable uses of other

riparians. Reasonableness is determined by present conditions and a new
user on a watercourse may supplant an existing user if his use is considered more reasonable. Of course, the question of replacing prior
users will only arise where water supplies are not adequate to meet all
needs. Thus, if a new coal facility will consume more than the available
amount of unused waters, it may be able to displace prior rights on the
strength of the assertion that new coal-fired energy supplies are more
important than the displaced function.

Energy facilities will not necessarily have an advantage under the
riparian law, however, because some energy facilities, such as coal
slurry pipelines, draw water from remote rather than abutting sources.
Because the riparian system prohibits the use of water on non-riparian
land, obtaining water to make the coal slurry could pose severe legal
restraints in riparian jurisdictions.

States west of the Mississippi River allocate water under what is
known as a prior appropriation system that bases water rights not on
land ownership but on seniority of use. Thus, users who established
their uses earlier in time are able to satisfy their needs before a subsequent user is entitled to water. The legal protection of prior uses
prevents displacement in favor of "more reasonable" uses as is possible
in the eastern system. Nevertheless, a western "prior appropriation"
can be rendered worthless in times of short supply when superior appropriations consume all the water. Furthermore, because the western system embraces the principle that public waters must be used for a beneficial purpose, a water-short western state might be disposed to block the
use of water for producing coal or electricity for use outside the state
on grounds of not being beneficial to the state. The Montana legislature
assumed just this posture when it enacted a law stating that the use of
water to slurry coal out of the state was not a beneficial use.

How the laws allocating water are ultimately applied will vitally
effect coal utilization, especially in the western states that have large
coal reserves. Resolution may require federal intervention in this field
of law.

E.   Solid Waste Disposal

Air and water pollution control programs generate large accumulations of solid waste in the form of scrubber and sewage sludge and other residues. Hence, air and water pollution control regulations create more solid wastes, and the disposal of those wastes then creates new sources of air and water pollution. Congress enacted the Resource Conservation and Recovery Act of 1976 (RCRA) to close the regulatory cycle.

RCRA requires a comprehensive study of the polluting effects of solid wastes from active and abandoned coal mines and of the adequacy of current disposal measures, and requires a study of coal sludge, including especially methods for the use of sludge.

RCRA encompasses all forms of solid waste and includes special hazardous waste provisions that may have the greatest impact on coal utilization. Treatment, storage, transportation and disposal of hazardous wastes that have adverse effects on health and the environment are all regulated. The Act authorizes state permit programs similar to those described in connection with air and water pollution control.

The principal solid wastes generated in coal-fired power plants; namely, flue-gas desulfurization wastes, bottom ash waste and fly ash waste, are not categorically denominated as hazardous. Instead, the wastes from each plant will be examined on a case-by-case basis. Arsenic, barium, cadmium, chromium, lead, mercury, selenium and silver are hazardous contaminants if they appear in excess of permitted limits. Because each of these contaminants may be found in coal, RCRA regulations will have to be followed by coal-fired utilities when any one of the elements exceeds federal standards.

If the coal wastes will be hazardous, RCRA mandates that new power-plants be constructed to avoid endangering underwater drinking sources that lie beyond the property line of the plant and to prevent jeopardizing sole or principal source aquifers as designated under the Safe Drinking Water Act. Measures must be taken to prevent unauthorized people or domestic livestock from entering the disposal site. In addition, a groundwater and leachate monitoring system must be installed, an EPA approved sampling program must be employed, and designated recordkeeping and inspecting systems must be implemented.

The hazardous waste regulations as applied to coal use show a greater inclination to encourage coal utilization at the expense of threats to environmental safety than do air and water regulations. As noted, the hazardous waste regulations apply only if a hazardous condition has been shown to exist at a specific site. This is to be contrasted with the other kinds of regulations that apply comprehensively unless a specific exemption is authorized by statute and granted by the administration. Perhaps the different approach taken with hazardous waste regulations was designed to prevent the arousal of public opposition to coal conversion at a time when other alternative energy sources are not available.

F.   Surface Mining Control and Reclamation Act of 1977

Strip mining is the least expensive and most favored-by-the-industry method of extracting coal, but it creates substantial threats to the environment both in the mining process and in land damage that is not repaired. The Midwest and Appalachian regions of the United States have suffered substantial damages from surface mining and the arid West is threatened by the presence of large reserves of strippable coal.

These factors prompted Congress to pass the Surface Mining Control and Reclamation Act of 1977 (Surface Mining Act). This law follows the model of delegation of federal authority to the states that has been seen before. The states were given 18 months to devise a permitting plan that incorporated the standards of the federal law and Department of Interior regulations. The state plans may also contain land use and environmental controls that are more stringent than the federal counterparts. An approved comprehensive state permit system must include environmental assessment procedures, a reclamation plan, and requirements for performance bonds. Irreparable land degradation, water pollution, and destruction of historic and archaeological sites are specific topics that must be addressed in the permitting process. In addition, soil surveys of potential prime farm land are required.

The reclamation plan must describe the condition, productivity, uses and potential uses of mine lands prior to mining, and proposed post-mining uses of the land must be explained. The post-mining condition of the land must meet specific performance standards pertaining to restoration of the land, and revegetation. Stringent operational standards apply to mining in critical areas including those with slopes greater than 20 degrees and alluvial valleys. The land must be restored to a condition capable of supporting any prior use including possible uses to which the land might have been put if it had not been mined at all. Prime farm and timber lands are given further special consideration.

The Surface Mining Act proscribes all surface mining in designated areas. These include historical areas likely to suffer significant damage, renewable resource lands whose long-range productivity would be reduced by mining and areas in which an existing land use plan indicates mining to be an incompatible use.

The National Coal Association has charged that the Surface Mining Act will unnecessarily increase the costs of coal production, decrease production, and render coal reserves unusable. So far, Congress has not yielded to pressures to loosen the strip mining regulations.

IV. OTHER FEDERAL LEGISLATION AFFECTING COAL UTILIZATION

Although environmental laws are often singled out as the cause of increasing the cost of coal use, many other federal laws also have an effect. Laws with respect to transportation, mine safety, and coal leasing all directly and indirectly play some role.

A. Transportation

Coal is moved by four major modes of transportation in the United States: rail, barge, pipeline and truck. Each mode is regulated by the federal government.

1. Railroad

Rail transportation is perhaps the most regulated mode of transportation in the United States. Regulation of interstate coal shipments by the railroads falls under the jurisdiction of the Interstate Commerce Commission (ICC), which exerts its major influence in the rates set for hauling commodities. Although the ICC law requires that rates be "just and reasonable," reasonableness is determined on a "value of service" consideration rather than on the fair-rate-of-return-on-investment criteria usually associated with the public utility regulation. The following factors are to be considered in determining value of service:

(1) a comparison of the proposed rate with established rates for similar shipments in that region;

(2) the economic effect of the rate on the shippers and their customers;

(3) the relationship between the proposed rate and the cost of providing service; and

(4) the probability that the rate will compensate for costs involved.

The value of service consideration is sometimes used by carriers in a monopoly situation to exploit that monopoly by charging excessive rates. Originally, the value of service consideration was developed to permit lower-than-average freight rates for low-priced, high-volume commodities so as to encourage their movement. Higher valued commodities were then charged higher rates because the cost would be absorbed more willingly by the final customers. Although this sometimes translates to a "what the traffic will bear" rate, the ICC has moderated this approach so value of service is actually the equivalent of "what the traffic can be reasonably required to bear."

ICC policies have a direct influence on competition between different modes of transportation, and in 1976 Congress passed the Railroad Revitalization and Regulatory Reform Act (4R Act) which prohibits the ICC from comparing rail rates to rate structures of competing modes as a basis for determining the reasonableness of a proposed rate. Thus the railroads, to the extent that marginal pricing is used, may have a competitive edge over other modes that must consider fully allocated costs in setting rates.

Other special constraints on railroads include: common carrier status, requiring them to provide some unprofitable service; a requirement that rate increases be negotiated annually (though a rate may be for

5 years if authorized in conjunction with a related facility improvement
investment of more than $1 million); and a prohibition on entering into
long-term shipping contracts. Singling out railroads for these special
regulations places them in an unfavorable competitive position vis-a-vis
other modes. Nevertheless, the Powerplant and Industrial Fuel Act provides some financial help to railroads by authorizing the appropriation
of $100 million for the maintenance, rehabilitation, improvement, and
acquisition of equipment and facilities used for the rail transportation
of coal.

Railroads can be expected to haul much of the coal that will be
burned in this country. How large their role turns out to be can be
strongly affected by whether or not current railroad regulations are
modified or retained.

2.   *Water Transportation*

Some coal is imported from and exported to foreign countries on
ocean-going vessels, but barges handle most domestic over-water transportation of coal. These movements take place mainly on the federal
waterways system which is constructed, operated, and maintained by the
US Army Corps of Engineers. Because of the federal subsidy of waterways
and for other reasons, barge shipping costs are relatively low. Although
legislative bills that would impose user charges on waterways users have
been introduced in Congress, none has passed. Such a change would reduce
the artificial competitive advantages now enjoyed by barge traffic vis-
a-vis rail and other modes of transportation.

By contrast, barge transportation has been hindered by the pricing
of intermodal (e.g., railroad-to-barge) transactions. At one time, railroads set rates that made it cheaper to transport coal the "long way
round" than to use a shorter route involving partial barge shipment. In
fact, the rate was set so high that it was better to use the longer rail
route even if barge rates were set at zero. Although the ICC supported
this rate structure, it was ultimately invalidated by a federal court.
Current ICC regulations require that:

(1) rates set for volume service on all-rail routes must be comparable to rates on a connecting segment of a rail-water route and

(2) efficiencies of volume service must be afforded to all who qualify for it.

3. *Pipelines*

Coal slurry pipelines are thought not to be susceptible to two major problems facing railroads: uncertain and increasing rates, and uncertainty about the reliability of labor. Nevertheless, pipelines face severe legal obstacles. Perhaps the most critical of them is whether pipeline companies will be afforded eminent domain which has not yet been authorized for coal slurry pipelines. In the absence of a grant of power from either the federal government or all of the pipeline states, no pipeline is likely to be built.

The Interstate Commerce Act prohibits railroads from hauling commodities that are owned by the railroad. Hence, railroads cannot both own mines and haul the coal. If such a clause were applied to coal slurry pipelines, the pipeline operator also could not own mines or coal-using facilities.

Pipelines might enjoy several advantages. If they are not designated to be common carriers, they will not be required by law to provide unprofitable service as railroads have been. Furthermore, pipelines have historically enjoyed higher rates-of-return than allowed to the rail industry, and have had the authority to enter into long-term shipping contracts that are denied the railroads.

4. *Trucks*

The trucking industry plays a small but important role in the transportation of coal, especially in the hilly Appalachian states where trucks move coal from mines that are small or difficult to reach by other modes.

Most truck transportation of coal is intra-state in nature and is encumbered by few federal regulations. One federal constraint is imposed by the Federal-Aid Highway Act which authorizes the withholding of federal highway funds from states that permit overweight vehicles to use federal highways. Even though the highway law requires the states to certify compliance with the laws, actual enforcement is often lax.

B.   Mine Safety

Congress has passed several laws to protect the health and safety
of the miners. The Federal Metal and Non-Metallic Mine Safety Act of
1966 (Metal Act) imposed safety standards and the Federal Coal Mine
Health and Safety Act of 1967 (Coal Act) imposed additional regulations
to protect both safety and health. Concerned about the high frequency
of mining injuries and illnesses, Congress enacted the Federal Mine Safety and Health Act of 1977 (Mine Safety Act) thus creating a comprehensive
health and safety law that subsumed the best features of both earlier
statutes. These include operational standards for mines, such as: permitted concentrations of noise, toxic and explosive gases, and respirable
dust; underground roof supports; fire protection; and maintenance of electrical equipment. The Mine Safety Act specifically addresses a major
cause of miner accidents - inadequate training - by setting minimum
training standards, including the requirement of yearly refresher training. Because the Department of Interior is charged with the conflicting
responsibility of maximizing energy resource development, Congress
shifted enforcement of the Mine Safety Act to the Department of Labor.

The Coal Act has been criticized on the grounds that it added expense and lowered production of coal mining operations without making
mines safer.

C.   Mineral Leasing

About half of the coal resources of the United States is found west
of the Mississippi. Of this Western coal 65% is owned by the federal
government and 20% more is controlled by it. Only one-quarter of the
federal western resources is leased to operators, leaving a large reservoir under direct government ownership.

Federal coal leasing has been at a standstill since September, 1977.
This is the culmination of a series of actions that began in 1970 when
the Department of Interior placed a moratorium on leasing until it could
assess why then-current leasing policies had resulted in an increase in
lands leased but a decrease in actual production. Then in 1975 the Department issued a final Environmental Impact Statement on its long-term

leasing program, the adequacy of which was soon challenged in a lawsuit
filed by an environmental group. As a result of that suit, leasing has
been halted pending the completion of a court-approved EIS.

In the meantime, Congress passed the Federal Coal Leasing Amendments
Act of 1976. This Act is intended to discourage speculation in the federal leasing program, make it easier for small companies to compete for
leases, and to place limits on the size of holdings of any lessee.

Although no new federal coal leasing program may be realistically
evaluated until an acceptable EIS has been issued, it is nevertheless
true that the mineability of much of the federal reserves is questionable. Until these questions are cleared up, the real effect on coal production of federal leasing regulations remains uncertain.

D. Antitrust Laws

Who shall be permitted to own and exploit the coal reserves of this
nation is potentially one of the most important coal-use questions that
may be raised. Oil prices and oil company profits have escalated rapidly as the world's demand for oil has begun to exceed the supply. At the
same time, oil companies have begun to acquire other energy resources,
especially coal. Whether this horizontal expansion from one form of
energy into another is a good thing for the nation is not clear at this
writing. Surely, it would not be desirable to have monopolistic dominion
exerted over coal in the fashion that now holds sway over oil. On the
other hand, denying entry into the field to the well-financed and proven
high-technology companies could unnecessarily impede the development of
the resource.

The present arsenal of federal laws that outlaw monopolistic price
fixing and similar anti-competitive practices is comprised of the Clayton
Act (1914), the Federal Trade Commission Act (1914), the Sherman Act (1890)
and various amendments to them. Of these, the Clayton Act is most
directly on point. It outlaws inter-company acquisitions and mergers
when the effect "may be substantially to lessen competition, or to tend
to create a monopoly." The Federal Trade Commission Act outlaws "Unfair

methods of competition in or affecting commerce...," and the Sherman Act
outlaws combinations and conspiracies "in restraint of trade or commerce."
(Commerce in this context means interstate or foreign commerce.) None of
the statutes was enacted with energy monopolies specifically in mind and
neither the antitrust division of the United States Justice Department
nor the Federal Trade Commission, which are the agencies charged with
enforcement of them, has been aggressive about using them in the energy
field. No important case involving horizontal expansion of energy companies into coal production has been decided. Furthermore, although
several proposed legislative bills were introduced into both houses of
Congress during the decade of the 1970s, none has come close to passage
and none now appears ripe for enactment.

   How is this inactivity to be explained? One possible answer is that
the energy companies dominate Congress and have the raw power to stop
such a measure. Another is that fear of more governmental regulation is
even greater than the fear of the energy companies. Still another is
that the facts of coal ownership and exploitation simply do not justify
the enactment of such measures at present. In support of this last argu-
ment, it can be said that much of the coal resources of the United States
is in the ownership of the federal government and the portion in private
ownership is widely dispersed. Moreover, the number of coal producing
companies is large and the oil companies do not now dominate the field.
Hence, coal monopolization may be only a future threat in the mind of
Congress and not enough of a present danger to reach the threshold of
Congressional action.

   As coal use increases in the coming decades, consolidation of ownership and markets seems almost certain to occur, if what has happened in
other industries is a guide. Sooner or later the reach of the Clayton
and Federal Trade Commission Acts into this field will probably be tested.
In the meantime, because of ever growing public knowledge about how
vitally important the issue is to the well-being of the nation, some of
the special legislative bills may be passed. Nevertheless, too little
is known at the moment to be able to make useful predictions about how
antitrust laws will effect coal use in the future.

V. SUMMARY

This chapter has demonstrated that federal laws and regulations both encourage and constrain increased coal use. Coal is a dirty fuel and both Congress and the states have imposed regulations to abate air and water pollution. Whether or not pollution is ultimately controlled will depend both on how effectively the laws are enforced by regulatory agencies and how willing Congress will be to adhere to strong anti-pollution policies in the face of demands for more coal at less cost. It would be erroneous to assume, however, that environmental laws are the sole source of legal constraint on increase coal use. Regulation of integral auxiliary operations, such as mining, transportation, and power generation also has an important effect. Any comprehensive examination of the laws affecting the utilization of coal must include careful scrutiny of each of these numerous laws individually, and, perhaps more important, how they work together. Finally, who will be permitted to exploit the nation's coal resources could importantly effect the efficiency of coal production. Forceful application of federal antitrust laws could deny entry into the field to the rich and experienced oil producing companies, whereas the present hands-off approach could ultimately allow a few giant companies to gain monopolistic control, despite the impediment posed by widely dispersed ownership. At present the topic is little developed and no clear trends in the law are discernible.